Biomass-Derived Carbon Materials

Production and Applications

Edited by
Alagarsamy Pandikumar, Perumal Rameshkumar, and
Pitchaimani Veerakumar

WILEY-VCH

Editors

Dr. Alagarsamy Pandikumar
Electro Organic and Materials
Electrochemisry Division
(CSIR)-Central Electrochemical
Research Institute
Karaikudi-630003, Tamil Nadu
India

Dr. Perumal Rameshkumar
Department of Chemistry
Kalasalingam Academy of Research and
Education, Krishnankoil-626126
Tamil Nadu
India

Dr. Pitchaimani Veerakumar
Institute of Atomic and Molecular
Sciences Academia Sinica (IAMS)
National Taiwan University, 10617
Taipei
Taiwan

Cover: © From "Recent Advances on Porous Carbon Materials for Electrochemical Energy Storage" by Libin Wang and Xianluo Hu, Chem. Asian J. 10.1002/asia.201800553, Copyright Wiley-VCH GmbH. Reproduced with permission.

Library of Congress Card No.: applied for

British Library Cataloguing-in-Publication Data
A catalogue record for this book is available from the British Library.

Bibliographic information published by the Deutsche Nationalbibliothek
The Deutsche Nationalbibliothek lists this publication in the Deutsche Nationalbibliografie; detailed bibliographic data are available on the Internet at <http://dnb.d-nb.de>.

Print ISBN: 978-3-527-34926-5
ePDF ISBN: 978-3-527-83289-7
ePub ISBN: 978-3-527-83291-0
oBook ISBN: 978-3-527-83290-3

Typesetting Straive, Chennai, India
Printing and Binding CPI Group (UK) Ltd, Croydon, CR0 4YY

C114234_041122

Contents

Preface *xi*
Acknowledgments *xiii*

1 Introduction to Biomass-Derived Carbon Materials *1*
A. Sivakami, R. Sarankumar, S. Vinodha, and L. Vidhya
1.1 Introduction *1*
1.2 Biomass Resources and Composition *3*
1.2.1 Plant-Based Biomass *4*
1.2.2 Fruit-Based Biomass *5*
1.2.3 Microorganism-Based Biomass *7*
1.2.4 Animal-Based Biomass *7*
1.3 Condition for Precursor Selection of Biomass-Derived Carbon *8*
1.4 Production Methods of Biomass-Derived Carbon *8*
1.4.1 Carbonization *9*
1.4.1.1 Hydrothermal Carbonization *9*
1.4.1.2 Pyrolysis *10*
1.5 Biomass-Derived Carbons (B-d-CMs) Activation Methods *11*
1.5.1 Physical Activation *11*
1.5.2 Chemical Activation *13*
1.5.3 Combination of Physical and Chemical Activation *14*
1.5.4 Modification and Structural Control of B-d-CMs *14*
1.5.4.1 Surface Modification and Heteroatom Doping of B-d-CMs *15*
1.5.4.2 B-d-CMs Surface Loading of Metal Oxides or Hydroxides *15*
1.5.4.3 Surface Incorporation with Different Nanostructures *17*
1.6 Production Process Description *17*
1.7 Cost Analysis *19*
1.8 Summary *19*
References *20*

2 Introduction to Biowaste-Derived Materials *27*
Thangavelu Kokulnathan, Balasubramanian Sriram, Sabarison Pandiyarajan, Subramanian Ramanathan, and Thangavelu Sakthi Priya
2.1 Introduction *27*

2.2 Synthesis 28
2.2.1 Activation Mechanism of BW-AC by Physical Activation 28
2.2.2 Activation Mechanism of BW-ACs by Chemical Activation 29
2.2.2.1 Influence of Alkaline Activating Agents 30
2.2.2.2 Influence of Acidic Activating Agents 31
2.2.2.3 Influence of Neutral Activating Agents 31
2.2.2.4 Influence of Self-Activating Agents 32
2.3 Characterization 32
2.3.1 Electron Microscopes 32
2.3.2 HR-TEM Analysis 34
2.3.3 FTIR Spectroscopy 35
2.3.4 Raman Spectroscopy 36
2.3.5 XPS Analysis 38
2.3.6 XRD Patterns 39
2.3.7 BET Analysis 41
2.4 Properties 43
2.4.1 Surface Defects in BW-AC 43
2.4.2 Characterizations of Carbon Defects 46
2.4.3 Intrinsic Carbon Defects Activity 47
2.4.4 Heteroatom Doping Defects (or) Extrinsic Carbon Defects Activity 48
2.4.5 Electronic Band Structure Properties 48
2.5 Summary 50
References 50

3 Biomass-derived Carbon-based Materials for Microbicidal Applications *63*
Selvamuthu Preethi, Arunachalam Arulraj, Ramalinga Viswanathan Mangalaraja, Velayutham Ravichandran, and Natesan Subramanian
3.1 Introduction 63
3.2 Biomass Materials 64
3.2.1 Carbon and Its Derivatives 65
3.3 Microbicidal 66
3.3.1 Mechanism of Action 67
3.3.2 Microbicidal Resistance 68
3.3.3 Factors Affecting Microbicidal Resistance 68
3.4 Microbicidal Performance of Biomass-Derived Carbonaceous Materials 69
3.4.1 Role of Material Physicochemical Properties 70
3.4.1.1 Structural Destruction 70
3.4.1.2 Oxidative Stress 73
3.4.1.3 Wrapping Effect 76
3.4.1.4 Photothermal Effect 77
3.4.1.5 Extraction of Lipid 78
3.4.1.6 Metabolic Inhibitory Effect 79
3.5 Bioengineering Prospective Toward Carbonaceous Materials 79

3.5.1 Wound Dressing *80*
3.5.2 Surface Modifications (Coating) on Medical Devices *81*
3.5.3 Nanoantibiotic Formulations *82*
3.6 Biosafety *83*
3.7 Conclusion and Future Perspectives *84*
Acknowledgment *85*
References *85*

4 Carbon-Based Nanomaterials Prepared from Biomass for Catalysis *93*
A. Rajeswari, E. Jackcina Stobel Christy, and Anitha Pius
4.1 Introduction *93*
4.2 Preparation of Biomass-Derived Carbon-Based Nanomaterials *94*
4.3 Graphene *95*
4.3.1 Preparation of Graphene *95*
4.3.2 Graphene from Different Sources *95*
4.4 Carbon Nanotubes (CNTs) *99*
4.4.1 Synthesis of CNTs *99*
4.4.2 Synthesis of CNTs Using Biomass Materials *99*
4.5 Carbon Quantum Dots (CQDs) *102*
4.5.1 CQDs from Biomass *102*
4.6 Catalytic Applications of Carbon-Based Nanomaterials *104*
4.6.1 Potential Advantages in Using Carbon-Based Nanomaterials for Advanced Catalysts *104*
4.6.2 Photocatalysts *105*
4.6.3 Electro Catalysts *107*
4.7 Conclusions, Future Outlook, and Challenges *107*
Acknowledgments *107*
References *108*

5 Biomass-Derived Carbon Quantum Dots for Fluorescence Sensors *113*
Somasundaram Anbu Anjugam Vandarkuzhali, Jeyabalan Shanmugapriya, Chinna Ayya Swamy P, Subramanian Singaravadivel, and Gandhi Sivaraman
5.1 Introduction *113*
5.2 Characterization of CDs *114*
5.3 Optical Properties *115*
5.3.1 Absorbance *115*
5.3.2 Fluorescence *115*
5.4 Methods for the Synthesis of CDs *115*
5.4.1 Hydrothermal Carbonization Method *116*
5.4.2 Microwave Method *116*
5.4.3 Chemical Oxidation Method *116*
5.4.4 Pyrolysis *117*
5.5 Application of CDs *117*

5.5.1 Metal Ion Sensing 117
5.5.1.1 Mercury (Hg^{2+}) Sensor 118
5.5.1.2 Iron (Fe^{3+}) Sensor 119
5.5.1.3 Lead (Pb^{2+}) Sensor 120
5.5.1.4 Copper (Cu^{2+}) Sensor 120
5.5.1.5 Miscellaneous Metal Ions 122
5.5.2 Anion Sensors 122
5.5.3 Miscellaneous Molecules 123
5.6 Conclusion and Future Perspectives 123
References 124

6 Biomass-Derived Mesoporous Carbon Nanomaterials for Drug Delivery and Imaging Applications *129*
Balaji Maddiboyina, Ramya Krishna Nakkala, and Gandhi Sivaraman
6.1 Introduction 129
6.2 Drug Delivery Systems Based on MCNs 130
6.2.1 Immediate-release DDS 130
6.2.2 Sustained-release DDS 130
6.2.3 Controlled/Targeted DDS 131
6.3 Photothermal Therapy 131
6.3.1 Synergistic Therapy 135
6.3.2 Cell Labeling 135
6.3.3 Removal of Toxic Substances 139
6.3.4 Transmembrane Delivery 139
6.3.5 Photoacoustic Imaging 139
6.3.6 Therapeutic Biomolecule Delivery 140
6.3.7 Biosensing 140
6.3.8 Magnetic Resonance (MR) Imaging 142
6.4 Conclusion and Future Perspectives 143
References 143

7 Mesoporous Carbon Synthesized from Biomass as Adsorbent for Toxic Chemical Removal *147*
Babu Cadiam Mohan, Srinivasan Vinju Vasudevan, Ramkumar Vanaraj, Sundaravel Balachandran, and Selvamani Arumugam
7.1 Introduction 147
7.2 Synthesized Methods of Mesoporous Carbons from Biowaste or Biomass 148
7.3 Application of Mesoporous Activated Carbons 149
7.3.1 Removal of Dyes 149
7.3.1.1 GWAC as an Adsorbent for Methylene Blue and Metanil Yellow 150
7.3.1.2 Rice Husk (RH)-Derived Mesoporous Activated Carbon (AC) for Methylene Blue (MB) Dye Removal 151
7.3.1.3 Activated Carbon from Rattan Waste for Methylene Blue (MB) Removal 152

7.3.1.4 Activated Carbon from Cattail Biomass (CAC) for Malachite Green (MG) Removal *152*
7.3.1.5 Wood Sawdust Waste Activated Carbon (WACF-P) for Xylenol Orange (XO) Removal *152*
7.3.1.6 Mesoporous Activated Carbon from Agricultural Waste for Methylene Blue Removal *153*
7.3.1.7 Mesoporous Activated Carbon from Edible Fungi Residue (EFR-AC) for Reactive Black 5 Removal *153*
7.3.1.8 Mesoporous Activated Carbon from Plant Wastes for Methylene Blue (MB) Removal *153*
7.3.1.9 Mesoporous Activated Carbon from *Corozo oleifera* Shell for Methylene Blue (MB) Removal *154*
7.3.1.10 Mesoporous Activated Carbon from Coconut Coir Dust for Methylene Blue (MB) and Remazol Yellow (RY) Removal *154*
7.3.1.11 Mesoporous Activated Carbon from Macadamia Nut Shell (MNS) Waste for Methylene Blue (MB) Removal *155*
7.3.1.12 Mesoporous Activated Carbon from Neobalanocarpus Heimii Wood Sawdust (WSAC) for Methylene Blue (MB) Removal *155*
7.3.2 Removal of Metal Ions *155*
7.3.2.1 Use of Chicken Feather and Eggshell to Synthesize a Novel Magnetized Activated Carbon for Sorption of Heavy Metal Ions *157*
7.3.2.2 Meso/micropore-Controlled Hierarchical Porous Carbon Derived from Activated Biochar as a High-Performance Adsorbent for Copper Removal *158*
7.3.3 Removal of Phenolic Compounds *158*
7.4 Conclusion and Future Outlooks *165*
References *165*

8 Biomass-derived Carbon as Electrode Materials for Batteries *171*
P. Vengatesh, C. Karthik Kumar, T.S. Shyju, and M. Paulraj
8.1 Introduction *171*
8.1.1 Batteries *172*
8.1.2 Classification of Batteries *172*
8.1.3 Characteristics of Batteries *172*
8.2 Role of Carbon with Mechanism of Rechargeable Batteries (RBs) *174*
8.2.1 Li-Ion Batteries (LIBs) *174*
8.2.2 Li-S Batteries (Li-S) *175*
8.2.3 Na-Ion Batteries (SIBs) *176*
8.2.4 Zn-Air Batteries (ZABs) *178*
8.3 Biomass-derived Carbonaceous Materials *179*
8.4 Electrochemical Performances of RBs using Biomass-derived Carbon Electrodes *181*
8.4.1 Li-Ion Batteries (LIBs) *181*
8.4.1.1 Biomass-derived Undoped Carbon Electrodes *181*

8.4.1.2 Metal Oxides @ Biomass-derived Carbon Nanocomposite Electrodes *186*
8.4.1.3 Metal Sulfides @ Biomass-derived Carbon Nanocomposite Electrodes *188*
8.4.2 Na-Ion Batteries (SIBs) *189*
8.4.2.1 Biomass-derived Undoped Carbon Electrodes *190*
8.4.3 Li-S batteries *195*
8.4.3.1 Biomass-derived Carbon Hosts *198*
8.4.4 Zn-Air Batteries *199*
8.5 Biomass-derived Heteroatom-Doped Carbon Electrodes for RBs *201*
8.5.1 Single-Heteroatom-Doped Carbon Electrodes *202*
8.5.2 Dual-Heteroatom-Doped Carbon Electrodes *204*
8.6 Summary and Future Prospectives *206*
References *207*

9 Recent Advances in Bio-derived Nanostructured Carbon-based Materials for Electrochemical Sensor Applications *215*
Akshat Mathur, Jayashankar Das, and Sushma Dave
9.1 Introduction *215*
9.2 Conclusion and Future Perspectives *224*
References *225*

10 Porous Carbon Derived From Biomass for Fuel Cells *229*
A. Sivakami, Aristatil Ganesan, P. Sakthivel, Kishore Sridharan, Sabarinathan Venkatachalam, and Sudhagar Pitchaimuthu
10.1 Introduction *229*
10.2 Fuel Cells – Theory and Fundamentals *233*
10.3 Catalyst Support Materials *234*
10.3.1 As a Catalyst *236*
10.3.2 Synthesis Methods of Porous Carbon from Biomass *236*
10.4 Porous Carbon Synthesis from Different Biomass *237*
10.4.1 Oxygen Reduction Reaction (ORR) *237*
10.5 Synthesis of Biomass-Derived ORR Catalyst for Fuel Cell *238*
10.6 Future Outlook *245*
10.7 Summary *245*
References *246*

11 Biomass-Derived Carbon-Based Materials for Supercapacitor Applications *253*
G. Murugadoss, M. Rajaboopathi, M. Rajesh Kumar, and A. M. Kamalan Kirubaharan
11.1 Introduction *253*
11.1.1 Capacitor *253*
11.1.2 Battery *254*
11.2 Supercapacitor *255*

11.2.1 Types of Supercapacitors 255
11.2.2 Electrical Double-Layer Capacitors (EDLC) 256
11.2.3 Pseudocapacitor 257
11.2.4 Hybrid Capacitors 258
11.3 Activated Carbon Obtained from Biomass for Supercapacitor Application 259
11.3.1 Essential for Carbon-based Electrodes 259
11.4 Electrochemical Measurements 262
11.5 Structural Diversities of Biomass-Derived Carbon for Supercapacitor Applications 262
11.5.1 Spherical Structure 263
11.5.2 Fibrous Structure 263
11.5.3 Tubular Structure 263
11.5.4 Sheet Structure 263
11.5.5 Porous Structure 265
11.5.6 Mesocrystal Structure 268
11.6 Conclusion and Future Perspectives 269
References 269

12 Biomass-Derived Carbon for Dye-Sensitized and Perovskite Solar Cells *275*
N. Santhosh, P. Vijayakumar, M. Senthil Pandian, and P. Ramasamy
12.1 Introduction 275
12.2 DSSC Working Principle 276
12.3 DSSC Components 277
12.3.1 Transparent Conducting Substrate (TCO) 277
12.3.2 Photoanode 277
12.3.3 Dye Sensitizer 277
12.3.4 Electrolyte 278
12.3.5 Counter Electrode 278
12.4 Perovskite Solar Cells 278
12.5 Tunability of Bandgap Energy 280
12.6 Development of Perovskite Solar Cells from Dye-Sensitized Solar Cells 280
12.6.1 Working Principle of PSC 281
12.6.2 Perovskite Solar Cells Architecture 281
12.6.3 Hole Transport Material 282
12.7 Biomass-Derived Carbon Counter Electrode for DSSC 283
12.7.1 Performance of DSSC with Counter Electrode via Bio-derived Carbon 284
12.7.2 Biomass-Derived Carbon as a Counter Electrode for Perovskite Solar Cells 285
12.8 Conclusion and Future Perspectives 287
References 287

13 Recent Advances of Biomass-Derived Porous Carbon Materials in Catalytic Conversion of Organic Compounds *293*
N. Mahendar Reddy, D. Saritha, Naveen K. Dandu, Ch.G. Chandaluri, and Gubbala V. Ramesh
13.1 Introduction *293*
13.2 Synthesis Procedures *295*
13.2.1 Carbonization *295*
13.2.1.1 Hydrothermal Carbonization (HTC) *296*
13.2.1.2 Pyrolysis *297*
13.2.2 Activation *297*
13.2.2.1 Physical Activation *297*
13.2.2.2 Chemical Activation *298*
13.2.3 Physicochemical Activation *299*
13.2.4 Microwave-based synthesis *299*
13.2.5 Functionalization/Doping/Composites of ACs *300*
13.3 Applications *302*
13.3.1 Heterogeneous Catalysis *302*
13.4 Conclusion and Future Challenges *308*
References *309*

14 Summary on Properties of Bio-Derived Carbon Materials and their Relation with Applications *317*
S. Vinodha, L. Vidhya, and T. Ramya
14.1 Removal of Toxic Chemicals *321*
14.2 Electrode Materials for Batteries *322*
14.3 Electrochemical Sensor Applications *323*
14.4 Fuel Cell Applications *324*
References *329*

Index *331*

Preface

Biomass-derived carbon-based materials are of great interest because of abundant and easy availability of bio-precursors, and the materials with low dimensions possess large surface area and porosity that allow for a variety of applications. Biomass-derived carbon is increasingly popular in making composite materials because of its continuity, interconnection, and porous and hierarchical structure. To date, a wide variety of composite materials involving biomass-derived carbon have been prepared and used for interesting applications. This book aims to provide a deep insight on the preparation and activation processes of biomass-derived carbon, synthesis of composites, and future opportunities on the exploration of these materials. The introductory chapters deal the possible sources, synthesis, properties, characterization, activation, and cost analysis of biomass-derived carbon-based materials. The remaining chapters elaborately discuss the applications of biomass-derived carbon-based materials, including catalysis, sensors, microbicidal activity, toxic chemicals removal, drug delivery, and electrochemical energy conversion and storage applications. The final chapters give an overview of properties of biomass-derived carbon materials and their relation with applications.

Hence, this book gathers and reviews multidisciplinary aspects of biomass-derived carbon-based materials research performed by chemists, physicists, materials scientists, biologists, and engineers. Readers can easily understand the fundamentals of biomass-derived carbon materials synthesis, activation processes, properties, characteristics, and their role in the current scenario of application-oriented research and development. This book will be helpful for researchers to establish their own research in the area of biomass-derived carbon materials.

Dr. Alagarsamy Pandikumar
Scientist
Electro Organic and Materials Electrochemisry Division
CSIR-Central Electrochemical Research Institute
Karaikudi-630003, Tamil Nadu, India

Dr. Perumal Rameshkumar
Assistant Professor
Department of Chemistry
School of Advanced Sciences
Kalasalingam Academy of Research and Education
Krishnankoil-626126, Tamil Nadu, India
and
Dr. Pitchaimani Veerakumar
Department of Chemistry
Institute of Atomic and Molecular Sciences Academia Sinica (IAMS)
National Taiwan University
Taipei-10617, Taiwan

Acknowledgments

We are grateful to all the authors who contributed their chapters to make this a valuable book and for the successful completion of the process. We are thankful to the publishing editor, Wiley-VCH, for accepting our proposal and giving us an opportunity to edit this book, and their help toward the successful completion of the work is greatly acknowledged.

1

Introduction to Biomass-Derived Carbon Materials

Sources, Production, Activation, and Cost Analysis

A. Sivakami[1], R. Sarankumar[2], S. Vinodha[3], and L. Vidhya[3]

[1] *Malla Reddy University, School of Sciences, Department of Physics, Maisammaguda, Hyderabad, 500100, India*
[2] *QIS Institute of Technology, Department of Electronics and Communication Engineering, Ongole, 523272, India*
[3] *Sethu Institute of Technology, Department of Chemical Engineering, Virudhunagar, 626115, Tamil Nadu, India*

1.1 Introduction

The transformation in climate and demand for energy encourages us to concentrate and cultivate a new technique for generating and storing energy via clean and green technologies. In order to fulfill our requirements for both energy generation and storage, it is necessary to produce high-performance materials [1–4]. Due to their less cost, comparatively simple preparation, richness, eco-friendly credentials, decent conductivity, and stability in chemical and thermal environments, carbon materials are placed at the topmost position in the list among other materials. The supply of different waste biomasses is growing globally, and the disposal of this biomass is an issue. It is mostly disposed of by burning materials in open spaces, which, by releasing greenhouse gases, directly affects our atmosphere. The conversion of biomass into useful materials is a beneficial alternative for waste management.

Because of the waste to richness principles, the carbons extracted from biomass attracted additional interest. In addition, the carbon materials derived from biomass demonstrated beneficial properties that broadened their applications. At present, by enhancing benefit of their physical and functional diversity, carbon materials, such as graphene, carbon nanotubes, activated carbon (AC), and porous carbon, are used in the energy storing area. However, demands for green and renewable energy storage materials have been spurred by the growth of advanced science and technology. As a source of electrode materials, biomass-derived carbon gained considerable attention due to its structural diversity, modifiable physical/chemical properties, environmental friendliness, and significant worth in trade and industry [5–10].

Since nature adds bizarre microstructures to biomass, the carbon materials derived from biomass also exhibit natural structural diversities, such as 0D spherical, 1D fibrous, 2D lamellar, and 3D spatial structures. In addition, it is important to

Biomass-Derived Carbon Materials: Production and Applications, First Edition.
Edited by Alagarsamy Pandikumar, Perumal Rameshkumar, and Pitchaimani Veerakumar.

mold the characteristics of the bio-derived carbon materials (B-d-CMs) according to their application. Different approaches for manufacturing and modifying carbon materials are pursued.

There are several general properties of carbon materials that make them attractive in different applications. Carbon is a common commodity, indicating that it has long been used by individuals. Graphite, carbon black (CB), and AC materials are included in these long-used traditional carbon materials [11–14]. New carbon materials with customized properties have been developed in the past century. This included carbon fibers, graphite that was strongly focused, and several others.

Much more sophisticated nanosized or nanostructured carbon materials have been developed in recent decades. Carbon materials are currently being intensively researched, in particular the newest nanocarbons, but also macroscopic carbons such as carbon fibers [15, 16]. Due to their distinguished physicochemical properties, innovative carbon materials namely graphene and its derivatives, fullerene, and carbon nanotubes have gained significant interest in the area of energy storing nowadays. They have good conductivity, outstanding chemical stability, porosity that can be tuned, large specific surface area, and enriched electroactive sites enriched. Porous carbon nanomaterials have gained significant attention because of their physiochemical properties and high surface area. The carbon with different

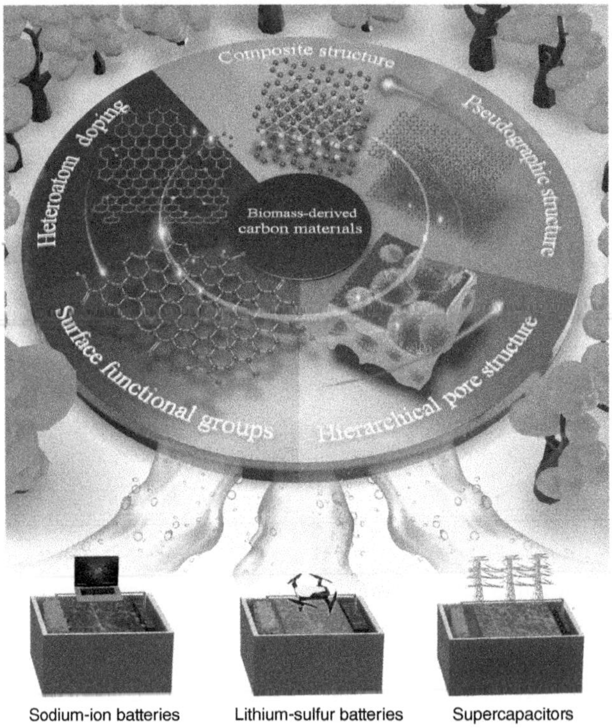

Figure 1.1 B-d-CMs structure strategies for different EES applications. Source: Ref. [19] / American Association for the Advancement of Science / CC BY 4.0.

pore sizes has attracted significant attention for highly efficient electrochemical storage applications [17, 18].

However, these carbon materials depend on precursors based on fossil fuels using energy-consuming synthetic methods (e.g. chemical vapor deposition, discharge of electric arcs, and laser ablation) that are toxic and expensive to the atmosphere. While these synthetic methods are advanced technology on a bench scale, due to complex synthetic processes, they are not yet ready for commercialization. Consequently, the establishment of more effective, environmentally sustainable and economic approaches to the processing of carbon materials is important.

B-d-CMs are showing great importance, and efforts have been devoted for enhancing the performance of electrochemical storage applications [19]. It is essentially to know how the structure design and diffusion kinetics of B-d-CMs are affecting the performance of electrochemical energy storage (EES) devices. It is shown in Figure 1.1.

1.2 Biomass Resources and Composition

Biomass refers to animal- and plant-based materials or by-products that may serve as a potential energy source. Protein, carbohydrates, starch, lignin, and lipids constitute biomass and are such components that differ dependent on the geographic situation and source. Proximate and final studies have shown that biomass is abundant in carbon, hydrogen, oxygen, and nitrogen, and traces of chlorine and sulfur are also shown. Biomass-derived carbon has many crucial advantages associated with additional electrode materials for energy and ecological applications, such as cheap and plentiful supply, environmentally safe, in situ nanoporous structure establishment, and processing elasticity [20–24]. As the source affects the finishing carbon return and its structural features, which are mandatory for energy storing and ecological applications, the option of a biomass precursor is crucial.

Agro-residues from crop production, solid waste from municipal, and further agro-based manufacturing units are key sources of precursors. Owing to the large availability and less cost, these precursors have gained a lot of attention. They do, however, have various chemical functionalities, creating them an ideal choice for a wide range of morphologies for the proposal of carbon materials. Biomass derivative carbons are deliberated to be favorable electrode materials for different forms of electrochemical energy storing and transformation systems due to the aforementioned advantages, including lithium batteries, supercapacitors, potassium batteries, sodium batteries, and fuel cells [25–29]. Given the quick growth in this sector, a thorough analysis and comparison of their manufacturing approaches, features, applications, and performance in these electrochemical energy storing applications are not only necessary but also urgent.

The carbon powder that comprises egg white, bacterial cellulose, mushrooms, peels of orange, human hair, dry elm samara, chitin, catkin, etc. has been used to manufacture a wide variety of biomaterials. These biomass products, however, can be classified into four main groupings, i.e. biomass based on microorganisms,

Table 1.1 Difference between biochar, activated carbon, and carbon black.

	Biochar	Activated carbons	Carbon black
Precursors	Biomass	Coal, asphalt, and biomass.	Petroleum, coal tar, and asphalt.
Carbon content	40–90%	80–95%	>95%
structural features	Amorphous and porous carbon with enriched surface functionalities	Amorphous carbon and highly porous	Microcrystal or amorphous carbon particles
Preparation method	Medium temperature pyrolysis (400–600 °C), followed by physical or chemical activations	High-temperature carbonization (700–1000 °C) with physical or chemical treatments	Combustion process with little or without air

Source: Ref. [30] / with permission of American Chemical Society.

animal based, plant based, and food based. It becomes more difficult to predict the concluding arrangement and construction of a derived biochar. Generally speaking, it is difficult to understand the awareness of the elemental and chemical composition of biomass as the responses can happen in the phases of carbonization and stimulation. The morphology and structure of resulting carbon could eventually be changed by these induced reactions. Table 1.1 lists the difference between activated, biochar, and CB [30].

Significant research studies are done in pursuit of biomass that at the same time receives functional groups containing oxygen or nitrogen, interrelated micro or mesoporous arrangements, and similarly has a large carbon content production that amplifies the application of the environment and resources. Most advanced precursors, however, end up with low yields. The biochar is a derivative of willow catkins, for example, proved excellent presentation in performance, capacitance, and cycling as energy storage devices. The final carbon outcome was significantly lesser (5.5% wt.) relative to similar biomass precursors including rice straw, considering these benefits [31–34]. The biochar outcome, heteroatom, and doping of the biochar depend greatly on the precursors' basic composition and chemical structure. It is thus important to discover and understand different precursor-related properties in order to enhance biochar outcome to create it appropriately in energy storing and ecological applications. The different biomass resources are given in Figure 1.2.

1.2.1 Plant-Based Biomass

The numerical chemical composition of plant-based biomass varies according to geographic factors, categorizations of organisms, and organ dependency. Still, cellulose, lignin, hemicellulose, and extractives consist of the qualitative chemical configurations of plant-based biomass [35, 36]. For instance, seed shells, palm, areca, etc. consist of substantial lignin and 83% of cellulose. Whereas, for

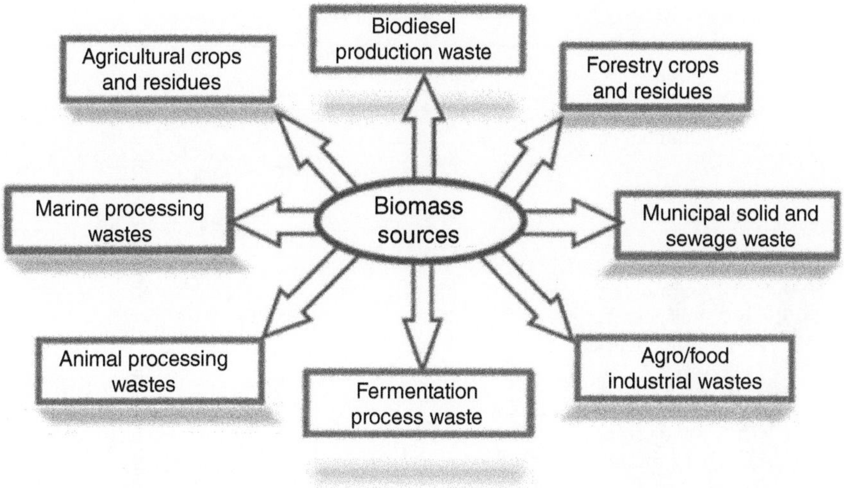

Figure 1.2 Popular biomass resources.

instance, plant's bast fibers are good in cellulose, and in cotton, jute, and fax, bast comprises 67, 64, and 56% comparatively higher cellulose with respect to other parts of the plant. Figure 1.3 shows the chemical composition of lignin, cellulose, and hemicellulose. Investigations on the basic composition of precursors, such as the involvement of microstructure, capacitance, and conductivity in precursors of oxygen and nitrogen content.

High oxygen content in precursors has been determined to yield low crystallinity and larger defects, and more unstable mixes are produced in the process of pyrolysis and thermal decay; however, higher levels of nitrogen content could yield enhanced electrochemical featured nitrogen-doped carbon.

Compared to lignin, cellulose and hemicellulose offer less stability in thermal decomposition and lead to considerably less carbon yields. Nevertheless, lignin hemicellulose and cellulose components contribute to biochar yield porosity. It was noted in the process of thermal pyrolysis at 500 °C that no noteworthy relations between hemicellulose and cellulose were seen, but obvious interactions between the components of lignin and cellulose were observed.

Therefore, in order to get biochar outcome with worthy conductivity, controllable faults, and a high degree of graphitization to make it suitable for application in ecological and energy storing, it is important to choose plant biomass precursors rich in nitrogen content and less in oxygen content, cellulose fraction, and lignin fraction.

1.2.2 Fruit-Based Biomass

The quantifiable elemental and chemical composition of fruit-based biomass can vary with regard to geographical parameters and species. The main fruit-based biomass components are sugars, lipids, ash, crude proteins, and fibers. The lipid quantity in the peels and pulp varies from 0.7 to 9.96% and 1.4 to 28.6%, while the

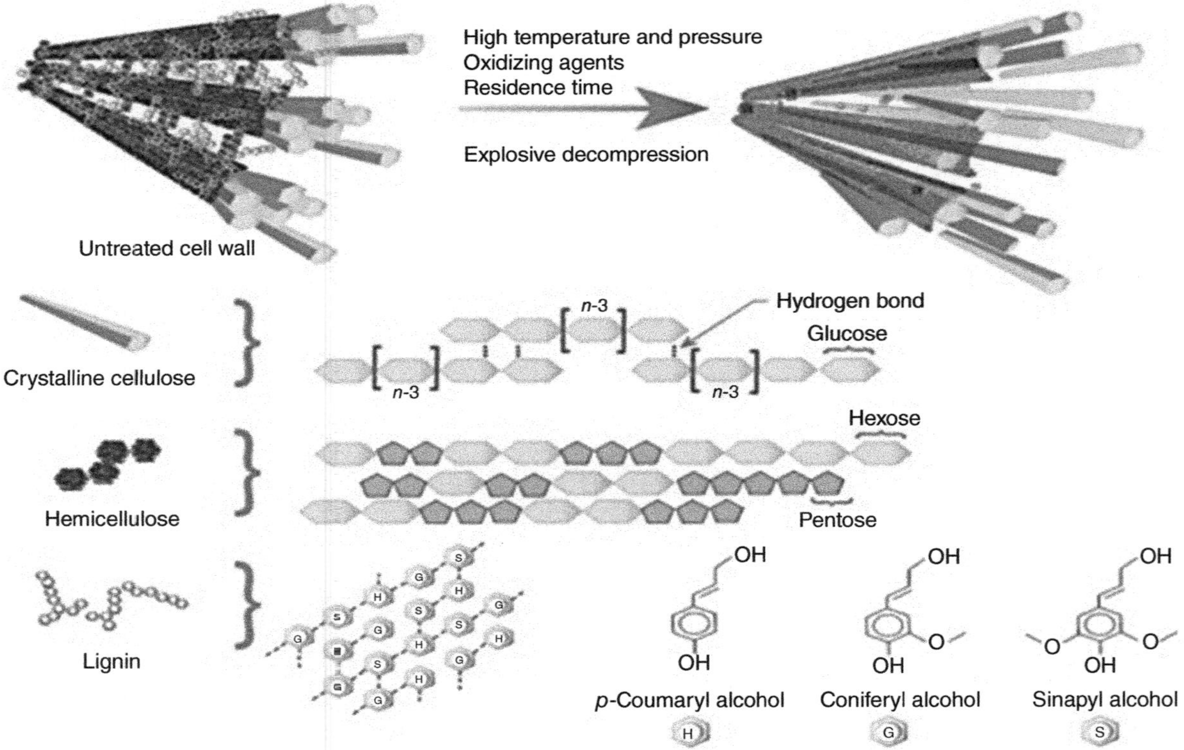

Figure 1.3 Plant-based biomass constituents. Source: Ref. [36] / with permission of Elsevier.

crude protein yields 3.5 to 28.6% and 5.8 to 43.4%, respectively. Larger levels of proteins and crude lipids, however, lead to disadvantages in the ultimate biochar outcome, as these proteins and crude lipids start degrading when temperature is low, exempting unstable compounds namely water vapors, methyl esters, olefins, carbon dioxide, and ammonia fumes. On the other hand, the presence of nitrogen and phosphorous content in proteins and crude lipids can lead to heteroatom-doped carbon production [25, 37, 38]. The crude fibers of lignin, hemicellulose, and cellulose are the main providers to the processing of carbon. The mass portions of crude fibers, however, are considerably less and typically good in cellulose, which affects the graphic structure and biochar yield.

1.2.3 Microorganism-Based Biomass

New measurements open up the prospect of using microorganisms derived from biochar, such as bacterial cellulose and fungi. The fungi, such as mushrooms and yeasts, have evidenced to be improved precursors of reformative biomass for biochar derivation, attributing their rapidly increasing capacity and their ease of use in environment to bulk. Carbohydrates, crude proteins, fibers, and fats are the main elements of microorganism-based biomass. Plant- and fruit-derived biomass are the key components found in the microorganism-based biochar. But the individual compounds and elements from these modules are considerably different.

The carbohydrates existing in the microorganism include chitins that establish a glucan cross-link, serving as the main source of carbon in the process of pyrolysis, while sucrose and starch are the source of carbohydrates in non-cross-linked plants with the lowest thermal stability. In microorganisms, crude fibers are primarily made of cellulose, which shows the same plant or fruit biomass carbonization behavior.

The main precursors are mushrooms based on microorganism-based precursors. Compared to other elements such as mycelium, the fruiting section of the mushroom is most commonly used for processing as it goes into depth. Owing to the existence of large nitrogen content, mushrooms are an attractive choice. The mushroom's nitrogen content varies from 3 to 10% and 17% nitrogen content is observed for some species. As they are able to generate carbon derived from nitrogen-doped biomass, the mushroom precursors are promising [39–41]. Geographical factors, however, have a major effect on the composition of the elements and can differ from region to region.

1.2.4 Animal-Based Biomass

Chitin is an alternative positive bio-precursor for a widespread variety of applications owing to the existence of larger nitrogen concentrations, chemical stability and large existence in the environment. Chitin is capable of creating chitin-catecholamine and chitinglucan complex cross-link networks and is able to form intermolecular hydrogen bonds. Compared with cellulose, chitin has larger thermal stability and carbon outcome. Chitin extraction can be used for the popular animal species containing mollusks, pests, and crustaceans. The crustaceans consist of a substantially high amount of chitin content ranging from 17 to 72%, i.e. *Carcinus*, *Pandalus*, *Carangon*, and *Cancer* [42, 43].

Likewise, the cuticles and sloughs of many types of insects, such as, butterflies *Holotrichia parallela*, and silkworms, have extraordinary chitin content concentrations ranging from 18.4 to 64%.

Chemical demineralization, deproteinization, and mechanical crushing are used to remove the chitin content from the biomass. The final extraction yield depends on biomass precursors, and it ranges from 4 to 40% in the processing system. Chitin's nitrogen content is extracted from animal biomass and matches with that of microorganism-based biomass with nitrogen concentrations, capable of its appropriateness in energy and ecological applications [44, 45].

1.3 Condition for Precursor Selection of Biomass-Derived Carbon

The following conditions must be addressed in order to synthesize good-quality biochar by means of superior conductivity and porosity and to satisfy the needs for huge ecological and energy applications.

The existence of nitrogen content improves nitrogen-doped carbon production with greater conductivity and enhanced cycling stability. Still, it is important to select precursors having less oxygen content, or else the aromatic carbon formation would be obstructed. The existence of strongly cross-linked, large molecular weight with thermal stability biomacromolecules including lignin, keratin, and chitin enables the formation of aromatic carbon and in the process of carbonization provides superior biochar. Aliphatic compounds must be prevented by the existence of little contents of noncrosslinked and molecular weight, otherwise they hinder aromatic carbon formations.

1.4 Production Methods of Biomass-Derived Carbon

Numerous activation and various methods of carbonization can be used to turn biomass into carbon. In order to turn biomass into value-added carbon goods, physical, chemical, and a combination may be used [16].

In carbon materials, many factors including surface properties, temperature, time, reagents, and availability cause an effect. The key processes used to extract carbon from biomass are pyrolysis and hydrothermal carbonization (HTC). Pyrolysis is performed at a defined temperature level in a restricted oxygen or inert atmosphere environment, whereas a thermochemical mechanism is used for transforming biomass into carbon. Temperature, temperature ramping rate, catalyst, and particle size are the products obtained from biomass pyrolysis. HTC is done with or without the use of a catalyst in a pressurized aqueous atmosphere at less temperature range from 120 to 250 °C. Compared to natural biomass coalification, the HTC procedure

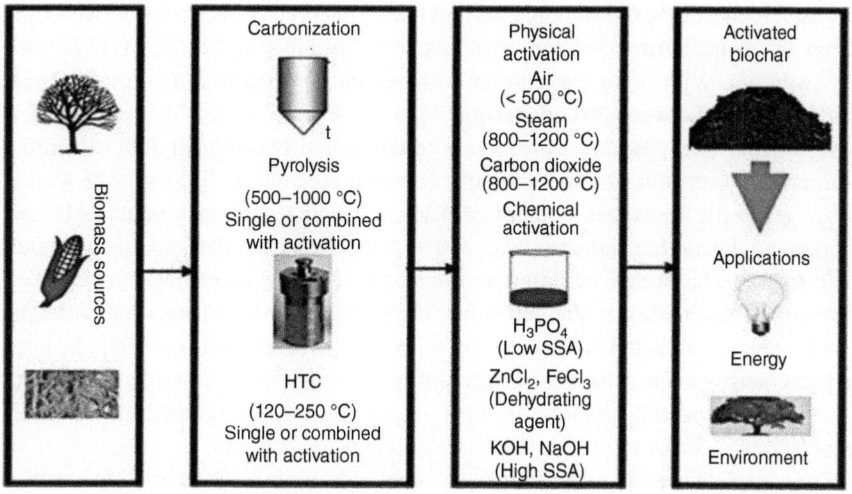

Figure 1.4 Overview of overall production methods of B-d-CMs. Source: Ref. [36] / with permission of Elsevier.

is done at a rate of higher reaction added with a smaller reaction length. Various publications have studied hydrothermal conversion in recent research [46–48].

The overview of different production methods of B-d-CMs is shown in Figure 1.4.

HTC is a thermochemical conversion method that requires a variety of components including precursor concentration, catalyst, residence period, and temperature. It uses subcritical waters to transform biomass to carbon products for successful dehydration and hydrolysis of hydrochar precursors with high-oxygen-rich functional groups. Through the use of additives or doping-containing precursors, other functional groups, including nitrogen groups, may also be added to hydrochars. Recovered carbon products have attracted interest in a wide range of uses, including energy harvesting, catalytic, and trap technologies.

The method used for transforming carbon materials into AC is activation. For activation, chemical and physical methods may be introduced. Physical activation by pyrolysis is achieved at 1200 °C in the presence of carbon dioxide. In the presence of a chemical agent, chemical activation takes place at temperatures between 450 and 900 °C. The most used chemical activators are NaOH, KOH, K_2CO_3, $FeCl_3$, H_3PO_4, and $ZnCl_2$ [48].

1.4.1 Carbonization

1.4.1.1 Hydrothermal Carbonization

The hydrochar material produced by the method of HTC is partly carbonized and contains oxygen groups of large density. The final yield, however, depends on the precursor features employed. For energy storage applications, hydrochars are straightaway used as electrodes. Poor porosity and small specific surface area are naturally present in the hydrochar formed by HTC. Consequent initiation or carbonization is compulsory to change its chemical and physical properties.

For supercapacitor applications, Zhu and his team created electrodes derived from fungi by hydrothermal-assisted pyrolysis [49]. Initially, at 120 °C, HTC is done for six hours. With large oxygen composition and less surface area, the particles obtained differ between 50 and 200 nm.

Hydrochar is exposed for three hours to additional pyrolysis at 700 °C in order to increase electronic conductivity and enhance porosity. A fixed surface area of 80 m^2 g^{-1} with an oxygen content of 5 wt. is the final carbon product obtained. Compared with other industrial AC, the content displays a particular capacitance of 196 F g^{-1}. The porous carbon materials dependent on hydrochar exhibit a high number of heteroatoms. Via activation using KOH, Salinas-Torres and research work group established AC by doping with O and N heteroatoms [50]. At large current densities, the existence of both nitrogen and oxygen groups increases charge transmission and capacitance. Uniform oxygen has been developed by Wei et al., which shows a large specific surface area and porous structure.

The complete overview of carbonization methods of producing B-d-CMs is shown in Figure 1.5.

1.4.1.2 Pyrolysis

In an oxygen-limited atmosphere, biomass pyrolysis is carried out when the temperature ranges from 300 to 1200 °C. Pyrolysis is categorized into fast pyrolysis and slow pyrolysis depending on the range of heating. The method of slow pyrolysis has a longer residence period and lower heating rates. It is carried out at the temperature of

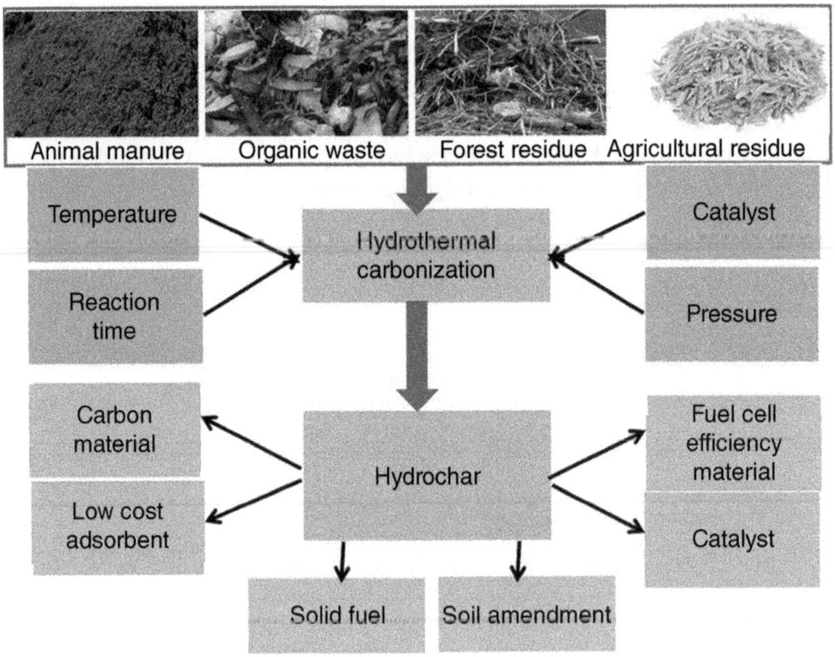

Figure 1.5 Complete overview of carbonization process. Source: Ref. [51] / with permission of Elsevier.

400–600 °C, resulting in maximized biochar yield and low bio-oil and syngas product yields [52]. Quick pyrolysis, on the other hand, possesses larger heating rate and lower residence period, providing the value of maximized bio-oil outcome upto 75%. Temperature is a main parameter in regulating the pyrolysis process and then affecting the behavior and biochar outcome, relative to rate of heating, reaction time, and particle size of feedstock.

Growing pyrolysis temperature decreases the yield of biochar, the ability of cation exchange, and the quality of nutrients, but increases its degree of aromatization, specific surface area, larger range of heating value, and solution pH. In addition, biochar generated at a lower temperature of pyrolysis has a lower stable fraction ratio than biochar produced at higher temperatures [53]. Due to its less conductivity, poorer pore characteristics, and less specific surface area, biochar developed at lower temperature pyrolysis is not appropriate for use as energy storing and conversion materials. Therefore, surface alteration and activation procedures are required prior to their implementation [54].

1.5 Biomass-Derived Carbons (B-d-CMs) Activation Methods

The two basic methods that are applied to achieve AC based on biomass are physical and chemical activation. Physical activation is easier and more environmentally friendly than the chemical activation, which is normally done at higher temperatures.

1.5.1 Physical Activation

Overall, various suitability and acceptable activation methods could be altered and material features could be different. It is a famous fact that, as opposed to physical activation, chemical activation needs less activation period and temperature. Chemical activation, however, has shown many significant drawbacks comprising of water washing step after activation that is needed and needed to eradicate≪Revise as "...that is needed to eradicate..." impurities. Normally, physical activation deploys a two-step procedure. Biomass content is initially pyrolyzed to create biochar, which is then triggered using gases, namely steam, CO_2, air, or their mixture by controlled gasification [23, 56]. Disorganized carbons are made in the process of pyrolysis from tar decomposition, blocking biochar pores, and minimizing their particular surface area.

The complete overview of different activation methods is shown in Figure 1.6.

The successively regulated gasification is able to promote further decomposition of the as-prepared biochar and get a fully created, usable, and interrelated porous structure. Porosity production also yields from carbon calcination and generation of volatile substances, and depending heavily on the triggering gas. CO_2 is safe and easy for usage, so it is always used. Activation period, temperature, rate of gas flow, and furnace selection can influence the degree of carbon calcination. By activation

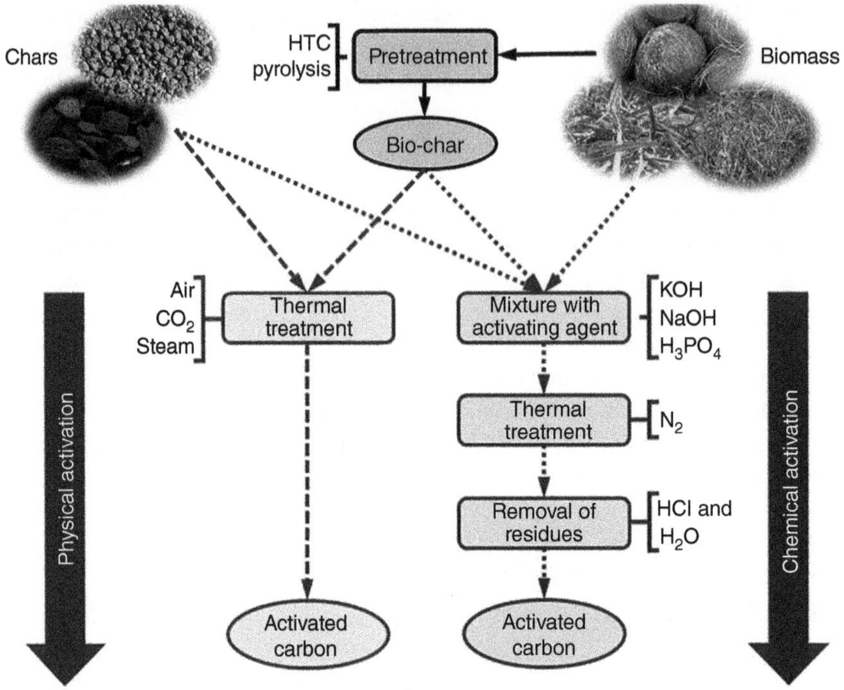

Figure 1.6 Overview of complete activation methods. Source: Ref. [55] / Frontiers Media S.A. / CC BY 4.0.

of CO_2, Zhang et al. [57] developed carbons derived from biomass from various forestry area and agricultural residue types. In general, higher surface areas and micropore volume can benefit from the higher activation temperature. Guo et al. [58] researched the activation of carbon-based coconut shell by CO_2 and methodically observed the effects on the particular surface region, micropore volume, and total volume of activation temperature flow rate and time.

Their findings showed that increasing activation temperature aided the formation of pores, expanded pore diameter, and increased generation of mesopores. In addition, the generation of micropores and mesopores was favored by increasing activation time, but excessively lengthy activation period led to pores collapsing and deteriorating. Taer et al. [59] investigated the CO_2 activation of carbon extracted from rubber wood and tested the capacitive and electrochemical characteristics of the AC content. With the rise in activation temperatures, the conductivity and basic capacitance of their findings increased. The increased specific surface area induced by the increasing temperature of activation was straightway connected to the rise in the volume of micropores in the carbon precursor. Though, pore deformation resulted from a larger temperature above 900 °C [60, 61]. Second, due to its cheap price and ease, steam is often used in biomass materials as an activation agent.

Demiral et al. [62] examined olive bagasse steam activation, and their findings exhibited that rise in activation period (30–45 minutes) and temperature (750–900 °C) could increase the total pore volume of the specific surface region.

Still, additional enhancement in activation period (60 minutes) decreased total volume and the real surface area of the pore, as the creation and expansion of micropores were less successful than the worsening of high porosity. Chang et al. [63] and Okada et al. [60] published similar findings. The pore structures were mainly contributed by micropores, close to CO_2 activation, and the ratio of micropore volume to total pore volume (Vmicro/VT) was between 0.63 and 0.84. Thus, all methods of activation of CO_2 and steam led to well-built microporosity. Compared to CO_2 activation, however, steam activation favors the widening of microporosity and increases the growth of mesopores at the cost of microporosity.

1.5.2 Chemical Activation

Air activation needs less temperature compared to the CO_2 and steam activation mentioned earlier. Nevertheless, since they commonly grow a hierarchical porosity, steam and CO_2 activations are often used. Research on biomass physical activation is still incomplete, in comparison to chemical activation. There is a need for further methodical research and evaluations of CO_2, air, and steam activations of carbons derived from biomass, particularly in the application scenarios in the field of electrochemical energy storing. Thermal preparation of the biomass carbon precursor and the triggering agent in the 450–900 °C temperature range is part of the process of chemical activation. In contrast to physical activation, chemical activation needs lesser pyrolysis temperature, improved carbon outcome, carbon yielded with large surface area, and properly arranged and defined microporous structure [64].

The large surface area and properly arranged and defined microporous structure of the carbon generated show a vital role in ecological and energy storing applications. KOH is widely used as an agent for chemical activation, performing as an oxidant and forming oxygen functional groups on biochar. Therefore, not only electrochemical double-layer capacitance but also pseudo-capacitance can be contributed by KOH-ACs. KOH first dehydrates into K_2O at 400 °C in the activation phase. Then, with the development of H_2, carbon reacts with H_2O, followed by CO_2 formation. K_2O reacts and forms K_2CO_3 with CO_2. Once the temperature is increased beyond 600 °C, it responds completely to KOH. K_2CO_3 begins to decompose above 700 °C and is absent from the system at 800 °C [65]. Simultaneously, metallic potassium is formed. To generate CO at higher temperatures, the CO_2 produced will react with carbon.

The three key mechanisms for porosity production by activation of KOH are generally accepted. A redox reaction decomposes the carbon matrix with KOH, leading to the formation of abundant micro- and mesopores. The formation of H_2O and CO_2 helps in the production of porosity. After extracting the metallic potassium and other potassium substances by rinsing, as shown in the figure, expanded carbon lattices integrated with intermediate metallic K are unable to restore their original structure. Therefore, on the basis of the synergistic effects of physical and chemical activation, microporosity, carbon lattice expansion, and high specific surface area are formed. Various precursors are developed into porous carbon material by means of distinguishing morphological characteristics, pore texture, and surface functional

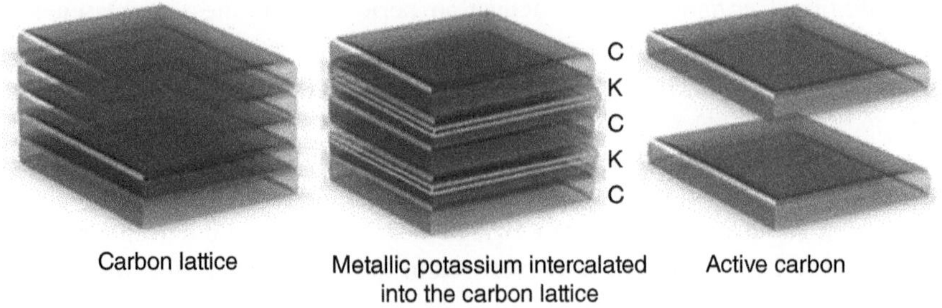

Figure 1.7 KOH activation mechanism. Source: Ref. [66] / with permission of Elsevier.

groups by controlling the activation parameters, further influencing the behavior of KOH-ACs in their electrochemical energy storing system applications [66]. The KOH activation mechanism is shown in Figure 1.7.

Peng et al. [67] developed carbons through leftover tea leaves via large temperature carbonization and activation process with KOH. The carbon materials demonstrated a porous structure with a large specific surface area ranging from 2245 to 2841, and when measured in a three-electrode device in a 2 M KOH electrolyte solution it showed perfect capacitive behaviors with a maximum specific capacitance of 330 F at a current density of 1 A. To enhance specific surface area and pore volume further, the KOH activation method is combined with CO_2 activation. The arrangement of CO_2 gasification, however, leads to a decrease in functional groups of oxygen, decreasing the contribution of pseudo-capacitance to the overall real capacitance (up to 30% loss). Therefore, the grouping of activation of KOH and CO_2 yields very complicated effect on the electrochemical features of biomass-derived carbon.

1.5.3 Combination of Physical and Chemical Activation

The combination of both physical and chemical activation shows better pore size distribution and porosity development of AC than the individual one. This two-stage mechanism comprises of initial chemical activation using different agents then followed by physical activation. The high mesopore content and surface-area-modified AC are obtained and reported by several authors, which is utilized in both chemical and physical activation processes [68–70].

1.5.4 Modification and Structural Control of B-d-CMs

The electrochemical performance of B-d-CMs depends upon their consistency of surface functionalities. The micro-, meso-, and macropore-sized ACs are produced based on their applications with choosing suitable activation process that is already known. The double-layer properties such as capacitance, self-charge characteristics, electrical conductivity, and wettability can be altered by surface functionalities of B-d-CMs with oxygen-containing carboxyl, hydroxyl, and phenolic groups [71–74]. There are several methods available to improve the electrochemical performance of B-d-CMs via surface functionalities modification, heteroatom doping, surface

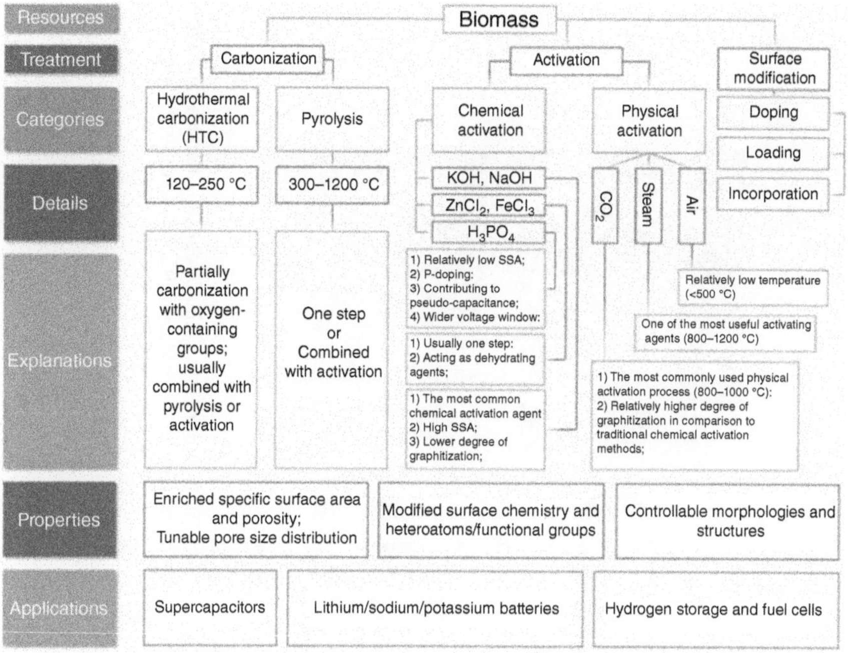

Figure 1.8 Complete overview of production methods converting biomass into carbon materials for EES devices. Source: Ref. [66] / with permission of Elsevier.

incorporation with nanostructures, and surface loading with oxides including physical and chemical activation. The complete production methods and structural modification of B-d-CMs are shown in Figure 1.8.

1.5.4.1 Surface Modification and Heteroatom Doping of B-d-CMs

Several authors investigated the surface modification of B-d-CMs using functional groups or doping different heteroatoms such as nitrogen, phosphorus, oxygen, sulfur, and boron. They have reported that the surface modification and heteroatom doping B-d-CMs are enhancing the double-layer properties, which include pseudo-capacitance, wettability, effective surface area, etc. for energy (supercapacitor electrodes, fuel cells, and batteries) and environmental (wastewater treatment) applications [75–78].

Instead of single heteroatom doping, codoping (two heteroatoms with B-d-CMs) is increasing the electrochemical performance of B-d-CMs because of synergistic effect. Among all heteroatoms, the nitrogen-containing groups (amide, pyrrolic, pyridinic, imide, and lactame) doped B-d-CMs show significance and excellent performance in CO_2 capture, catalyst, electrochemical storage, and environmental applications [79].

1.5.4.2 B-d-CMs Surface Loading of Metal Oxides or Hydroxides

The surface of bio-mass derived carbon (BM-DCs) is also modified by loading different transition metal oxides or hydroxides such as NiO, MnO_2, RuO_2, and

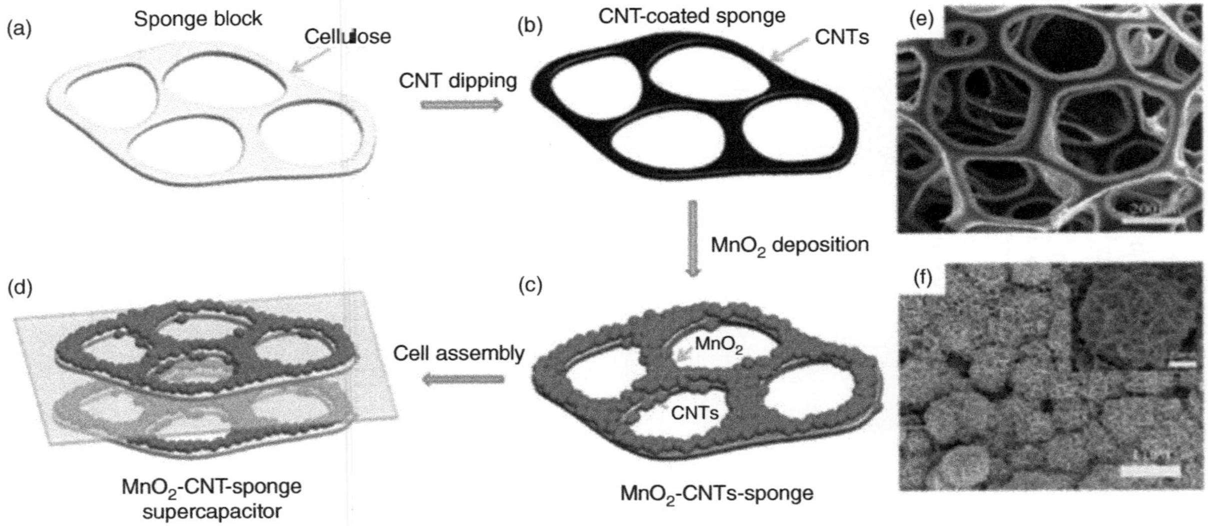

Figure 1.9 (a–d) The complete hybrid MnO_2/CNT sponge electrode fabrication (Source: From Ref. [83] / with permission of American Chemical Society) and (e and f) representation of the scanning electron microscope (SEM) and high resolution scanning electron microscope (HRSEM) images showing flower-like individual MnO_2 nanoparticles. Source: Ref. [83] / with permission of American Chemical Society.

Co_3O_4 [80, 81]. Due to loading of metal oxides into the BM-DCs, the agglomeration of metal particles is reduced, but the surface redox activity can be enhanced, which is useful to improve the electrochemical performance of B-d-CMs. Among all metal oxides, the NiO is a promising significant electrode material for energy applications because of its unique properties and availability.

1.5.4.3 Surface Incorporation with Different Nanostructures

The B-d-CMs are incorporated with different nanostructures, such as carbon nanotubes (CNTs) and graphene, which are important electrode materials for electrochemical storage applications. Deng et al. [82] investigated cellulose/carbon nanotube nanocomposite fibers followed by deacetylation and carbonization, and as-prepared active nanofibers were used as electrodes for supercapacitors. Chen et al. [83] studied that the MnO_2 nanoparticles are incorporated with CNT sponge hybrid electrode for supercapacitor applications. The complete diagrammatic representation of fabrication of hybrid MnO_2/CNT electrode is used for supercapacitor applications as shown in Figure 1.9.

1.6 Production Process Description

The abundant amount of AC, carbonaceous materials are prepared from various biomass wastes such as pecan shell, coconut shell, rice waste, bamboo cane, and peat [84–86] because these ACs are mainly used as odors, removal of colors, and metal ions. In general, the production cost of steam-AC is lower than that of acid-AC. The economic estimation of production of AC process depends upon the various raw materials is considered using the economic criteria namely return on investment, net present value, simply payback period, and internal rate of return. The profitable is attained in the stand-alone case indicated by economic indices and more attractive in the integrated plant. They found that cost estimation was sensitive to KOH cost, plant capacity, and product selling price.

In terms of product output and activation route, AC plant economies are mostly extremely sensitive. The economy is also significantly impacted by the price of the product, based on its quality (adsorption capacity) and not quantity (weight). The raw material costs and plant capability are also factoring that affect the costs of the raw materials. Calculations showed that pet coke complies with all these requirements and will be the best option for the manufacture of AC [87].

Large capacity manufacturing facilities are necessary to produce appealing economics for the other materials examined and this might offer challenges in small markets. The study might start with a plant capacity determined at the moment of breakeven. The conclusion is that the highest raw material yield should be chosen, a chemical activation scheme should be adopted, and the product price based on product-surfaced areas (or more generally on product adsorption capacity of adsorbate) be considered for attractive investment in AC production.

Weight-based cost calculation, large output of CB, which is reflected in substantially improved economic indices near pet coke and charcoals, makes chemical activation more appealing than physical activation. The utilization of wood and tires is the last to be evaluated economically owing to their poor yield. When expenses are referred to on a productive surface basis, the picture is again substantially changed. The most appealing raw materials are pet coke, followed by wood and lignite, as they have large surface areas and yield compared to black carbon, wood, and used tires, with lower surface areas despite strong yields.

It may be stated that the price of the carbon products on the basis of surface area, such as pet coke, charcoal, and lignite, is better compared to the economic indices. In contrast to others for which these characteristics have values that result in worse economic results, this material mix results in high yields with large surface area. Economic improvement in chemical activation is significantly more obvious than in physical activation because carbon development with extremely high surface areas uses zero expenses for start-up material to manufacture ACs [88, 89].

Here the various product yields and surface areas of the raw material used in physical and chemical activation are listed in Tables 1.2 and 1.3 [90].

Table 1.2 Raw materials used, product yield, and surface areas in physical activation (chemical activation).

Raw material	Pyrolysis yield, wt.%	Activation yield, wt.%	Total carbon yield, wt.%	BET surface area, m^2/g
Wood	25	50	12.5 (22)	800 (800)
Used tires	30	50	15 (20)	500 (700)
Pet coke	90	70	63 (45)	1000 (3000)
Carbon black	95	50	47.5 (60)	500 (500)
Charcoal	90	50	45 (44)	900 (2000)
Lignite	52	30	15.6 (25)	800 (2500)

Source: Ref. [90] / with permission of Elsevier.

Table 1.3 The raw materials and their production cost calculation based on heating value.

Raw material	Heating value, kcal/kg	Raw material cost, $/kg	Product price, $/kg (physical activation)	Product price, $/kg (chemical activation)
Wood	4541	0.808	1.54	1.54
Used tires	8760	1.558	0.96	1.34
Pet coke	7833	1.394	1.92	5.76
Carbon black	8202	1.459	0.96	0.96
Charcoal	6667	1.186	1.72	3.84
Lignite	4156	0.739	1.54	4.22
Diesel oil	10 250	1.26	—	—

Source: Ref. [90] / with permission of Elsevier.

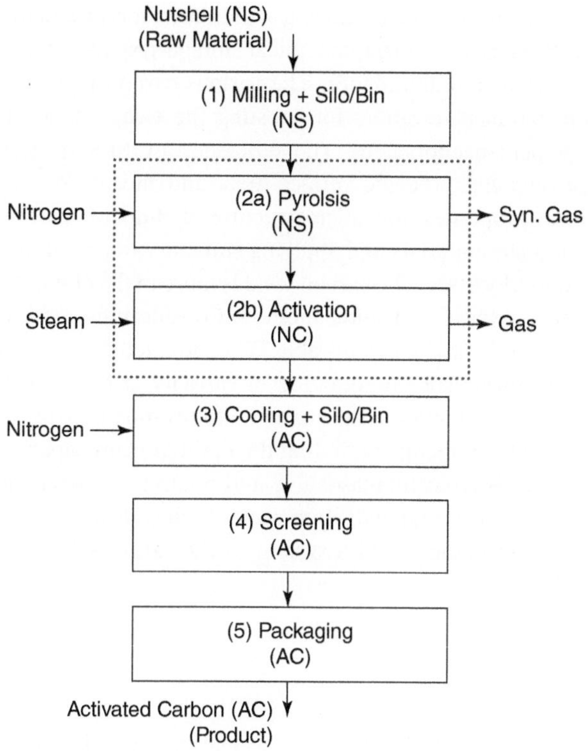

Figure 1.10 Proposed design equipment for the production of activated carbon from industrial waste nutshells. Source: From Ref. [91] / MDPI / CC BY 4.0.

1.7 Cost Analysis

Cost can be measured in a number of different ways, and each way of accounting for the cost of producing ACs from biomass in its own insights. The cost can be estimated for including equipment and its installation, financing cost, and maintenance cost for focusing particular applications such as batteries, supercapacitors, and fuel cells.

Leon et al. [91] developed the conceptual design of the industrial production plant of AC (6.6 ton/day) from waste nutshells (31.15 tons/day) by physical activation method in terms of different stages. The cost of the design included the cost of main equipment, nutshells price, basic operations, and services. Several authors designed different plants based on the yield of AC, low-cost production, and high surface area for EES devices.

The complete production process and the cost estimation of AC from industrial waste nutshells are shown in Figure 1.10.

1.8 Summary

Carbon materials are the most used material in the electrochemical storage devices such as fuel cells, lithium batteries, and especially supercapacitors.

Among others, the biomass is one of the promising candidates for producing carbonaceous materials, which is used for electrochemical storage system because of its microstructure, cheap price, and abundance. The various carbon materials are produced using different biomass resources for boosting the electrochemical performance of batteries, supercapacitors, etc. The efficiency of EES devices completely depends on porosity, high specific surface area, and morphology of B-d-CMs. The physiochemical properties and microstructure of B-d-CMs can be enhanced by selecting the biomass materials and applying suitable carbonization–activation–surface modification processes. The carbon yield is more in the chemical activation method than in the physical activation method. Considering practical applications of B-d-CMs in EES is still challengeable. The product description and cost analysis are also discussed. The computation of capacity at break-even points, i.e. at the capacity, where production costs are equal to revenue may be an alternate approach of handling the challenge of finding the optimal plant capacity. It may be observed that capacity is typically reasonable and relative size depends upon the criteria such as product yield and activation route. Finally, B-d-CMs are undoubtfully increasing the performance of EES system in the future similar to graphite in lithium-ion batteries.

References

1 Gao, Y.P., Wu, X., Huang, K.J. et al. (2017). Two-dimensional transition metal diseleniums for energy storage application: a review of recent developments. *CrystEngComm* 19: 404–418.

2 Liu, X., Zhang, J.Z., Huang, K.J., and Hao, P. (2016). Net-like molybdenum selenide–acetylene black supported on Ni foam for high-performance supercapacitor electrodes and hydrogen evolution reaction. *Chem. Eng. J.* 302: 437–445.

3 Zhao, Y.Q., Lu, M., Tao, P.Y. et al. (2016). Hierarchically porous and heteroatom doped carbon derived from tobacco rods for supercapacitors. *J. Power Sources* 307: 391–400.

4 Gao, Y.P., Zhai, Z.B., Huang, K.J., and Zhang, Y.Y. (2017). Energy storage applications of biomass-derived carbon materials: batteries and supercapacitors. *New J. Chem.* 41: 11456–11470.

5 Tang, W., Zhang, Y., Zhong, Y. et al. (2017). Natural biomass-derived carbons for electrochemical energy storage. *Mater. Res. Bull.* 88: 234–241.

6 Damen, L., Hassoun, J., Mastragostino, M., and Scrosati, B. (2010). Solid-state, rechargeable Li/$LiFePO_4$ polymer battery for electric vehicle application. *J. Power Sources* 195: 6902–6904.

7 Yahya, M.A., Baharuddin, M., Ngah, C.W.Z.C.W., and Hashim, M.A. (2015). Activated carbon electrode from desiccated coconut residue for high performance of supercapacitors. *J. Mater. Sci. Res.* 4: 45–48.

8 Chen, X.Y., Chen, C., Zhang, Z.J., and Xie, D.H. High performance porous carbon through hard-soft dual templates for supercapacitor electrodes. *J. Mater. Chem. A* 1: 7379–7383.

9 Wang, J., Zhang, X., Li, Z. et al. (2020). Recent progress of biomass-derived carbon materials for supercapacitors. *J. Power Sources* 451: 227794–227811.

10 Chu, S. and Majumdar, A. (2012). Opportunities and challenges for a sustainable energy future. *Nature* 488: 294–303.

11 Sivakami, A., Bagyalakshmi, S., Balamurugan, K.S., and Ikshan, N.I. Graphene-Carbon nanotubes nanocomposite modified electrochemical sensors for toxic chemicals, Graphene based electrochemical sensors for toxic chemicals. In: *Materials research foundations*, vol. 82 (ed. A. Pandikumar and P. Rameshkumar), 211–242.

12 Winter, M. and Brodd, R.J. What are batteries, fuel cells, and supercapacitors? *Chem. Rev.* 104: 4245–4270.

13 Goodenough, B. (2015). Energy storage materials: a perspective. *Energy Storage Mater.* 1: 158–161.

14 Wang, H., Yang, Y., and Guo, L. (2017). Renewable-biomolecule-based electrochemical energy-storage materials. *Adv. Energy Mater.* 7: 1700663.

15 Yu, Z., Tetard, L., Zhai, L., and Thomas, J. (2015). Supercapacitor electrode materials: nanostructures from 0 to 3 dimensions. *Energy Environ. Sci.* 8: 702–730.

16 Sevilla, M. and Mokaya, R. (2014). Energy storage applications of activated carbons: supercapacitors and hydrogen storage. *Energy Environ. Sci.* 7: 1250–1280.

17 Wang, Q., Yan, J., and Fan, Z. (2016). Carbon materials for high volumetric performance supercapacitors: design, progress, challenges and opportunities. *Energy Environ. Sci.* 9: 729–762.

18 Bai, Y., Shen, B., Zhang, S. et al. (2018). Storage of mechanical energy based on carbon nanotubes with high energy density and power density. *Adv. Mater.* 31 (9): 1800680.

19 Li, R., Zhou, Y., Li, W. et al. (2020). Structure engineering in biomass-derived carbon materials for electrochemical energy storage. *Research* 8685436–8685463.

20 Li, Y., Zhang, Q., Zhang, J. et al. (2015). A top-down approach for fabricating free-standing bio-carbon supercapacitor electrodes with a hierarchical structure. *Sci. Rep.* 5: 14155.

21 Seh, Z.W., Sun, Y., Zhang, Q., and Cui, Y. (2016). Designing high energy lithium–sulfur batteries. *Chem. Soc. Rev.* 45: 5605–5634.

22 Tian, J., Liu, Z., Li, Z. et al. (2017). Hierarchical S-doped porous carbon derived from by-product lignin for high-performance supercapacitors. *RSC Adv.* 7: 12089–12097.

23 Gao, Z., Zhang, Y., Song, N., and Li, X. (2017). Biomass-derived renewable carbon materials for electrochemical energy storage. *Mater. Res. Lett.* 5: 69–88.

24 De, S., Balu, A.M., Waal, J.C., and Luque, R. (2015). Biomass-derived porous carbon materials: synthesis and catalytic applications. *ChemCatChem* 7, 1608–1629.

25 Liu, Y., Chen, J., Cui, B. et al. (2018). Design and preparation of biomass-derived carbon materials for supercapacitors: a review. *C – J. Carbon Res.* 4 (4): 53.

26 Wang, R., Wang, P., Yan, X. et al. (2012). Promising porous carbon derived from celtuce leaves with outstanding supercapacitance and CO2 capture performance. *ACS Appl. Mater. Interfaces* 4: 5800–5806.

27 Chen, S., Tang, S., Sun, Y. et al. (2018). Preparation of a highly porous carbon material based on quinoa husk and its application for removal of dyes by adsorption. *Materials* 11: 1407.

28 Wang, J. and Kaskel, S. (2012). KOH activation of carbon-based materials for energy storage. *J. Mater. Chem.* 22: 23710–23725.

29 Madhu, R., Veeramani, V., and Chen, S. (2014). Heteroatom-enriched and renewable banana-stem-derived porous carbon for the electrochemical determination of nitrite in various water samples. *Sci. Rep.* 4: 4679.

30 Liu, W.J., Jiang, H., and Yu, H.Q. (2015). Development of biochar-based functional materials: toward a sustainable platform carbon material. *Chem. Rev.* 115: 12251–12285.

31 Sudhan, N., Subrmani, K., Karnan, M. et al. (2017). Biomass-derived activated porous carbon from rice straw for a high-energy symmetric supercapacitor in aqueous and non-aqueous electrolytes. *Energy Fuel* 31 (1): 977–985.

32 Chang, K.-L., Chen, C.-C., Lin, J.-H. et al. (2014). Rice straw-derived activated carbons for the removal of carbofuran from an aqueous solution. *New Carbon Mater.* 29: 47–54.

33 Bernard, A. (2020). Goodman, utilization of waste straw and husks from rice production: a review. *J. Bioresour. Bioproduct.* 5: 143–162.

34 Yan, Y., Shi, M., Wei, Y. et al. (2018). Process optimization for producing hierarchical porous bamboo-derived carbon materials with ultrahigh specific surface area for lithium-sulfur batteries. *J. Alloys Compd.* 738: 16–24.

35 Biswas, R., Uellendahl, H., and Ahring, B.K. (2015). Wet explosion: a universal and efficient pretreatment process for lignocellulosic biorefineries. *BioEnergy Res.* 8: 1101–1116.

36 Thomas, P., Lai, C.W., and Bin Johan, M.R. (2019). Recent developments in biomass-derived carbon as a potential sustainable material for supercapacitor-based energy storage and environmental applications. *J. Anal. Appl. Pyrolysis* 140: 54–85.

37 Morais, D.R., Rotta, E.M., Sargi, S.C. et al. (2017). Proximate composition, mineral contents and fatty acid composition of the different parts and dried peels of tropical fruits cultivated in Brazil, Artic. *J. Braz. Chem. Soc.* 28: 308–318.

38 Sanchez Orozco, R., Balderas Hernandez, P., Roa Morales, G. et al. (2014). Characterization of lignocellulosic fruit waste as an alternative feedstock for bioethanol production. *BioResources* 9: 1873–1885.

39 Arroyo, J., Farkaš, V., Sanz, A.B., and Cabib, E. (2016). Strengthening the fungal cell wall through chitin-glucan cross-links: effects on morphogenesis and cell integrity. *Cell. Microbiol.* 18: 1239–1250.

40 Sales-Campos, C., Araujo, L.M., de Minhoni, M.T., A., and de Andrade, M.C.N. (2011). Physiochemical analysis and centesimal composition of Pleurotus ostreatus mushroom grown in residues from the Amazon. *Cienc. Tecnol. Aliment.* 31: 456–461.

41 Moda, E.M., Horii, J., and Spoto, M.H.F. (2005). Edible mushroom Pleurotus sajor-caju production on washed and supplemented sugarcane bagasse. *Sci. Agric.* 62: 127–132.

42 Liu, S., Sun, J., Yu, L. et al. (2012). Extraction and characterization of chitin from the beetle holotrichia parallela motschulsky. *Molecules* 17: 4604–4611.

43 Ibitoye, E.B., Lokman, I.H., Hezmee, M.N.M. et al. (2018). Extraction and physicochemical characterization of chitin and chitosan isolated from house cricket. *Biomed. Mater.* 13: 025009.

44 Laith, A.A., Ambak, M., Abol-Munafi, A.B. et al. (2017). Metabolomic analysis of marine and mud crabs based on antibacterial activity. *Aquac. Rep.* 7: 7–15.

45 Shavandi, A., Silva, T.H., Bekhit, A.A., and Bekhit, A.E.-D.A. (2017). Keratin: dissolution, extraction and biomedical application. *Biomater. Sci.* 5: 1699–1735.

46 Peng, Z., Guo, Z., Chu, W., and Wei, M. (2016). Facile synthesis of high-surface-area activated carbon from coal for supercapacitors and high CO2 sorption. *RSC Adv.* 6: 42019–42028.

47 Radhakrishnan, L., Reboul, J., Furukawa, S. et al. (2011). Preparation of microporous carbon fibers through carbonization of al-based porous coordination polymer (Al-PCP) with furfuryl alcohol. *Chem. Mater.* 23: 1225–1231.

48 Chen, R., Li, L., Liu, Z. et al. (2017). Preparation and characterization of activated carbons from tobacco stem by chemical activation. *J. Air Waste Manage. Assoc.* 67: 713–724.

49 Zhu, H., Wang, X., Yang, F., and Yang, X. (2011). Promising carbons for supercapacitors derived from fungi. *Adv. Mater.* 23: 2745–2748.

50 Salinas-Torres, D., Ruiz-Rosas, R., Valero-Romero, M.J. et al. (2016). Asymmetric capacitors using lignin based hierarchical porous carbons. *J. Power Sources* 326: 641–651.

51 Nizamuddin, S., Baloch, H.A., Griffin, G.J. et al. (2017). An overview of effect of process parameters on hydrothermal carbonization of biomass. *Renewable and Sustainable Energy Reviews* 73: 1289–1299.

52 Biswal, M., Banerjee, A., Deo, M., and Ogale, S. (2013). From dead leaves to high energy density supercapacitors. *Energy Environ. Sci.* 6: 1249.

53 Raymundo-Pinero, E., Cadek, M., and Beguin, F. (2009). Tuning carbon materials for supercapacitors by direct pyrolysis of seaweeds. *Adv. Funct. Mater.* 19: 1032–1039.

54 Jeong, Y.W., Sang, K.C., Choi, Y.S., and Kim, S.J. (2014). Production of biocrude-oil from swine manure by fast pyrolysis and analysis of its characteristics. *Renew. Energy* 79: 14–19.

55 Jimena, C.-G., Alain, C., and Vanessa, F. (2020). Energy storage in supercapacitors: focus on tannin-derived carbon electrodes. *Front. Mater.* 7: 217. https://doi.org/10.3389/fmats.2020.00217.

56 Abioye, A.M. and Ani, F.N. (2015). Recent development in the production of activated carbon electrodes from agricultural waste biomass for supercapacitors: a review. *Renew. Sust. Energ. Rev.* 52: 1282–1293.

57 Zhang, T., Walawender, W.P., Fan, L.T. et al. (2004). Preparation of activated carbon from forest and agricultural residues through CO_2 activation. *Chem. Eng. J.* 105: 53–59.
58 Guo, S., Peng, J., Li, W. et al. (2009). Effects of CO_2 activation on porous structures of coconut shell-based activated carbons. *Appl. Surf. Sci.* 255: 8443–8449.
59 Taer, E., Deraman, M., Talib, I.A. et al. (2010). Physical, electrochemical and supercapacitive properties of activated carbon pellets from pre-carbonized rubber wood sawdust by CO_2 activation. *Curr. Appl. Phys.* 10: 1071–1075.
60 Okada, K., Yamamoto, N., Kameshima, Y., and Yasumori, A. (2003). Porous properties of activated carbons from waste newspaper prepared by chemical and physical activation. *J. Colloid Interface Sci.* 262: 179–193.
61 Pallarés, J., González-Cencerrado, A., and Arauzo, I. (2018). Production and characterization of activated carbon from barley straw by physical activation with carbon dioxide and steam. *Biomass Bioenergy* 115: 64–73.
62 Demiral, H., Demiral, İ., Karabacakoğlu, B., and Tümsek, F. (2011). Production of activated carbon from olive bagasse by physical activation. *Chem. Eng. Res. Des.* 89: 206–213.
63 Chang, C.F., Chang, C.Y., and Tsai, W.T. (2000). Effects of burn-off and activation temperature on preparation of activated carbon from corn cob agrowaste by CO_2 and steam. *J. Colloid Interface Sci.* 232: 45–49.
64 Dutta, S., Kim, J., Ide, Y. et al. (2017). 3D network of cellulose-based energy storage devices and related emerging applications. *Mater. Horiz.* 4: 522–545.
65 Lukatskaya, M.R., Dunn, B., and Gogotsi, Y. (2016). Multidimensional materials and device architectures for future hybrid energy storage. *Nat. Commun.* 7: 12647.
66 Zhu, Z. and Xu, Z. (2020). The rational design of biomass-derived carbon materials towards next-generation energy storage: a review. *Renew. Sust. Energ. Rev.* 134: 110308.
67 Peng, C., Xb, Y., Wang, R. et al. (2013). Promising activated carbons derived from waste tea-leaves and their application in high performance supercapacitors electrodes. *Electrochim. Acta* 87: 401–408.
68 Toso, J.P., Lopez, R.H., de Azevedo, D.C.S. et al. (2011). Evaluation of a mixed geometry model for the characterization of activated carbons. *Adsorption* 17: 551–560.
69 Sutrisno, B. and Hidayat, A. (2015). The effects of activation temperature on physico-chemical characteristics of activated carbons derived from biomass wastes. *AIP Conference Proceedings*, p. 060016. AIP Publishing LLC.
70 Rangabhashiyam, S. and Balasubramanian, P. (2019). The potential of lignocellulosic biomass precursors for biochar production: performance, mechanism and wastewater application – a review. *Ind. Crop. Prod.* 128: 405–423.
71 Li, Y., Ye, K., Cheng, K. et al. (2014). Anchoring CuO nanoparticles on nitrogen-doped reduced graphene oxide nanosheets as electrode material for supercapacitors. *J. Electroanal. Chem.* 727: 154–162.

72 Li, Y., Wang, G., Wei, T. et al. (2016). Nitrogen and sulfur co-doped porous carbon nanosheets derived from willow catkin for supercapacitors. *Nanomater Energy* 19: 165.

73 Cao, H., Peng, X., Zhao, M. et al. (2018). Oxygen functional groups improve the energy storage performances of graphene electrochemical supercapacitors. *RSC Adv.* 8: 2858–2865.

74 Wei, T., Wei, X., Gao, Y., and Li, H. (2015). Large scale production of biomass-derived nitrogendoped porous carbon materials for supercapacitors. *Electrochim. Acta* 169: 186–194.

75 Zheng, L., Wang, Y., Wang, X. et al. (2010). The effects of surface modification on the supercapacitive behaviors of carbon derived from calcium carbide. *J. Mater. Sci.* 45: 6030–6037.

76 Qu, D. (2002). Studies of the activated carbons used in double-layer supercapacitors. *J. Power Sources* 109: 403–411.

77 Zhang, S., Tian, K., Cheng, B.H., and Jiang, H. (2017). Preparation of N-doped supercapacitor materials by integrated salt templating and silicon hard templating by pyrolysis of biomass wastes. *ACS Sustain. Chem. Eng.* 5: 6682–6691.

78 Trung, N.B., Tam, T.V., Dang, D.K. et al. (2015). Facile synthesis of three-dimensional graphene/nickel oxide nanoparticles composites for high performance supercapacitor electrodes. *Chem. Eng. J.* 264: 603–609.

79 Wu, S., Hui, K.S., Hui, K.N., and Kim, K.H. (2016). Ultrathin porous NiO nanoflake arrays on nickel foam as an advanced electrode for high performance asymmetric supercapacitors. *J. Mater. Chem.* 4: 9113–9123.

80 Wang, Y., Zhang, Y., Pei, L. et al. (2017). Converting Ni-loaded biochars into supercapacitors: implication on the reuse of exhausted carbonaceous sorbents. *Sci. Rep.* 7: 41523.

81 Wang, Y., Song, Y., and Xia, Y. (2016). Electrochemical capacitors: mechanism, materials, systems, characterization and applications. *Chem. Soc. Rev.* 45: 5925–5950.

82 Deng, L., Young, R.J., Kinloch, I.A. et al. (2013). Supercapacitance from cellulose and carbon nanotube nanocomposite fibers. *ACS Appl. Mater. Interfaces* 5: 9983–9990.

83 Chen, W., Rakhi, R.B., Hu, L. et al. (2011). High-performance nanostructured supercapacitors on a sponge. *Nano Lett.* 11: 5165–5172.

84 Njoku, V., Islam, M., Asif, M., and Hameed, B. (2014). Preparation of mesoporous activated carbon from coconut frond for the adsorption of carbofuran insecticide. *J. Anal. Appl. Pyrolysis* 110: 172–180.

85 Alvarez, J., Lopez, G., Amutio, M. et al. (2014). Upgrading the rice husk char obtained by flash pyrolysis for the production of amorphous silica and high quality activated carbon. *Bioresour. Technol.* 170: 132–137.

86 Wu, M., Guo, Q., and Fu, G. (2013). Preparation and characteristics of medicinal activated carbon powders by CO_2activation of peanut shells. *Powder Technol.* 247: 188–196.

87 Ng, C., Marshall, W.E., Rao, R.M. et al. (2003). Activated carbon from pecan shell: process description and economic analysis. *Ind. Crop. Prod.* 17: 209–217.

88 Ko, D.C.K., Mui, E.L.K., Lau, K.S.T., and McKay, G. (2004). Production of activated carbons from waste tire-process design and economic analysis. *Waste Manag.* 24 (9): 875–888.
89 Nazem, M., Zare, M., and Shirazian, S. (2020). Preparation and optimization of activated nano-carbon production using physical activation by water steam from agricultural wastes. *RSC Adv.* 10: 1463–1475.
90 Stavropoulos, G.G. and Zabaniotou, A.A. (2009). Minimizing activated carbons production cost. *Fuel Process. Technol.* 90: 952–957.
91 Leon, M., Silva, J., Carrasco, S., and Barrientos, N. (2020). Design, cost estimation and sensitivity analysis for a production process of activated carbon from waste nutshells by physical activation. *Processes* 8: 945.

2

Introduction to Biowaste-Derived Materials

Synthesis, Characterization, and Its Properties

Thangavelu Kokulnathan[1], Balasubramanian Sriram[2], Sabarison Pandiyarajan[3,4], Subramanian Ramanathan[5], and Thangavelu Sakthi Priya[3]

[1] *Department of Electro-Optical Engineering, National Taipei University of Technology, Taipei, Taiwan*
[2] *Department of Materials and Mineral Resources Engineering, National Taipei University of Technology, Taipei, Taiwan*
[3] *Department of Chemical Engineering and Biotechnology, National Taipei University of Technology, Taipei, Taiwan*
[4] *Department of Mechanical Engineering, National Taipei University of Technology, Taipei, Taiwan*
[5] *Environmental and Energy Engineering, Kyungpook National University, Daegu, South Korea*

2.1 Introduction

The worldwide waste level is going on increasing as wastes are dumped into the environment as solid waste. There is environmental risk in the management of solid waste [1]. Most researchers can use renewable solid waste to conduct their research more efficiently. Renewable waste, such as biowaste-derived activated carbon (BW-AC), can be used as a starting material because it lowers the cost of raw materials while also lowering the risk associated with solid waste management. The importance of research into innovative production techniques and the usage of low-cost materials cannot be overstated. All-inexpensive residues with high carbon and low inorganic content can be employed as starting materials for the synthesis of BW-AC, one of the most well-known and widely used adsorbents according to Muñoz-Guillena et al. [2] and Arami-Niya et al. [3]. BWs such as carbon residues, including cannonball tree flower [4], rice husks (RH) [5], bean husks [6], olive stones [7], date pits [8], coconut shell [9], coffee bean husks [10], cassava peel [11], bagasse [12], grain sorghum [13], cotton stalk [14], rubberwood sawdust [15], jute fiber [16], chicken feather [17], fruit stones [18], nutshells [19], kola-nut testa [20], almond green hull powder [21], bovine bone [22], and golden shower fruit [23], have been reported in the literature as carbon materials. The preparation of AC from the biowaste involves two primary processes (Figure 2.1) [24–27]. (i) The procedure begins with the pre-carbonization of biowaste in air, carbon dioxide (CO_2), or steam at a temperature below 800 °C [28, 29]. (ii) The second stage is the chemical activation approach, and the precursor is treated with activating

Biomass-Derived Carbon Materials: Production and Applications, First Edition.
Edited by Alagarsamy Pandikumar, Perumal Rameshkumar, and Pitchaimani Veerakumar.

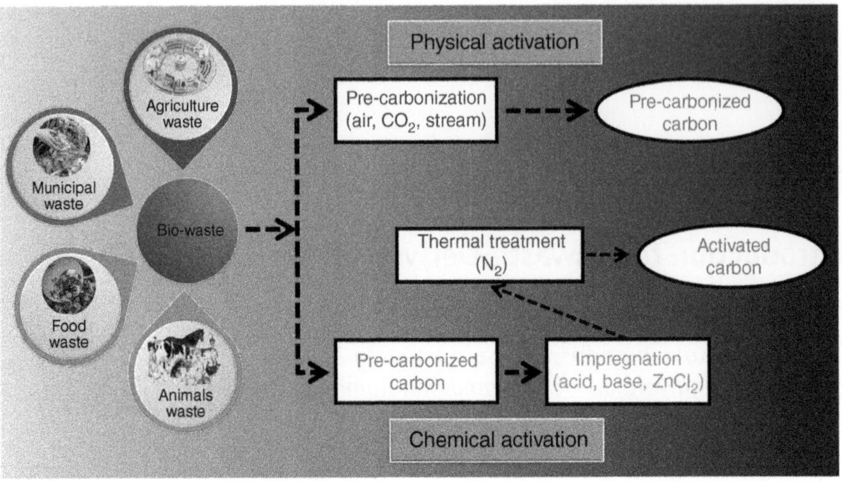

Figure 2.1 Preparation processes in physical and chemical methods of activated carbon synthesis. Sources: Pixavril / Adobe Stock; Eric Isselée / Adobe Stock.

agents to activate the first hydration, which leads to pyrolytic decomposition and tar materialization, with higher carbon yield [30, 31]. As a result, diverse porosity characteristics, such as micro, meso, and macrospores, are correlated to BW-AC, and their uniform distribution is essential to effective charge transfer. Furthermore, these activation processes are then used to improve the high surface area, an interior porous structure (composed of pores of varying sizes), and a wide range of oxygenated functional groups of the AC [32, 33].

2.2 Synthesis

2.2.1 Activation Mechanism of BW-AC by Physical Activation

The most notable advantages of physical activation are pristine, green engineering and elimination of secondary waste management. The physical activation of BW-AC involves two types of thermolysis processes [34]. Initially, the combustion on the carbonaceous matter must be initiated after the pyrolysis or carbonization of the carbonaceous precursor. Generally, it is carried out at different temperatures up to 1200 °C in the presence of air, CO_2, and steam. Here, heteroatoms are eliminated and volatiles are released in this process, resulting in chars with high carbon content [35]. As the carbonization temperature rises, the carbon content rises as well, but porosity development remains minimal. Gasification is the second stage in the physical activation process. Through the second stage, it involves in selective removal of most reactive carbon atoms by controlled gasification processes, which result in the BW-AC increased porosity characteristics. As shown in the Eqs. (2.1) and (2.2), this gasification process is the endothermic reaction as it can remove carbon atoms from the network within solid carbon areas [36, 37].

$$C_{(s)} + CO_{2(g)} \rightarrow 2CO_{(g)}, \quad \Delta H = +159\ \mathrm{kJ\ mol^{-1}} \tag{2.1}$$

$$C_{(s)} + H_2O_{(g)} \rightarrow CO_{(g)} + H_{2(g)}, \quad \Delta H = +117\ \mathrm{kJ\ mol^{-1}} \tag{2.2}$$

The temperature during the gasification is influenced by the gasification agent used, which is generally water vapor, CO_2, or O_2 (air). With water vapor or CO_2 gasification, a temperature of about 700–900 °C is used. CO_2 is generally preferred because of its lower reactivity at high temperatures, which makes the activation process easier to control. Besides, the gasification temperatures should be much lower (around 300–450 °C) when using pure O_2 or air. Moreover, because of the higher reactivity and the exothermicity of the reaction with O_2, the use of O_2 makes it more difficult to control gasification and thus porosity development [35]. Therefore, depending on the precursor material and process conditions, both CO_2 and steam can be suitable gasifying agents [38].

2.2.2 Activation Mechanism of BW-ACs by Chemical Activation

During the chemical activation process, numerous chemical agents such as potassium hydroxide (KOH), sodium hydroxide (NaOH), sodium carbonate (Na_2CO_3), potassium carbonate (K_2CO_3), phosphoric acid (H_3PO_4), zinc chloride ($ZnCl_2$), ferric chloride ($FeCl_3$), and even more agents have been used as activating agents [39, 40]. Activating agents are classified into four types based on the acid–basic theory and activation mechanism: alkaline, acidic, neutral, and self-activating agents (Figure 2.2) [24]. The synergistic impact of pore generation, pore expansion, pore combination, and pore compression lead to increasing the porous structure of

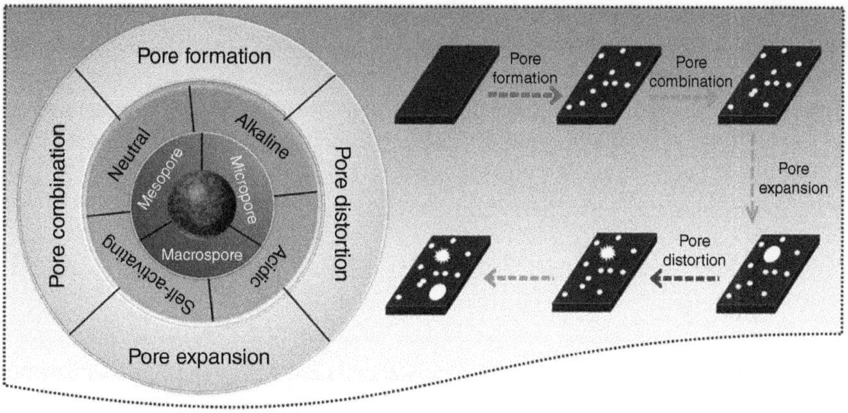

Figure 2.2 Schematic illustration of activation mechanism by acidic, alkaline, neutral, and self-activating agents with formation pathway of the porous structure by chemical activation.

BW-AC. The dehydrating effect of the active agents used prevents the development of tar, thus enhancing the productivity of porous carbon. Moreover, the objective of activation is to increase the adsorptive capacity of the carbon material by creating and developing (volume and size) porosity. Besides, chemical activation produces higher carbon yields than physical activation [41]. Chemical activation involves impregnating BW-ACs with an activating reagent as well as heat treating the impregnated material in an inert atmosphere. The chemical activation is progressing at the same time as the carbonization and activation steps. Based on the weight of BW-AC, the yield was determined and the chemical recovery was estimated using Eq. (2.3):

$$\text{Chemical recovery} = \frac{\begin{pmatrix}\text{weight of the sample}\\ \text{before washing}\end{pmatrix} - \begin{pmatrix}\text{weight of the sample}\\ \text{after washing}\end{pmatrix}}{\text{weight of impregnated chemical}} \tag{2.3}$$

2.2.2.1 Influence of Alkaline Activating Agents

Alkaline hydroxides can be utilized to produce BW-ACs with vastly high Brunauer–Emmett–Teller (BET)-specific surface areas without environmental issues. Three types of alkaline activating agents are commonly used such as strong, medium, and weak alkaline activating agents. The strong alkaline activating agents, such as KOH [4] and NaOH [12], are the first. The middle alkaline activating agents, such as potassium carbonate (K_2CO_3) [19] and sodium carbonate (Na_2CO_3) [42], represent the second category. The weak alkaline activating agents, often known as strong alkali and weak acid salts, include potassium phosphate (K_3PO_4) [43], potassium oxalate ($K_2C_2O_4$) [44], potassium acetate (CH_3COOK) [45], sodium silicate (Na_2SiO_3) [46], sodium aluminate ($Na_2Al_2O_4$) [47], and others. Among these alkaline activating agents, a higher potential to produce a large number of micropores, and a high specific surface area ($S_{BET} = 3000\,m^2\,g^{-1}$) at the '70s [35], KOH has been widely used for the synthesis of "super ACs." It has long been considered to reduce the formation of tar, reduce the temperature of the activating reaction, higher rate of elimination of non-carbon components, and increase the pyrolysis reaction rate [14–16]. Further, the KOH activation mechanism is followed by the following steps (Eqs. (2.4)–(2.8)) [35, 48]:

$$2\text{KOH}\Delta \rightarrow K_2O + H_2O\ \text{(dehydration)} \tag{2.4}$$

$$C + H_2O\Delta \rightarrow H_2 + CO\ \text{(water gas reaction)} \tag{2.5}$$

$$H_2O + C\Delta \rightarrow H_2 + CO_2\ \text{(watergas shift reaction)} \tag{2.6}$$

$$K_2O + H_2\Delta \rightarrow 2K + H_2O\ \text{(reduction by } H_2\text{)} \tag{2.7}$$

$$K_2O + C\Delta \rightarrow 2K + CO\ \text{(reduction by C)} \tag{2.8}$$

The involvement of metallic potassium vapor is distinct and is essential for the initiation of pores in BW-ACs; this superior role does not exist in the pyrolysis by acidic or neutral activating agents. For this reason, the specific surface areas of AC activated by strong alkalis are much larger than those activated by acidic or neutral activating agents.

2.2.2.2 Influence of Acidic Activating Agents

Acidic activating agents include H_3PO_4, phosphorous acid (H_3PO_3), metaphosphoric acid (HPO_3), pyrophosphoric acid ($H_4P_2O_7$), tetraphosphoric acid ($H_6P_4O_{13}$), nitric acid (HNO_3), sulfuric acid (H_2SO_4), phytic acid ($C_6H_{18}O_{24}P_6$), hydrogen peroxide (H_2O_2), ammonium phosphate ($[(NH_4)_3PO_4]_n$), isopropyl dihydrogen phosphate ($C_3H_9O_4P$), triisobutyl phosphate ($C_{12}H_{27}O_4P$), isopropylated triaryl phosphate ($C_{18}H_{15}R_3O_4P$), diisopropyl phosphate ($C_6H_{15}O_4P$), etc. For the synthesis of AC in an acidic medium, H_3PO_4 is the most widely utilized acidic medium agent. After heating, the phosphate esters dissociate into H_3PO_4 and smaller P molecules, i.e. phytic acid decomposes into H_3PO_4 and $C_6H_{12}O_6$ [49]. As a result, the activation role of H_3PO_4 can be attributed to the pore-forming mechanism of phosphate esters, as shown in the following seven factors (Eqs. (2.9)–(2.14)): (i) dehydration, (ii) elimination, (iii) swelling role, (iv) accelerating carbonization, (v) framework role, (vi) oxidation, and (vii) aromatic condensation [24]. Örkün et al. investigated the effect of H_3PO_4 concentration modifications on various precursors and determined that it had been effective in enhancing the formation of pores and optimizing the morphology, surface chemistry, and size distribution [50].

$$nH_3PO_4 \rightarrow H_{n+2}P_nO_{3n+1} + (n-1)H_2O \tag{2.9}$$

$$2H_3PO_4 \rightarrow P_2O_5 + 3H_2O \tag{2.10}$$

$$2H_3PO_4 \rightarrow H_2P_2O_7^{2-} + H_2O + 2H^+ \tag{2.11}$$

$$2H_2P_2O_7^{2-} \rightarrow P_4 + 6O_2 + 2H_2O \tag{2.12}$$

$$2P_2O_5 + 5C \rightarrow P_4 + 5CO_2 \tag{2.13}$$

$$P_2O_5(\text{liquid}) \rightarrow P_2O_5(\text{gas}) \tag{2.14}$$

2.2.2.3 Influence of Neutral Activating Agents

Neutral activating agents include $ZnCl_2$, $FeCl_3$, calcium chloride ($CaCl_2$), magnesium chloride ($MgCl_2$), potassium chloride (KCl), ammonium chloride (NH_4Cl), potassium sulfate (K_2SO_4), potassium permanganate ($KMnO_4$), potassium bitartrate ($KHC_4H_4O_6$), sodium amide ($NaNH_2$), monosodium phosphate (NaH_2PO_4), monopotassium phosphate (KH_2PO_4), etc. The key pore-forming strategies of neutral activating agents include oxidation, templating, and gasification. At their oxidation states, positive ions (e.g. Zn^{2+}, Fe^{3+}, Mn^{7+}, Mg^{2+}, Na^+, etc.) could react with carbon atoms via reduction processes, removing sections of the carbon atoms and leaving channels behind. Meanwhile, the construction of carbon-bonded compounds of differing sizes and morphologies might occupy space in the carbon matrix, which can be eliminated during acid washing. According to Qiu et al., the pores were produced due to a synergistic impact of a chemical etching by K_2MnO_4,

MnO_2, K_2CO_3, MnO, etc., and physical activation by low-molecular gases such as CO and CO_2 [51], i.e. the $KMnO_4$ activation mechanism is followed by the following six steps (Eqs. (2.15)—(2.20)):

$$2KMnO_4 \rightarrow K_2MnO_4 + MnO_2 + O_2 \quad (2.15)$$

$$K_2MnO_4 + C \rightarrow MnO + K_2CO_3 \quad (2.16)$$

$$2MnO_2 + C \rightarrow 2MnO + CO_2 \quad (2.17)$$

$$K_2CO_3 + 2C \rightarrow 2K_2O + 3CO \quad (2.18)$$

$$CO_2 + C \rightarrow 2CO \quad (2.19)$$

$$K_2O + C \rightarrow 2K + CO \quad (2.20)$$

2.2.2.4 Influence of Self-Activating Agents

Although the activating agents mentioned earlier have been widely used in industry, they do have some limitations, such as equipment corrosion, consumption of washing water, and inevitable secondary pollution [52]. Another approach that has been widely mentioned is the direct pyrolysis of BW without activation. This approach does not require additional activating chemicals and requires less energy. However, the performance of BW-ACs constructed by using the approach is highly dependent on the characteristics of BW carbon precursors, which limits the application applicability and portability [53]. Besides, the aforementioned method used for activation is "self-activation," which uses gases released during the pyrolysis process or non-additional metals in BW to activate the carbon source, and has gained significant attention. There are three major types of self-activating agents, including (i) organic acid salts, (ii) certain natural components in some carbon precursors, and (iii) in situ pyrolysis gases. Activating the device using a self-activating chemical could simplify the method, reduce expense, minimize the chances of environmental consequences, and allow for even more scalable production [23, 54].

2.3 Characterization

Characterization is an important process of measuring and determining the properties of BW-AC material. Various types of characterization methods such as scanning electron microscope (SEM), high-resolution transmission electron microscopes (HR-TEM), Fourier-transform infrared spectroscopy (FTIR), Raman spectroscopy, X-ray photoelectron spectroscopy (XPS), X-ray diffraction (XRD), and BET are used to analyze the BW-AC.

2.3.1 Electron Microscopes

Electron microscopes (EMs) like SEM and TEM are commonly used to study the topography of AC-based materials. EMs act basically like a light microscope but instead of using visible light, it uses very high accelerated electrons as a source of illumination. Similarly, the optical light microscope is limited by its magnifications and resolution, related to the wavelength of the visible light photons, but the wavelength of accelerated electrons is very short and produces a very high resolution in FE-SEM and HR-TEM. Likewise, a light microscope exhibits only a small part of the overall sample, but electron microscopy images provide valuable evidence about size, shape, composition, and so on. The conversion of BW precursors into their derived AC is carried out through two-step procedure [59]. Pre-carbonization of BW precursors is done in the first step using one of the two techniques: pyrolysis or hydrothermal pre-carbonization [60]. In the second step, the activation of the BW-derived carbonaceous materials is through the physical and chemical methods to improve their specific surface area and porosity [61]. As a result, diverse porosity characteristics, such as micro, meso, and macropores, are connected with AC, and their uniform distribution is essential for improving physicochemical properties [25]. Various types of morphologies including 0D spherical structure, 1D fibrous structure, 2D lamellar structure, and 3D skeleton structure are generated through a different synthesis [62].

Nguyen et al. [55] reported that 0D structures like AC spheres (ACSs) were obtained from raw sawdust, as shown in detail in Figure 2.3a. The obtained ACSs

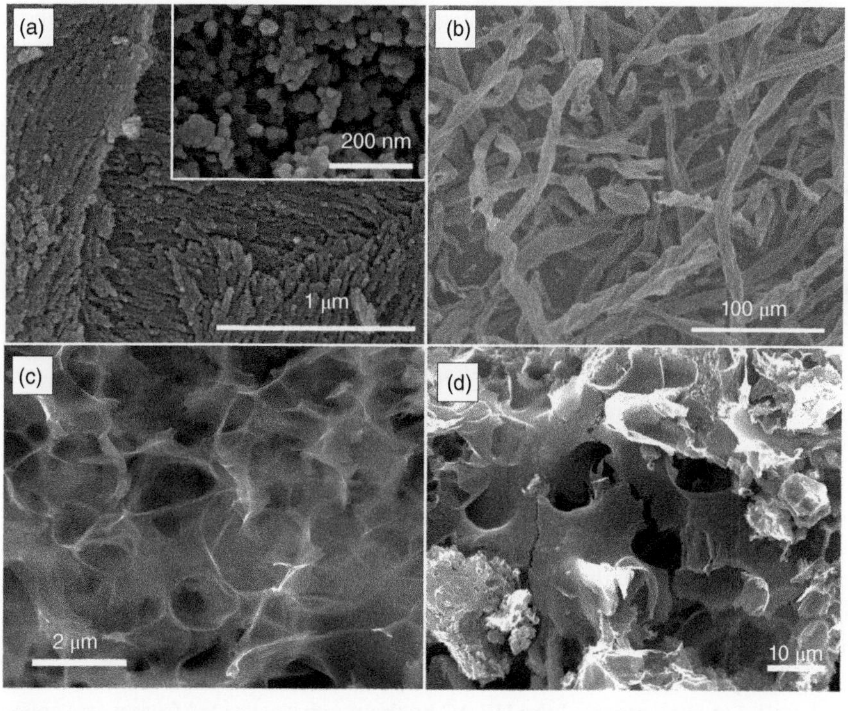

Figure 2.3 SEM images of (a) 0D–ACSs (Source: Ref. [55] / with permission of Elsevier), (b) 1D–ACFs (Source: Ref. [56] / with permission of American Chemical Society), (c) 2D–ACNSs (Source: Ref. [57] / with permission of American Chemical Society), and (d) 3D–ACNSs (Source: Ref. [58] / with permission of Elsevier).

show particle size in the range of 10–20 nm. Similarly, Wang et al. [63] reported that the hierarchical porous (HP)-ACSs were synthesized by a simple method, exhibiting smooth surface and sphere-like morphology with an average size of about 1.5–2 μm. Cheng et al. [56] reported a 1D-activated helical and tubular carbon fiber (ACFs) activated carbon fiber like morphology from cotton by KOH activation. Figure 2.3b shows that the ACFs have a wall thickness ranging from 1.5 to 2.0 μm and a tubular diameter varying from 5.0 to 7.0 μm. This type of structure is more favorable for rapid ion diffusion and electron transfer. Wang et al. [57] have reported highly interconnected 2D-activated carbon nanosheets (ACNS)-like structures with the thickness of 10–30 nm fabricated from hemp bast fiber (Figure 2.3c). It entails hydrothermal pre-carbonization of hemp bast fiber, which yields hydrochar, which was then activated with KOH at temperatures of 700, 750, and 800 °C in an inert atmosphere. The NS-like structure is achieved by activating pre-carbonized hemp bast fiber with KOH. Because of the hierarchical nature of hemp fibers, KOH activation results in a high degree of graphitization in this study. Thirumal et al. converted tamarind fruit shells (TFSs) into 3D-porous ACNSs for its application in supercapacitor applications [64]. The TFSs were mixed with KOH in the ratio of 1 : 5 in deionized (DI) water. Then, the solution of TFSs-KOH samples was evaporated at 100 °C. The advantage of using KOH is that it acts as a carbonization catalyst, and it promotes P-ACNSs like morphology. In another study, Selvaraj et al. prepared 3-ACNSs like morphology (Figure 2.3d) by template-free pyrolysis method from *Prosopis juliflora* wood waste carbon blocks (WCB) [58]. In their approach, they have pre-carbonized material at 300 °C for three hours. After that, the pre-carbonized sample was activated with KOH with a 3 : 1 wt% ratio mixture and then was carbonized at different temperatures at 700, 800, and 900 °C for two hours under argon gas atmosphere.

2.3.2 HR-TEM Analysis

HR-TEM is a powerful spectroscopic technique that provides information at an atomic-scale level and generates a higher resolution image. The HR-TEM images of an RH prepared AC were collected by Vu et al. [65]. The HR-TEM of this as-prepared AC showed HP-AC-*like* structure and confirms its amorphous character, which contains a large number of mesopores and micropores as shown in Figure 2.4a and b. The significant number of white spots between the disordered carbon layers shows that the AC from RH contains a lot of micropores and mesopores. Luo and coworkers synthesized AC derived from forestry fir bark (FB) waste and followed by KOH activation [66]. The as-prepared AC exhibited a great number of pores in the micrometer and nanometer size (Figure 2.4c), which could provide a large surface area. This observation is well-matched with the literature of Song et al. [27]. Liu et al. reported the HP-AC by the pyrolysis derived from waste coffee grounds (WCGs) [67]. The HP-AC exhibited fragmentized and thinner-*like* NSs, indicating the formation of abundant micropores (Figure 2.4d–f). These findings showed that the $FeCl_3$ catalyst could produce a mesopores-dominant structure

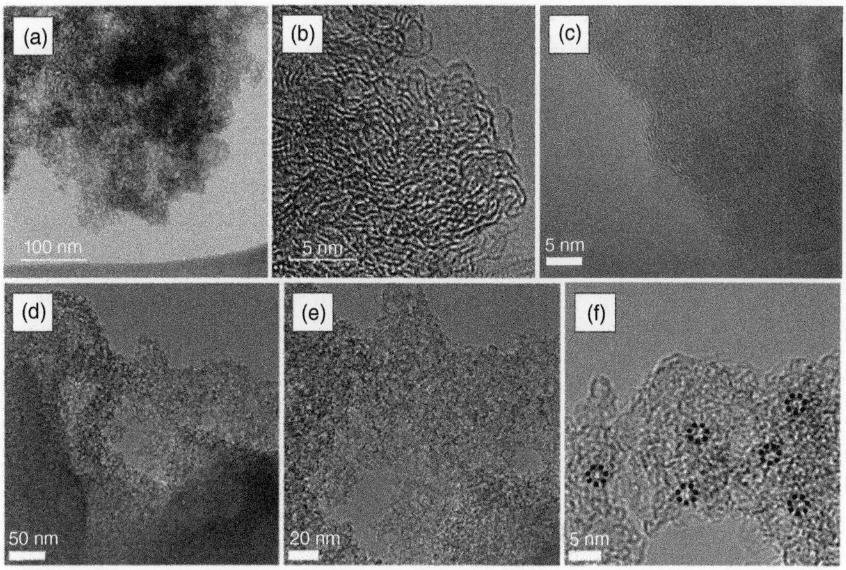

Figure 2.4 HR-TEM images (a and b) HP-AC (Source: Ref. [65] / With permission of Royal Society of Chemistry), (c) AC-FB (Source: Ref. [66] / The Royal Society /CC BY 4.0.), and (d–f) HP-AC (Source: Ref. [67] / Springer Nature/CC BY 4.0).

and that further KOH activation added even more micropores to build an HP-AC morphology.

2.3.3 FTIR Spectroscopy

FTIR spectroscopy is one of the most powerful qualitative techniques for the evaluation of the chemical structure of carbon materials. It has several advantages than the typical infrared spectroscopy, such as easy to interpret the entire interactions of different types of surface functional groups, improved intrinsic signal-to-noise ratio, enhanced resolution, higher precision level of the frequency scale, and offering quantitative information of individual functionalities present on the carbon surface. The surface-functional groups formed or consumed during treatment can be identified by comparing the FTIR spectra of the sample pre- and post-treatment. The recorded spectrum is a plot of the measured IR intensity (units in transmittance) contrasted with wave number of light (units in an inverse of wavelength) [69–73]. The possible functional groups present on carbon surface are listed in Table 2.1 along with FTIR wave numbers. Princy et al. [68] prepared walnut shell derived mesoporous AC (Wn) at different temperatures (600, 700, 800 °C, respectively), and its FTIR analysis results can be seen in Figure 2.5. Based on the outcomes, the authors confirmed the existence of functional groups such as carbonyl C—O stretching, C=O stretching of carboxylic acid, C=C stretching, O—H stretching mode at 1047, 1605, 1413, and 3405 cm^{-1}, respectively. Remarkably, the peak intensities are decreased with increasing activation temperature indicating the decomposition of C—H bonds

Table 2.1 Possible FTIR peak assignments of functional groups on carbon surfaces.

Functional group	Assignment of wave number (cm^{-1})	References
Carboxylic acids		
C=O/stretching	1800–1600, 1750–1720, 1717, 1712, 1710	[70, 72, 74–76]
C—H/stretching	3000–2600, 2924	[77, 78]
O—H/stretching	3530, 3500	[79, 80]
Quinones	1680–1550, 1620–1580, 1650, 1645, 1635	[70, 72, 74, 77]
Phenolic groups		
C=O (lactones)/stretching	1760, 1750, 1720, 1710	[74, 81–83]
C=O (ketones)/stretching	1707, 1705, 1700, 1570, 1560	[70, 74, 83, 84]
C—O (ethers)/stretching	1400–1100, 1300–1000, 1400–1000	[74, 77]
C—OH/stretching	1400–1000, 1220–1000, 1300–1200	[74, 77]
O—H/stretching	3620–2500, 3605, 3393	[70]
C=N	1600, 1570	[69, 71, 85]
Others		
Pyridine-like groups	1610–1480, 1570–1512	[70]
C—N	1190, 1250, 1250	[69–71]
Cyclic amides	1676, 1461–1685, 1670	[71, 73, 83]
N—H	1480, 1560	[69, 73]
Nitro groups	1330–1530, 1574	[86, 87]

to form more stable aromatic C=C bonds [88]. Besides, the oxygen-containing functional groups could enhance the reversible adsorption/desorption of working ions in the electrochemistry, which is beneficial for improving the total specific capacitance in energy storage applications. However, excess of an oxygen-containing functional group will lead to compromise the electronic conductivity [89].

2.3.4 Raman Spectroscopy

Raman spectroscopy is a non-destructive standard technique to characterize the adequate structural information of crystalline and amorphous carbons. In this sense, two fundamental peaks usually observed in the region around ~1300–1350 and ~1580–1600 cm^{-1} are recognized as the D-band (defects and disorder) and G-band (graphite band) owing to C—C graphitic lattice vibration mode, respectively [90–92]. The G-band is arising due to the E_{2g} in-plane symmetry stretching vibration of sp^2 carbon atoms (Figure 2.6a), which is Raman active. Moreover, the G-band not only occurs in the sixfold rings but also occurs in all sp^2 carbons. The D-band is arising due to the A_{1g} mode (Figure 2.6b), which is comparable to the in-plane breathing vibration mode of sp^2 atoms in rings [91, 92]. This mode is forbidden

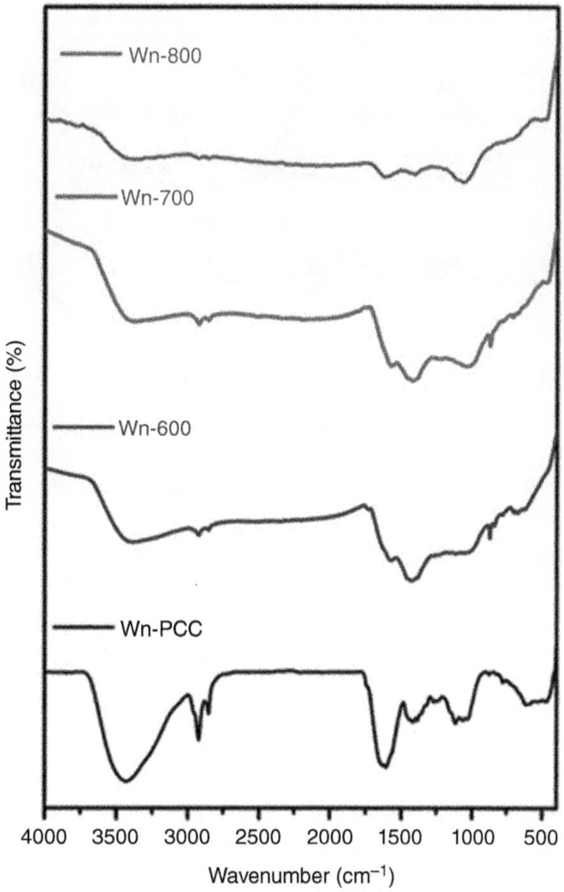

Figure 2.5 FTIR spectrum of AC prepared at different temperatures. Source: Ref. [68] / with permission of Elsevier.

to large solitary grains like highly ordered pyrolytic graphite and results from a relaxation of the symmetry rule for Raman scattering due to the prevalence of disorders like crystal edges. The position, ratio, and intensities of G- and D-bands provide the information of orientation and grain or domain size. The D-band is dispersive, and the intensity is varying with the excitation energy of the photon, whereas the G peak is not dispersive and its intensity is strictly associated with the presence of aromatic carbons. Those bands are fitted to Gaussian, symmetric Lorentzian, and asymmetric Breit–Wigner Fano functions, respectively. The ratio of intensity between D-band (I_D) and G-band (I_G) is interconnected to the in-plane length (L_a) of the graphene-like layers, lattice defect, and degree of structural disorder in carbonaceous materials as is described by Ferrari and Robertson (F–R) model and Tuinstra–Koenig (T–K) formula [92] (Eq. (2.21))

$$I_D/I_G = C(\lambda)/L_a \tag{2.21}$$

where C (515.5 nm) ~44 Å. Besides, the 2D-band is also the main feature in the Raman spectra that is related to two phonons near the K point. This band appears at ~2700 cm^{-1} and occurs in all sp^2 hybridized carbon materials accompanied with the

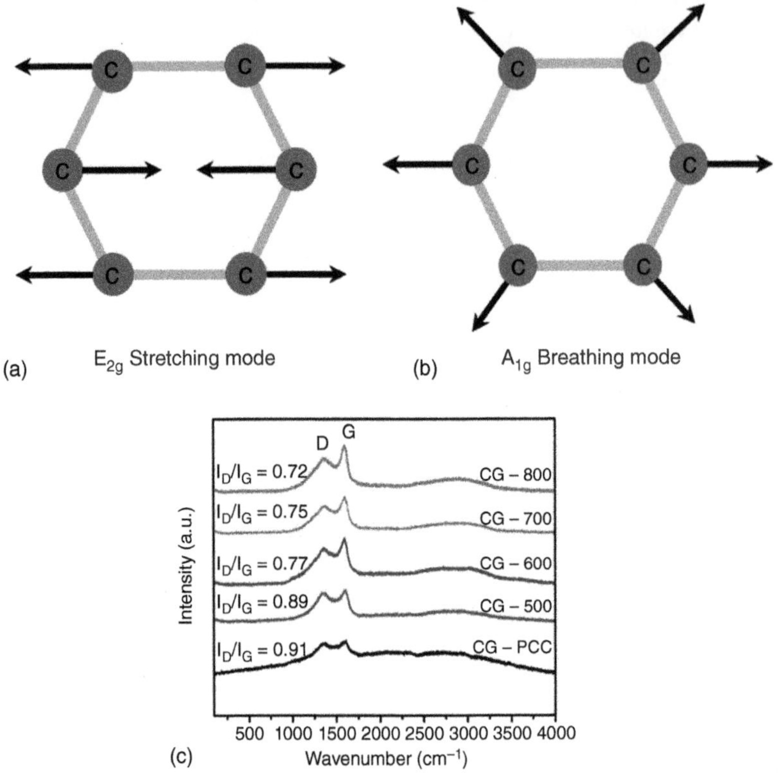

Figure 2.6 Carbon vibration in (a) G-band mode (Source: Adapted from Refs. [92, 93]), (b) D-band mode (Source: Adapted from Refs. [92, 93]), and (c) Raman spectra of CG ACs at various temperatures (Source: Ref. [4] / with permission of Elsevier). (Note: G mode is just due to the relative motion of sp^2 carbon atoms and can be found in chains as well.)

double-resonance Raman scattering effect. Elanthamilan et al. prepared *Couroupita guianansis* (CG)-derived AC at different temperatures (500, 600, 700, 800, 900 °C), and its Raman analysis results can be seen in Figure 2.6c [4]. In that outcome, two prominent peaks are observed at 1360 and 1594 cm^{-1}, which correspond to the D-band and G-band, due to its C—C graphitic lattice vibration mode. A small hump that appears around ~2700 cm^{-1} is ascribed to 2D–band due to its minor existence of sp^2 hybridized carbon network associated with the double resonance. In addition, the I_D/I_G value for CG-pre-carbonized carbon (PCC), CG-500, CG-600, CG-700, and CG-800 is 0.91, 0.89, 0.77, 0.75, and 0.72, respectively. The authors concluded that the results revealed the graphitization of CG ACs enhanced with an increase of activation temperature [4].

2.3.5 XPS Analysis

XPS analysis was used to identify the oxidation states and surface chemistry of various ACs [94]. This technique has been widely employed to study the

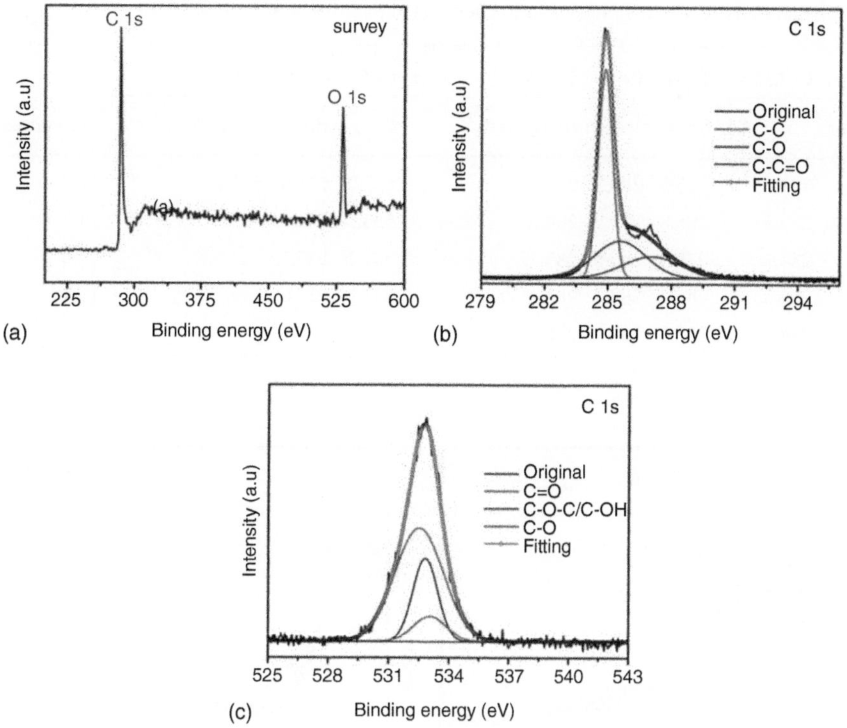

Figure 2.7 (a) Survey XPS spectra, (b) C 1s, and (c) O 1s of AC (Source: Ref. [68] / with permission of Elsevier).

oxygen-containing surface groups. The XPS core spectra of AC show the only existence of C 1s and O 1s elements in the survey [95]. Generally, the XPS spectrum of C 1s exhibited four functional groups such as C—C or C=C, C—O, C=O, and O—C=O [70]. Likewise, the spectrum of O 1s showed three major functional groups including C=O, C—O, and C—O—C/C—OH [96]. Figure 2.7a shows the survey spectra C 1s and O 1s of walnut-derived AC [68]. As shown in Figure 2.7a, the high-resolution spectrum of C 1s has three peaks at 284.8, 285.6, and 287.1 eV, which corresponded to the satellite, C—C, C—O, and C=O bonds, respectively. Likewise, the analysis of the O 1s region (Figure 2.7b) revealed that the presence of three functional groups at the binding energy of 532.5, 533.0, and 532.8 eV was assigned to C=O, C—O, and C—O—C/C—OH bonds, respectively. Finally, it should be noted that, in the XPS spectra of AC, the peaks of functional groups appear at ~284–290 and ~532–534 for C 1s and O 1s elements, respectively. From the previous literature, the various physical and chemical activators of AC have been listed in Table 2.2.

2.3.6 XRD Patterns

XRD technique is an important characterization tool for studying the crystalline nature and phase purity of synthesized materials in a non-destructive manner.

Table 2.2 Summary of various physical and chemical activators of AC.

	C–C/C=C	C–O	C=O	COOH	π–π	C=O	C–O	–OH	
Activator	**C1s fitting binding energy (eV)**				**O 1 s fitting binding energy (eV)**				**References**
KOH	284.6	285.5	286.5	289.3	—	532.7	531.6	533.7	[97]
$ZnCl_2$	284.5	285.8	287.7	290.8	—	533.5	532.3	—	[98]
H_3PO_4	284.6	285.3	286.5	289.1	—	531.1	533.5	—	[99]
H_2SO_4	283.8	285.4	288.7	285.4–288.7	—	531.7	532.6	535.7	[100]
CO_2	284.8	286.1	287.7	288.9	—	—	—	—	[101]

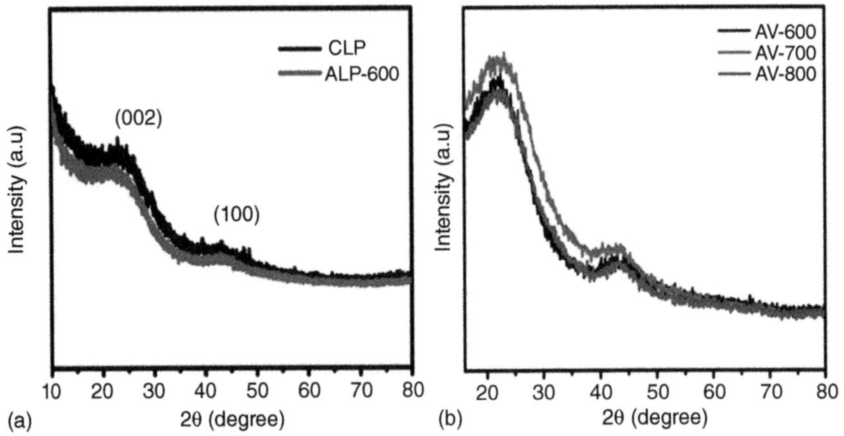

Figure 2.8 XRD pattern of (a) amorphous (Source: Ref. [107] / with permission of Elsevier) and (b) crystalline for BW-AC (Source: Ref. [108] / with permission of American Chemical Society).

The XRD pattern of as-prepared AC shows broad peaks, indicating amorphous with more graphitic nature and crystal disorders [102, 103] as well as crystalline nature in some of the carbon materials [104]. The mean crystallite size (D) can be estimated from the full width at half-maximum (FWHM) of the XRD peak, using the Debye–Scherrer equation [105]. The interlayer spacing (d) was calculated from the Bragg equation [106]. The XRD pattern of AC shown in Figure 2.8a indicates two broad peaks at ~23° and ~43° corresponding to (002) and (100) reflections, respectively. The broad peaks suggest that the AC is amorphous in nature [107]. In another study, the XRD patterns (Figure 2.8b) for AC that exhibit two diffraction peaks at ~25° and ~43° correspond to the reflection from the (002) and (100) planes [108], respectively. The observed broad lines indicate that the ACs are moderately crystalline in nature. From the previous literature, the nature of AC from different BW has been listed in Table 2.3.

Table 2.3 The XRD pattern of synthesized various BW-based ACs.

Starting material	Amorphous	Crystalline	Structure	References
Banana peel	—	✓	Crystalline due to the presence of silica	[109]
Walnut shell	✓	—	—	[110]
Potato waste	✓	—	Graphitic carbon	[111]
Soybean pods	✓	—	Low degree of graphitization and low crystallinity	[104]
Onion leaves	—	✓	Crystalline graphite	[112]
Tamarind shell	✓	—	Amorphous	[113]
Bagasse	✓	—	Graphitic carbon with disordered structure	[114]
Orange peel	✓	—	Graphitic carbon with disordered nature	[115]
Garlic skin	✓	—	Disordered carbon layer	[116]
Nori	—	✓	—	[117]
Rotten carrot	✓	—	Graphitic carbon structure	[118]

Source: Ref. [1] / with permission of Elsevier.

2.3.7 BET Analysis

Porosity is one of the crucial surface properties of the AC, resulting from the activation processes. Compared to the physical activation, the chemical activation process has achieved the nature of pore sizes and shapes, which are greatly dependent on the activating agents, such as KOH, NaOH, KCl, NaCl, H_3PO_4, and $ZnCl_2$, that produce a wide range of pore size distribution with large channels as shown in Figures 2.9 and 2.10 [35]. In general, KOH and NaOH are acting as oxidizing agents, whereas KCl, NaCl, H_3PO_4, and $ZnCl_2$ are acting as dehydrating agents that promote the dehydration effect in the carbon matrix, resulting in the formation of cyclic compounds and condensation process. This cross-linking reaction facilitates grain refinement and porosity. The BET analysis is used to determine the surface area and porosity of the prepared AC from various BWs. According to the International Union of Pure and Applied Chemistry (IUPAC) classification by Dubinin, the varieties of pore sizes can be classified into three kinds. Commonly, AC pores are divided into different categories such as macropores (>50 nm), mesopores (2–50 nm), and micropores (<2 nm) following the IUPAC classification [35]. Besides, the pore width less than 0.7 nm is ascribed as ultra-micropores classification [122].

The exemplification of the micro, meso, and macropores in an AC can be seen in Figure 2.8. The conventional ACs are typically consisting all three types of pores in their microstructure. The surface pores that exist in the AC are accountable for the adsorption of a large number of organic molecules and gases. During the activation process, the macropores are primarily fashioned through the oxidation on

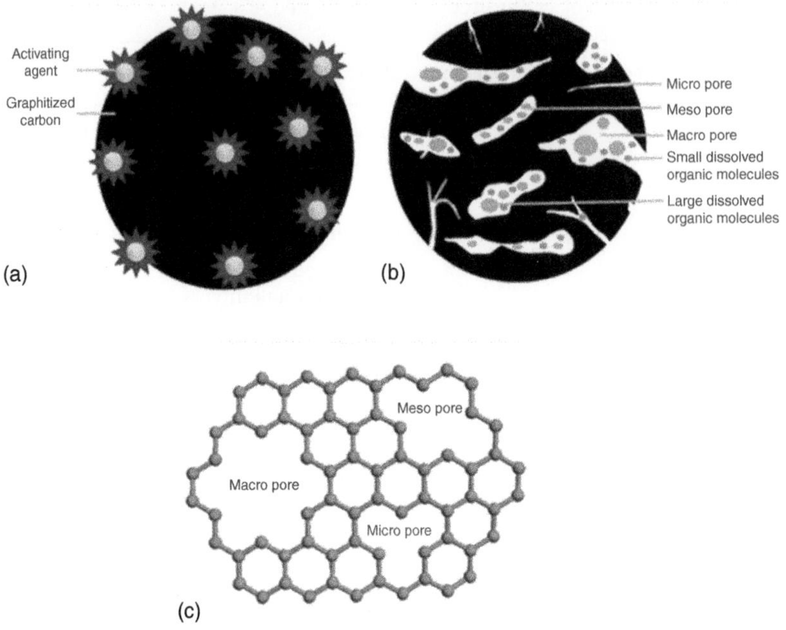

Figure 2.9 Formation and types of porous nature in activated carbon (Source: Adapted from Refs. [119–121]).

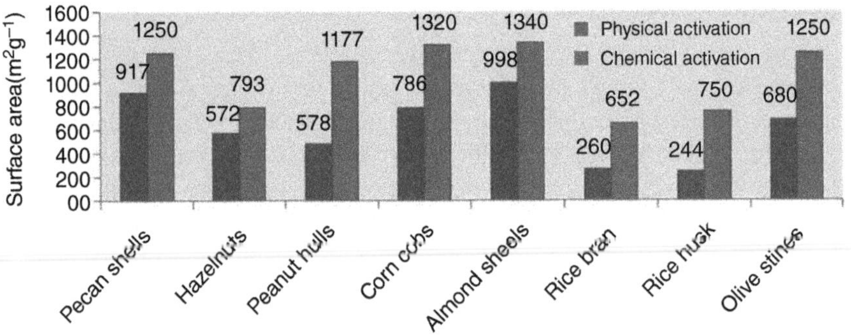

Figure 2.10 Comparison of the surface area for the carbon material by chemical activation and physical activation processes (Source: Ref. [35] / with permission of Elsevier).

the basal plane of the external surface area of raw material. Afterward, the mesopores and micropores are formed by an outbreak of organic molecules within the structure of the raw material [123]. Compared to all three types of pores, micropores are widely accounted for above 95% of the total internal surface area with volumes ranging from 0.15 up to 0.6 $cm^3\ g^{-1}$, whereas mesopores and macropores are much smaller and might vary between 10 and 200 $m^2\ g^{-1}$, respectively [124, 125]. Based on the internal specific surface area with expanded volume, micropores are more favorable for the adsorption of pollutant dyes and polar compounds [126]. In contrast, the mobility of ions in the micropores is 10 times slower than the mesopores in the bulk electrolyte. Thus, the suitable pore size distribution of micro to meso can provide high mobility of ions and high capacitance for the energy storage applications [1, 59]. The BW-derived AC materials and their surface characterization are summarized in Table 2.4.

2.4 Properties

2.4.1 Surface Defects in BW-AC

The BW-ACs are highly eco-friendly materials. It is naturally diverse and has intricate microarchitectures, extensive and low-cost source, feasibility, and large interlayer distances, disordered structures, and also abundant functional groups on their structure. It can provide more potential charge storage sites, which would be benefitted in energy storage applications. Among these properties, structural defect of the carbon skeleton affects the charge state of the carbon atoms due to the structural defect density and activity of increasing potential active sites and overall electrocatalytic performance of carbon structures [127–129]. The structural defect can be classified into intrinsic carbon defects, which are formed by thermal vibration of lattice atoms without any dopants, including edges, vacancies, holes, or topological defects, and extrinsic carbon defects are mainly caused by heteroatom doping approach [130, 131]. The schematic representation of a type of defect in BW-AC is shown in Figure 2.11.

The *intrinsic carbon defect* is extensively affecting the carbon skeleton and has mainly occurred at the edge and topological place of the carbon atoms. In *edge defects*, a large number of unpaired π-electrons on the edges have effectively

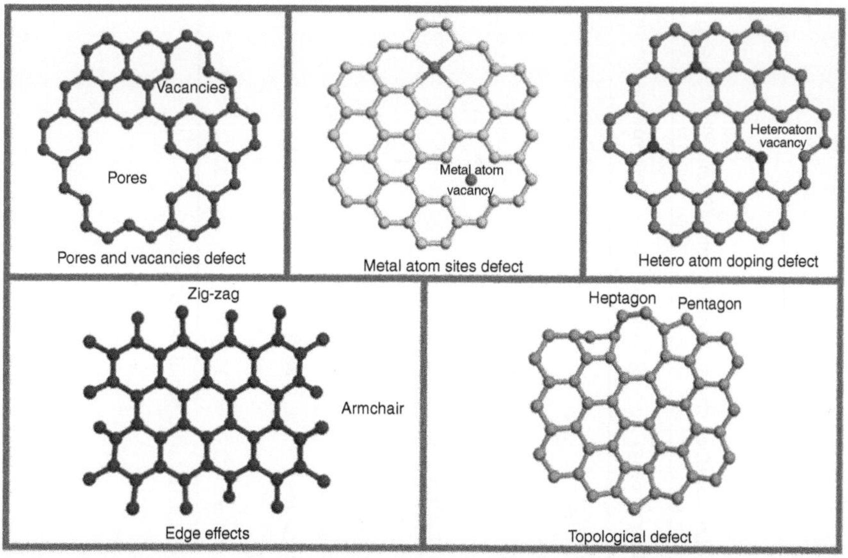

Figure 2.11 Type of the defect in BW-AC (Source: Adapted from Refs. [130–132]).

Table 2.4 Surface characterization of different BW-AC using BET analysis.

Starting material	BET ($m^2\ g^{-1}$)	Pore volume ($cm^3\ g^{-1}$)	Pore size (nm)	Nature of pore	References
Banana peel	1004–2521	0.52–1.30	1.69–1.04	Micro and mesopores	[109]
Walnut shell	62	—	—	—	[110]
Potato waste	1052	0.61	2.0–5.0	Micro, meso, and macropores	[111]
Soybean pods	2245	1.35	3.03	Micro, meso, and interconnected macropores	[104]
Onion leaves	555.1	2.608	0.6–0.8	Micro and mesopores	[112]
Tamarind shell	1040.31	0.61	~1.4–4	Micro and mesopores	[113]
Bagasse	2296	1.34	2.63	Meso, micro, and macropores	[114]
Orange peel	1004–2521	0.52–1.30	1.69–1.04	Micro and mesopore	[115]
Garlic skin	2818	—	0.8–1.0; 2.0–4.0	Meso, micro, and macropores	[116]
Nori	832.4	0.54	>0.5	Micro and mesopores	[117]
Rotten carrot	1154.99	0.92	3.22	Micro and mesopores	[118]

Source: Ref. [1]/ with permission of Elsevier.

accelerated, transferred, and reduced the formation of intermediates, which are key energy sources. Meanwhile, the edges of the carbon atoms possess higher charge density that causes higher active sites on their structure. In addition, the edges of the hexagonal carbon network are divided into two categories: (i) zigzag and (ii) armchair edges [132]. During the preparation of AC, it can produce more edges effect, for example, the mechanical ball milling method, which is used universally for making graphite carbon skeleton, produces more edge defects on their carbon structure. This method effectively minimizes the size of the carbon materials and exposed more edges sites on their carbon skeleton, and also this method introduces a large number of high-energy mechanical impact forces on intrinsic defects carbon [133]. Apart from this method, the chemical method is also used for the preparation of AC with rich edge defects. In these methods, the chemicals, such as KOH, NaOH, KCl, NaCl, H_3PO_4, and $ZnCl_2$ break the sp^3-hybridized C—C bond from the carbon skeleton, and an inert atmosphere with high temperature also produces zigzag-edged sp^2 hybridized carbon nanoribbons. The zigzag-edged sp^2 carbon has higher stability and excellent activity than the other materials with doping carbon defects. In addition, some other methods are also used to produce defective carbon atoms without destroying the carbon walls, such as plasma etching. Some researchers have produced dopant-free intrinsic defects on the carbon atom, for example Dai et al. produce carbon materials with radio frequency argon plasma for dopant-free carbon production, and it has rich intrinsic defects on their carbon structure; it clearly shows in TEM characterization with a large number of holes and edge defects on the graphene [134, 135].

In *topological defects*, some interference in the electronic symmetry of aromatic rings causes local charge redistribution; therefore, the adjacent carbon atoms are optimized to produce electrocatalyst with high active sites on their structure. Topological defects are normally occurring in the carbon skeleton with non-hexagonal structures such as pentagons, heptagons, and octagons. Usually, the carbon materials with topological defects are produced by in situ etching, chemical vapor deposition (CVD), and nitrogen removal process. Mu et al. used some chemicals for these materials' production, for example, in the fullerene framework, KOH acts as an etchant to produce intrinsic defects pentagons carbon from the in situ etching method. The carbon materials with intrinsic defects pentagons are observed by aberration-corrected scanning transmission electron microscopy (AC-STEM). Another nitrogen removal method is the high-temperature treatment process; in this method, the original N atoms are removed from the nanomaterials, which causes the edge of vacancy defects in the carbon skeleton to form dangling carbon bonds resulting in topological defects. And the X-ray absorption near-edge structure (XANES) measurements were also used to confirm the processing of temperature increases, and N content decrease causes a continuous increase of topological defect density, so it improved the catalytic performance of CO_2 reduction [136–138].

The *extrinsic carbon defects* are mainly caused by heteroatoms or single metal atoms doping. For example, the heteroatom with different electronegativities such as B, N, P, and F doped with carbon network produces an asymmetric charge, redistributes the spin density, breaks the electrical neutrality of carbon matrix, and also optimizes the carbon electron properties so it will generate active site on the charged carbon [139]. Among these heteroatom doping, commonly N-doping is used due to these lone electron pairs on pyridinic-N configuration; it enhances the binding ability of active sites of the catalyst with other molecules, and it forms reaction intermediates that act as a key molecule. Apart from this N-doping, the B-doped carbon-doped boron nitrides (CBN) also causes spin density and asymmetric charge introduction, and it also stabilizes key reaction intermediates, so it minimizes some activation barriers on CO_2 reduction reaction. In addition, P-doping can also introduce distortion of the carbon lattice due to the larger size of P, and also it produces large structural defects; therefore, it enhances the catalytic performance on electrocatalyst. The F-doping with carbon forms C—F bond with different properties like covalent or ionic or semi-ionic. Because of the high electronegativity of F to make carbon atoms with a high positive charge, it became more active sites of electrocatalytic carbon dioxide reduction (ECR) reaction and this more positive carbon atom with more active sites will enhance the ECR application [127, 140, 141].

The construction of extrinsic carbon defects or heteroatom-doped carbon structure is from two main methods such as pyrolysis and CVD. Some of the researchers synthesized this heteroatoms-doped carbon structure from the following methods – the N-doped three-dimensional (3D) graphene foam (NG) synthesized by Ajayan and his groups from the CVD method. In this method, they used methane as the precursor for N doping and grew graphene foam (NG) by using a CVD method, and the resultant NG product was obtained by hydrochloric acid etching nickel skeleton followed by repeated washing and also freeze-drying. For B-doping, Einaga et al. used $B(OCH_3)$ as the B source and used microwave plasma-assisted CVD method for the synthesis of B-doped diamond (BDD) on Si (111) wafers. It has good stability toward electrocatalytic reduction of CO_2 to formaldehyde. In the P-doping, Li et al. used onion-like carbon (OLC) for the host, and they synthesized P-OLC by the CVD method. Finally, the F-doped carbon (FC) was constructed by Wang et al. from the pyrolyzing method, and they used mixed commercial BP 2000 and polytetrafluoroethylene for pyrolyzing to produce FC catalyst [142–145].

2.4.2 Characterizations of Carbon Defects

In recent decades, more advanced characterization techniques were developed to detect and analyze the types of defects in ECR, which are used to understand the internal catalytic mechanism of different defect sites. These characterization methods can be divided into two types such as direct observation by using electron microscopy and indirect analysis by using spectral characterization and physical structural analysis. Electron microscopy is used to direct focus on catalysts imaging, its resolution is in nanoscale, and also an atomic level of observation was occurring. Surface defects of the catalyst were analyzed by most commonly used characterization techniques such as scanning transmission electron microscopy (STEM) and high-angle annular dark field scanning transmission electron microscopy (HAADF-STEM). When the observation of defects in the dark field images is more conductive than the bright field images, it is used to observe the honeycomb lattices of carbon-based catalysts, and also the type of defect is identified by sub-nanometer or even atomic-scale observation. In addition, some of the spectral characterization techniques also include for catalyst defects indirect analysis such as Raman spectrum, XPS, and X-ray absorption spectroscopy (XAS). From the Raman spectrum, defects and disorder induction are represented by the D-band, the graphitization characteristics of the sp^2 network are signified by the G-band, and also the ratio of these bands (I_D/I_G) used to analyze the degree of defects in the carbon structure. In XPS analysis, apart from the C-sp^2 bond some other bonds such as C-sp^3, C—N, C—O, and C=O can be referred to by the C1s spectrum that is used to analyze the 3D structure of carbon atom and intrinsic defects of the catalyst. The geometry of carbon atoms is revealed by XAS, which effectively characterizes the defects; it also includes near edge from X-ray absorption structure (XANES) and fine structure from extended X-ray absorption (EXAFS). Near the adsorption boundary of the carbon structures is found by XANES, which give most useful information such as oxidation state, binding environment, and local geometry of the absorbed atoms. The oscillating part of the spectrum is referred to as EXAFS, and it provides valuable information on the coordination number and chemical bond length around the adsorbed atom. The BET is used to analyze the specific surface area and pore size distribution of the catalyst, and it is also used to differentiate the morphology and structure of different catalysts [145–148].

2.4.3 Intrinsic Carbon Defects Activity

The intrinsic defects can affect the charge state of carbon nanomaterials in the carbon skeleton and also increase the density of the active site and improve the performance of an electrocatalytic activity. For example, the undoped carbon nanocages (CNC) are synthesized by Hu et al.; they used 700 °C for pyrolysis, and due to this high temperature of pyrolysis, it has a high specific surface area and large I_D/I_G ratio. It indicates the abundant defects that occur at corners, fringes, and holes, and due to these defects, it shows the best activity on oxygen reduction reaction (ORR). In recent years, the intrinsic defects in pure carbon nanomaterials are effectively applied for electrocatalyst research of ECR. Zhang et al. synthesized N-doped carbon spheres with intrinsic carbon defects that improve the catalytic activity of ECR, and also it was increasing the ECR performance with increased defect concentration but it decreases ECR performance with increased N content on dopant. The NEXAFS spectra can analyze the sp^2 defects in octagonal and pentagonal structures, then the sp^3 edges in an armchair and zigzag structures, and also it positively increases the ECR activity. The number of carbon defects directly affects the activity and selectivity of the catalysts. This carbon defect mechanism is verified by density functional theory (DFT) calculations; one perfect sp^2 carbon needs higher armchair and zigzag edge defects for *COOH formation, and also some free energy needs to decrease pentagonal defects. The *COOH adsorption optimization causes octagonal defects, and also it causes the conduction of electrons by positive C atoms promotion and the further promoting causing *COOH–*CO reduction [132, 149].

2.4.4 Heteroatom Doping Defects (or) Extrinsic Carbon Defects Activity

In the extrinsic carbon defects, some heteroatoms with high electronegativity are incorporated into the carbon skeleton, which will form the carbon structure, and also these heteroatoms attached to the sp^2 hybridized network will replace some of the carbon atoms. The structure of carbon materials can be optimized by these heteroatoms, and these heteroatoms will improve the electrocatalytic activity of carbon materials. Some of the heteroatoms are incorporated in the carbon skeleton such as B, N, O, F, P, and S. Among these heteroatoms N atom is the most commonly doped atom due to its high electronegativity than a carbon atom. This N-doped carbon atom will have different catalytic activity effects. In the N-doped carbon atom such as pyridinic, pyrrolic, and graphitic N atom is the important active center in the ECR [150–153]. The effective defect in the carbon atom will change the physical and chemical states of the carbon structure, and it will enhance the ECR performance.

2.4.5 Electronic Band Structure Properties

When we come to discuss the electrical conductivity or property of the material, we should remember about the conductor, insulator, and semiconductor. Figure 2.12 shows how electrical conduction occurs in the material. In conductor materials, the conduction and valence bands are overlapped with zero bandgaps, facilitating faster electron transportation. In semiconductor materials, there is a small Fermi-level bandgap less than 5 eV, and some kind of energy (i.e. thermal or photo) accelerates the electrons to reach the conduction band. The insulator materials are a large Fermi-level bandgap greater than 5 eV, which is a forbidden transition of an electron to reach the conduction band. Several attempts have been employed over the past 75 years to label the electronic properties of carbon materials including AC. The major problem regarding the conductivity of AC is neither metal nor a semi-conductor. Interesting to know is that the carbon materials

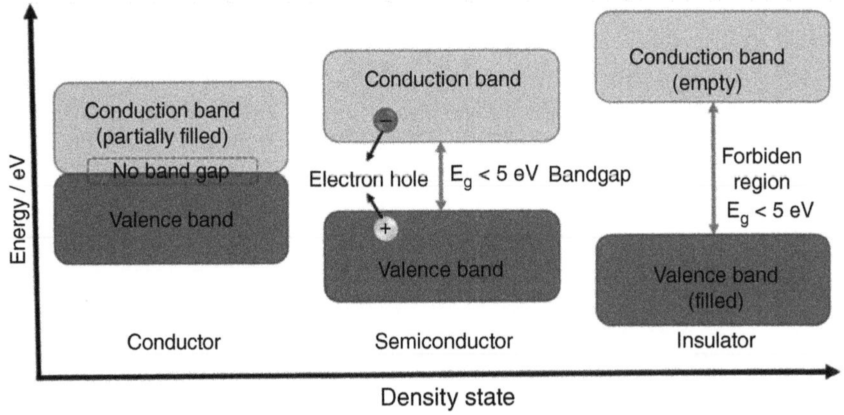

Figure 2.12 Schematic illustration of the energy band.

are considered semi-metal materials owing to their zero-bandgap energy with free flow of π-electrons. In the ground state, the electronic configuration of carbon atom stands $1s^2$, $2s^2$, $2p_x^{\ 1}$, $2p_y^{\ 1}$, $2p_z^{\ 0}$, which means that there are two filled electrons in 1s orbital, two filled electrons in 2s orbital, and the remaining two electrons are filled in $2p_x$ and $2p_y$, which are not filled orbitals. Besides, the remaining $2p_z$ atomic orbital is a vacant orbital owing to its energetically favorable phenomenon to occupy two electrons in 2s orbital and two of them only in $2p_x$ and $2p_y$ orbitals. Conversely, in the most typical situation, the valence electrons of carbon maintain the electronic configuration $2s^1$, $2p_x^{\ 1}$, $2p_y^{\ 1}$, and $2p_z^{\ 1}$ by contributing one of the 2s electrons into the vacant $2p_z$ orbital in the excited state. However, the different types of carbon allotropes and the flexibility of their bonding vary with their electrically conductive property. Remember that the conductive material should have free electrons to transport the electric current. Diamond, an allotrope of carbon, that did not have such free electrons indicates that it is an electrical insulator. Similar to graphite and other carbon-based materials, ACs are generally agreed to behave as semiconductors, which entails that their electrical conductivity increases with temperature. On the other hand, the conductivity of carbon greatly depends on its nature whether it is amorphous or crystalline. The amorphous nature of carbon conducts electricity at room temperature, but the crystalline state does not. At very high temperatures, all the states of carbons conduct electricity because of the release of bounded electrons in grain boundaries due to the high energy supplied to them. A peripheral disturbance, including the appearance of other functional groups such as carbonyl, carboxyl, and hydroxyl groups, can easily overcome the required energy for carbon atom excitation [154, 155]. Complex electronic band structures have been derived from quantum mechanics for the discussion of amorphous carbon. As a characteristic feature, the mobility of electrons in these structures is to take place by bouncing between less localized bandgap or higher energy state. According to Kastning's model, the graphitic nanocrystals are distinguished as three types of charge carriers: (i) delocalization of π-electrons through the entire particle of carbon; (ii) delocalization of π-electrons through graphitic nanocrystal, which is so-called single domain; and (iii) localized at either single domain or atomic group. Among the three types of charge carriers, (i) and (ii) are probable to contribute to the electrical conductivity of AC. However, the type (i) electronic states are seen to be entirely unoccupied, thus their involvement in the conductivity should be negligibly small. Besides, the graphene domains are charged with one and favorably two coherent π-electrons, being delocalized through the single domain. All these types of domains are in thermodynamic equilibrium. In this circumstance, the charge transport is considered to be the transport of charged domains, related to the conduction of protons in solvents with an enormous network of hydrogen bonds, such as water, and is known as the Grotthus mechanism. On the other hand, the uncharged domains do not contribute to electrical conductivity [37, 156].

2.5 Summary

In this chapter, we have provided a brief overview of AC from various BW. The various AC materials synthesized from various BW precursors have been discussed. The activation methods such as physical activation and chemical activation as well as activation agents are discussed. Among these methods, chemical activation displays important advantages, including high yield, large specific surface area, porous structural morphology, and conductivity. The characterization techniques of SEM, HR-TEM, FTIR, Raman spectra, XPS, XRD, and BET for various BW-AC materials have been well discussed. All the physiochemical properties of the BW-AC are explained.

References

1 Shanmuga Priya, M., Divya, P., and Rajalakshmi, R. (2020). A review status on characterization and electrochemical behaviour of biomass derived carbon materials for energy storage supercapacitors. *Sustain. Chem. Pharm.* 16: 100243. 10.1016/j.scp.2020.100243.

2 Muñoz-Guillena, M.J., Linares-Solano, A., and de Lecea, C.S.M. (1992). Determination of calorific values of coals by differential thermal analysis. *Fuel* 71: 579–583. 10.1016/0016-2361(92)90157-J.

3 Arami-Niya, A., Abnisa, F., Shafeeyan, M.S. et al. (2012). Optimization of synthesis and characterization of palm shell-based bio-char as a by-product of bio-oil production process. *Bioresources* 7: 246–264. 10.15376/biores.7.1.0246-0264.

4 Elanthamilan, E., Sriram, B., Rajkumar, S. et al. (2019). Couroupita guianansis dead flower derived porous activated carbon as efficient supercapacitor electrode material. *Mater. Res. Bull.* 112: 390–398. 10.1016/j.materresbull.2018.12.028.

5 Canh, V.D., Tabata, S., Yamanoi, S. et al. (2021). Evaluation of porous carbon adsorbents made from rice husks for virus removal in water. *Water (Switzerland)* 13: 1–10. 10.3390/w13091280.

6 Mondal, S., Sinha, K., Aikat, K., and Halder, G. (2015). Adsorption thermodynamics and kinetics of ranitidine hydrochloride onto superheated steam activated carbon derived from mung bean husk. *J. Environ. Chem. Eng.* 3: 187–195. 10.1016/j.jece.2014.11.021.

7 Jaouadi, M., Marzouki, M., Hamzaoui, A.H., and Ghodbane, O. (2021). Enhanced electrochemical performance of olive stones-derived activated carbon by silica coating for supercapacitor applications. *J. Appl. Electrochem.* 10.1007/s10800-021-01623-4.

8 Bouchelta, C., Medjram, M.S., Zoubida, M. et al. (2012). Effects of pyrolysis conditions on the porous structure development of date pits activated carbon. *J. Anal. Appl. Pyrolysis* 94: 215–222. 10.1016/j.jaap.2011.12.014.

9 Deng, Z., Sun, S., Li, H. et al. (2021). Modification of coconut shell-based activated carbon and purification of wastewater. *Adv. Compos. Hybrid Mater.* 4: 65–73. 10.1007/s42114-021-00205-4.

10 Gonçalves, M., Guerreiro, M.C., Oliveira, L.C.A. et al. (2013). Micro mesoporous activated carbon from coffee husk as biomass waste for environmental applications. *Waste Biomass Valoriz.* 4: 395–400. 10.1007/s12649-012-9163-1.

11 Sudaryanto, Y., Hartono, S.B., Irawaty, W. et al. (2006). High surface area activated carbon prepared from cassava peel by chemical activation. *Bioresour. Technol.* 97: 734–739. 10.1016/j.biortech.2005.04.029.

12 Kalderis, D., Bethanis, S., Paraskeva, P., and Diamadopoulos, E. (2008). Production of activated carbon from bagasse and rice husk by a single-stage chemical activation method at low retention times. *Bioresour. Technol.* 99: 6809–6816. 10.1016/j.biortech.2008.01.041.

13 Khalil, K.M.S., Allam, O.A.S., Khairy, M. et al. (2017). High surface area nanostructured activated carbons derived from sustainable sorghum stalk. *J. Mol. Liq.* 247: 386–396. 10.1016/j.molliq.2017.09.090.

14 Tian, X., Ma, H., Li, Z. et al. (2017). Flute type micropores activated carbon from cotton stalk for high performance supercapacitors. *J. Power Sources* 359: 88–96. 10.1016/j.jpowsour.2017.05.054.

15 Khasri, A., Jamir, M.R.M., Ahmad, A.A., and Ahmad, M.A. (2021). Adsorption of remazol brilliant violet 5r dye from aqueous solution onto melunak and rubberwood sawdust based activated carbon: interaction mechanism, isotherm, kinetic and thermodynamic properties. *Desalin. Water Treat.* 216: 401–411. 10.5004/dwt.2021.26852.

16 Manasa, P., Lei, Z.J., and Ran, F. (2020). Biomass waste derived low cost activated carbon from *Carchorus Olitorius* (Jute Fiber) as sustainable and novel electrode material. *J. Energy Storage* 30: 101494. 10.1016/j.est.2020.101494.

17 Rahmani-Sani, A., Singh, P., Raizada, P. et al. (2020). Use of chicken feather and eggshell to synthesize a novel magnetized activated carbon for sorption of heavy metal ions. *Bioresour. Technol.* 297: 122452. 10.1016/j.biortech.2019.122452.

18 Suhas, Gupta, V.K., Singh, L.P. et al. (2021). A novel approach to develop activated carbon by an ingenious hydrothermal treatment methodology using *Phyllanthus emblica* fruit stone. *J. Clean. Prod.* 288: 125643. 10.1016/j.jclepro.2020.125643.

19 Dao, T.M. and LeLuu, T. (2020). Synthesis of activated carbon from macadamia nutshells activated by H_2SO_4 and K_2CO_3 for methylene blue removal in water. *Bioresour. Technol. Rep.* 12: 100583. 10.1016/j.biteb.2020.100583.

20 Slyvanus, N.O. (2020). Chemical modification of Kola-Nut (Cola Nitida) testa for adsorption of Cu^{2+}, Fe^{2+}, Mg^{2+}, Pb^{2+}, and Zn^{2+} from aqueous solution. *Makara J. Sci.* 24: 10.7454/mss.v24i1.11727.

21 Zbair, M., Anfar, Z., Ait Ahsaine, H. et al. (2018). Acridine orange adsorption by zinc oxide/almond shell activated carbon composite: operational factors, mechanism and performance optimization using central composite design and surface modeling. *J. Environ. Manag.* 206: 383–397. 10.1016/j.jenvman.2017.10.058.

22 Cazetta, A.L., Azevedo, S.P., Pezoti, O. et al. (2014). Thermally activated carbon from bovine bone: optimization of synthesis conditions by response surface methodology. *J. Anal. Appl. Pyrolysis* 110: 455–462. 10.1016/j.jaap.2014.10.022.

23 Veerakumar, P., Sangili, A., Chen, S.M. et al. (2020). Fabrication of platinum-rhenium nanoparticle-decorated porous carbons: voltammetric sensing of furazolidone. *ACS Sustain. Chem. Eng.* 8: 3591–3605. 10.1021/acssuschemeng.9b06058.

24 Gao, Y., Yue, Q., Gao, B., and Li, A. (2020). Insight into activated carbon from different kinds of chemical activating agents: a review. *Sci. Total Environ.* 746: 141094. 10.1016/j.scitotenv.2020.141094.

25 Dubey, P., Shrivastav, V., Maheshwari, P.H., and Sundriyal, S. (2020). Recent advances in biomass derived activated carbon electrodes for hybrid electrochemical capacitor applications: challenges and opportunities. *Carbon N. Y.* 170: 1–29. 10.1016/j.carbon.2020.07.056.

26 Sivachidambaram, M., Vijaya, J.J., Kennedy, L.J. et al. (2017). Preparation and characterization of activated carbon derived from the: *Borassus flabellifer* flower as an electrode material for supercapacitor applications. *New J. Chem.* 41: 3939–3949. 10.1039/c6nj03867k.

27 Song, S., Ma, F., Wu, G. et al. (2015). Facile self-templating large scale preparation of biomass-derived 3D hierarchical porous carbon for advanced supercapacitors. *J. Mater. Chem. A* 3: 18154–18162. 10.1039/c5ta04721h.

28 Tsai, W.T., Chang, C.Y., and Lee, S.L. (1998). A low cost adsorbent from agricultural waste corn cob by zinc chloride activation. *Bioresour. Technol.* 64: 211–217. 10.1016/S0960-8524(97)00168-5.

29 Daud, W.M.A.W., Ali, W.S.W., and Sulaiman, M.Z. (2003). Effect of activation temperature on pore development in activated carbon produced from palm cell. *J. Chem. Technol. Biotechnol.* 78: 1–5. 10.1002/jctb.712.

30 Elaiyappillai, E., Srinivasan, R., Johnbosco, Y. et al. (2019). Low cost activated carbon derived from Cucumis melo fruit peel for electrochemical supercapacitor application. *Appl. Surf. Sci.* 486: 527–538. 10.1016/j.apsusc.2019.05.004.

31 Srinivasan, R., Elaiyappillai, E., Pandian, H.P. et al. (2019). Sustainable porous activated carbon from Polyalthia longifolia seeds as electrode material for supercapacitor application. *J. Electroanal. Chem.* 849: 113382. 10.1016/j.jelechem.2019.113382.

32 González-García, P. (2018). Activated carbon from lignocellulosics precursors: a review of the synthesis methods, characterization techniques and applications. *Renew. Sust. Energ. Rev.* 82: 1393–1414. 10.1016/j.rser.2017.04.117.

33 Mohamad Nor, N., Lau, L.C., Lee, K.T., and Mohamed, A.R. (2013). Synthesis of activated carbon from lignocellulosic biomass and its applications in air pollution control – a review. *J. Environ. Chem. Eng.* 1: 658–666. 10.1016/j.jece.2013.09.017.

34 Bouchelta, C., Medjram, M.S., Bertrand, O., and Bellat, J.P. (2008). Preparation and characterization of activated carbon from date stones by physical activation with steam. *J. Anal. Appl. Pyrolysis* 82: 70–77. 10.1016/j.jaap.2007.12.009.

35 Lewoyehu, M. (2021). Comprehensive review on synthesis and application of activated carbon from agricultural residues for the remediation of venomous pollutants in wastewater. *J. Anal. Appl. Pyrolysis* 159: 105279. 10.1016/j.jaap.2021.105279.

36 Aworn, A., Thiravetyan, P., and Nakbanpote, W. (2008). Preparation and characteristics of agricultural waste activated carbon by physical activation having micro- and mesopores. *J. Anal. Appl. Pyrolysis* 82: 279–285. 10.1016/j.jaap.2008.04.007.

37 Barroso Bogeat, A. (2021). Understanding and tuning the electrical conductivity of activated carbon: a state-of-the-art review. *Crit. Rev. Solid State Mater. Sci.* 46: 1–37. 10.1080/10408436.2019.1671800.

38 Rambabu, N., Rao, B.V.S.K., Surisetty, V.R. et al. (2015). Production, characterization, and evaluation of activated carbons from de-oiled canola meal for environmental applications. *Ind. Crop. Prod.* 65: 572–581. 10.1016/j.indcrop.2014.09.046.

39 Illán-Gómez, M.J., García-García, A., Salinas-Martinez De Lecea, C., and Linares-Solano, A. (1996). Activated carbons from Spanish coals. 2. Chemical activation. *Energy Fuel* 10: 1108–1114. 10.1021/ef950195+.

40 Ahmadpour, A. and Do, D.D. (1996). The preparation of active carbons from coal by chemical and physical activation. *Carbon N. Y.* 34: 471–479. 10.1016/0008-6223(95)00204-9.

41 Ceyhan, A.A., Şahin, Ö., Saka, C., and Yalçin, A. (2013). A novel thermal process for activated carbon production from the vetch biomass with air at low temperature by two-stage procedure. *J. Anal. Appl. Pyrolysis* 104: 170–175. 10.1016/j.jaap.2013.08.007.

42 Maulina, S., Handika, G., Irvan, I., and Iswanto, A.H. (2020). Quality comparison of activated carbon produced from oil palm fronds by chemical activation using sodium carbonate versus sodium chloride. *J. Korean Wood Sci. Technol.* 48: 503–512. 10.5658/WOOD.2020.48.4.503.

43 Lu, Q., Zhang, Z.X., Wang, X. et al. (2018). Catalytic fast pyrolysis of biomass impregnated with potassium phosphate in a hydrogen atmosphere for the production of phenol and activated carbon. *Front. Chem.* 6: 1–10. 10.3389/fchem.2018.00032.

44 Chen, Y.D., Chen, W.Q., Huang, B., and Huang, M.J. (2013). Process optimization of $K_2C_2O_4$-activated carbon from kenaf core using Box-Behnken design. *Chem. Eng. Res. Des.* 91: 1783–1789. 10.1016/j.cherd.2013.02.024.

45 Garba, Z.N., Hussin, M.H., Galadima, A., and Lawan, I. (2019). Potentials of Canarium schweinfurthii seed shell as a novel precursor for CH_3COOK activated carbon: statistical optimization, equilibrium and kinetic studies. *Appl Water Sci* 9: 1–13. 10.1007/s13201-019-0907-y.

46 Tong, K.T., Vinai, R., and Soutsos, M.N. (2018). Use of Vietnamese rice husk ash for the production of sodium silicate as the activator for alkali-activated binders. *J. Clean. Prod.* 201: 272–286. 10.1016/j.jclepro.2018.08.025.

47 Gao, Y., Yue, Q., Xu, S., and Gao, B. (2015). Activated carbons with well-developed mesoporosity prepared by activation with different alkali salts. *Mater. Lett.* 146: 34–36.

48 Park, Y.-T., Im, C.-G., Kim, Y.-T., and Rhee, B.-S. (2009). Thermostable adsorption filter immobilized with super activated carbons by quinoline soluble isotropic pitch binder (I-a novel adsorption filter). *Carbon Lett.* 10: 198–201. 10.5714/cl.2009.10.3.198.

49 Brito, M.J.P., Veloso, C.M., Santos, L.S. et al. (2018). Adsorption of the textile dye Dianix® royal blue CC onto carbons obtained from yellow mombin fruit stones and activated with KOH and H_3PO_4: kinetics, adsorption equilibrium and thermodynamic studies. *Powder Technol.* 339: 334–343. 10.1016/j.powtec.2018.08.017.

50 Örkün, Y., Karatepe, N., and Yavuz, R. (2012). Influence of temperature and impregnation ratio of H_3PO_4 on the production of activated carbon from hazelnut shell. *Acta Phys. Pol. A* 121: 277–280. 10.12693/APhysPolA.121.277.

51 Qiu, D., Guo, N., Gao, A. et al. (2019). Preparation of oxygen-enriched hierarchically porous carbon by $KMnO_4$ one-pot oxidation and activation: mechanism and capacitive energy storage. *Electrochim. Acta* 294: 398–405. 10.1016/j.electacta.2018.10.049.

52 Sun, K., Leng, C.Y., Jiang, J.C. et al. (2017). Microporous activated carbons from coconut shells produced by self-activation using the pyrolysis gases produced from them, that have an excellent electric double layer performance. *Xinxing Tan Cailiao/New Carbon Mater.* 32: 451–459. 10.1016/S1872-5805(17)60134-3.

53 Chen, Y.D., Huang, M.J., Huang, B., and Chen, X.R. (2012). Mesoporous activated carbon from inherently potassium-rich pokeweed by in situ self-activation and its use for phenol removal. *J. Anal. Appl. Pyrolysis* 98: 159–165. 10.1016/j.jaap.2012.09.011.

54 Wang, A., Sun, K., Xu, R. et al. (2021). Cleanly synthesizing rotten potato-based activated carbon for supercapacitor by self-catalytic activation. *J. Clean. Prod.* 283: 125385. 10.1016/j.jclepro.2020.125385.

55 Nguyen, V.H., Nguyen, D.T., Nguyen, T.T. et al. (2021). Activated carbon with ultrahigh surface area derived from sawdust biowaste for the removal of rhodamine B in water. *Environ. Technol. Innov.* 24: 101811. 10.1016/j.eti.2021.101811.

56 Cheng, P., Li, T., Yu, H. et al. (2016). Biomass-derived carbon fiber aerogel as a binder-free electrode for high-rate supercapacitors. *J. Phys. Chem. C* 120: 2079–2086. 10.1021/acs.jpcc.5b11280.

57 Wang, H., Xu, Z., Kohandehghan, A. et al. (2013). Interconnected carbon nanosheets derived from hemp for ultrafast supercapacitors with high energy. *ACS Nano* 7: 5131–5141. 10.1021/nn400731g.

58 Selvaraj, A.R., Muthusamy, A., Inho-Cho, H.J. et al. (2021). Ultrahigh surface area biomass derived 3D hierarchical porous carbon nanosheet electrodes for high energy density supercapacitors. *Carbon N. Y.* 174: 463–474. 10.1016/j.carbon.2020.12.052.

59 Bi, Z., Kong, Q., Cao, Y. et al. (2019). Biomass-derived porous carbon materials with different dimensions for supercapacitor electrodes: a review. *J. Mater. Chem. A* 7: 16028–16045. 10.1039/c9ta04436a.

60 Deng, J., Li, M., and Wang, Y. (2016). Biomass-derived carbon: synthesis and applications in energy storage and conversion. *Green Chem.* 18: 4824–4854. 10.1039/c6gc01172a.

61 Williams, P.T. and Reed, A.R. (2006). Development of activated carbon pore structure via physical and chemical activation of biomass fibre waste. *Biomass Bioenergy* 30: 144–152. 10.1016/j.biombioe.2005.11.006.

62 Jiang, L., Sheng, L., and Fan, Z. (2018). Biomass-derived carbon materials with structural diversities and their applications in energy storage. *Sci. China Mater.* 61: 133–158. 10.1007/s40843-017-9169-4.

63 Wang, L., Li, Y., Yang, K. et al. (2017). Hierarchical porous carbon microspheres derived from biomass-corncob as ultra-high performance supercapacitor electrode. *Int. J. Electrochem. Sci.* 12: 5604–5617. 10.20964/2017.06.16.

64 Thirumal, V., Dhamodharan, K., Yuvakkumar, R. et al. (2021). Cleaner production of tamarind fruit shell into bio-mass derived porous 3D-activated carbon nanosheets by CVD technique for supercapacitor applications. *Chemosphere* 282: 131033. 10.1016/j.chemosphere.2021.131033.

65 Vu, D.L., Seo, J.S., Lee, H.Y., and Lee, J.W. (2017). Activated carbon with hierarchical micro-mesoporous structure obtained from rice husk and its application for lithium-sulfur batteries. *RSC Adv.* 7: 4144–4151. 10.1039/C6RA26179E.

66 Luo, L., Wu, X., Li, Z. et al. (2019). Synthesis of activated carbon from biowaste of fir bark for methylene blue removal. *R. Soc. Open Sci.* 6: 10.1098/rsos.190523.

67 Liu, X., Zhang, S., Wen, X. et al. (2020). High yield conversion of biowaste coffee grounds into hierarchical porous carbon for superior capacitive energy storage. *Sci. Rep.* 10: 1–12. 10.1038/s41598-020-60625-y.

68 Elanthamilan, E., Catherin Meena, B., Renuka, N. et al. (2021). Walnut shell derived mesoporous activated carbon for high performance electrical double layer capacitors. *J. Electroanal. Chem.* 901: 115762. 10.1016/j.jelechem.2021.115762.

69 Przepiórski, J., Skrodzewicz, M., and Morawski, A.W. (2004). High temperature ammonia treatment of activated carbon for enhancement of CO_2 adsorption. *Appl. Surf. Sci.* 225: 235–242. 10.1016/j.apsusc.2003.10.006.

70 Biniak, S., Szymański, G., Siedlewski, J., and Świątkoski, A. (1997). The characterization of activated carbons with oxygen and nitrogen surface groups. *Carbon N. Y.* 35: 1799–1810. 10.1016/S0008-6223(97)00096-1.

71 Choma, J., Osuchowski, L., Marszewski, M. et al. (2016). Developing microporosity in Kevlar®-derived carbon fibers by CO_2 activation for CO_2 adsorption. *J. CO2 Util.* 16: 17–22. 10.1016/j.jcou.2016.05.004.

72 Jia, Y.F., Xiao, B., and Thomas, K.M. (2002). Adsorption of metal ions on nitrogen surface functional groups in activated carbons. *Langmuir* 18: 470–478. 10.1021/la011161z.

73 Przepiórski, J. (2006). Enhanced adsorption of phenol from water by ammonia-treated activated carbon. *J. Hazard. Mater.* 135: 453–456. 10.1016/j.jhazmat.2005.12.004.

74 Kaźmierczak, J., Biniak, S., Swiątkowski, A., and Radeke, K.H. (1991). Interdependence of different parameters characterizing the chemistry of an activated carbon surface. *J. Chem. Soc. Faraday Trans.* 87: 3557–3561. 10.1039/FT9918703557.

75 Meldrum, B.J. and Rochester, C.H. (1990). In situ infrared study of the surface oxidation of activated carbon dispersed in potassium bromide. *J. Chem. Soc. Faraday Trans.* 86: 2997–3002. 10.1039/FT9908602997.

76 Pradhan, B.K. and Sandle, N.K. (1999). Effect of different oxidizing agent treatments on the surface properties of activated carbons. *Carbon* 37: 1323–1332.

77 Fanning, P.E. and Vannice, M.A. (1993). A DRIFTS study of the formation of surface groups on carbon by oxidation. *Carbon N. Y.* 31: 721–730. 10.1016/0008-6223(93)90009-Y.

78 Sricharoenchaikul, V., Pechyen, C., Aht-Ong, D., and Atong, D. (2008). Preparation and characterization of activated carbon from the pyrolysis of physic nut (*Jatropha curcas* L.) waste. *Energy Fuel* 22: 31–37. 10.1021/ef700285u.

79 Gómez-Serrano, V., Piriz-Almeida, F., Durán-Valle, C.J., and Pastor-Villegas, J. (1999). Formation of oxygen structures by air activation. A study by FT-IR spectroscopy. *Carbon N. Y.* 37: 1517–1528. 10.1016/S0008-6223(99)00025-1.

80 Edwin Vasu, A. (2008). Surface modification of activated carbon for enhancement of nickel(ii) adsorption. *E-J. Chem.* 5: 814–819.

81 Nakahara, M. and Sanada, Y. (1995). FT-IR ATR spectroscopy of the edge surface of pyrolytic graphite and its surface/PVC interface. *J. Mater. Sci.* 30: 4363–4368. 10.1007/BF00361518.

82 Xu, T. and Liu, X. (2008). Peanut shell activated carbon: characterization, surface modification and adsorption of Pb^{2+} from aqueous solution, Chinese. *J. Chem. Eng.* 16: 401–406. 10.1016/S1004-9541(08)60096-8.

83 Cagniant, D. and Gruber, R. (1998). Structural characterization of nitrogen-enriched coals. *Energy Fuel* 12: 672–681.

84 Mawhinney, D.B. and Yates, J.T. (2001). FTIR study of the oxidation of amorphous carbon by ozone at 300 K-Direct COOH formation. *Carbon N. Y.* 39: 1167–1173. 10.1016/S0008-6223(00)00238-4.

85 Lazar, G. and Lazar, I. (2003). IR characterization of a-C:H:N films sputtered in $Ar/CH_4/N_2$ plasma. *J. Non-Cryst. Solids* 331: 70–78. 10.1016/j.jnoncrysol.2003.09.004.

86 Salame, I.I. and Bandosz, T.J. (2001). Surface chemistry of activated carbons: combining the results of temperature-programmed desorption, Boehm, and potentiometric titrations. *J. Colloid Interface Sci.* 240: 252–258. 10.1006/jcis.2001.7596.

87 Saha, B., Tai, M.H., and Streat, M. (2001). Metal sorption performance of an activated carbon after oxidation and subsequent treatment. *Process Saf. Environ. Prot.* 79: 345–351. 10.1205/095758201753373113.

88 Mopoung, S., Moonsri, P., Palas, W., and Khumpai, S. (2015). Characterization and properties of activated carbon prepared from tamarind seeds by KOH activation for Fe(III) adsorption from aqueous solution. *Sci. World J.* 2015: 10.1155/2015/415961.

89 Li, X., Tang, Y., Song, J. et al. (2018). Self-supporting activated carbon/carbon nanotube/reduced graphene oxide flexible electrode for high performance supercapacitor. *Carbon N. Y.* 129: 236–244. 10.1016/j.carbon.2017.11.099.

90 Li, B., Dai, F., Xiao, Q. et al. (2016). Activated carbon from biomass transfer for high-energy density lithium-ion supercapacitors. *Adv. Energy Mater.* 6: 1–6. 10.1002/aenm.201600802.

91 Blankenship, T.S., Balahmar, N., and Mokaya, R. (2017). Oxygen-rich microporous carbons with exceptional hydrogen storage capacity. *Nat. Commun.* 8: 10.1038/s41467-017-01633-x.

92 Björkman, Å. (1969). Thermische Klärschlammbehandlung. *Schweiz. Z. Hydrol.* 31: 632–645. 10.1007/BF02543692.

93 Redondo, E., Fevre, L.W.L., Fields, R. et al. (2020). Enhancing supercapacitor energy density by mass-balancing of graphene composite electrodes. *Electrochim. Acta* 360: 0–9. 10.1016/j.electacta.2020.136957.

94 Shafeeyan, M.S., Daud, W.M.A.W., Houshmand, A., and Shamiri, A. (2010). A review on surface modification of activated carbon for carbon dioxide adsorption. *J. Anal. Appl. Pyrolysis* 89: 143–151. 10.1016/j.jaap.2010.07.006.

95 Rao, K.V.V., Kishore, G., Rao, K.S. et al. (2010). X-ray photoelectron spectroscopy studies on activated carbon prepared from rind of Citrus nabilis. *Asian J. Chem.* 22: 4377–4381.

96 Sriramulu, D., Vafakhah, S., and Yang, H.Y. (2019). Activated: luffa derived biowaste carbon for enhanced desalination performance in brackish water. *RSC Adv.* 9: 14884–14892. 10.1039/c9ra01872g.

97 Yi, K., Iradukunda, Y., Luo, F. et al. (2021). Preparation of activated carbon derived from licorice residue and its electrochemical properties. *Int. J. Electrochem. Sci.* 16: 1–10. 10.20964/2021.03.33.

98 Hira, S.A., Yusuf, M., Annas, D. et al. (2020). Biomass-derived activated carbon as a catalyst for the effective degradation of Rhodamine B dye. *Processes* 8: 10.3390/PR8080926.

99 Gao, X., Wu, L., Li, Z. et al. (2018). Preparation and characterization of high surface area activated carbon from pine wood sawdust by fast activation with H_3PO_4 in a spouted bed. *J. Mater. Cycles Waste Manage.* 20: 925–936. 10.1007/s10163-017-0653-x.

100 Kigozi, M., Kali, R., Bello, A. et al. (2020). Modified activation process for supercapacitor electrode materials from African maize cob. *Materials (Basel)* 13: 1–20. 10.3390/ma13235412.

101 IlKim, M., Seo, S.W., Kwak, C.H. et al. (2021). The effect of oxidation on the physical activation of pitch: crystal structure of carbonized pitch and textural properties of activated carbon after pitch oxidation. *Mater. Chem. Phys.* 267: 124591. 10.1016/j.matchemphys.2021.124591.

102 Sudhan, N., Subramani, K., Karnan, M. et al. (2017). Biomass-derived activated porous carbon from rice straw for a high-energy symmetric supercapacitor in aqueous and nonaqueous electrolytes. *Energy Fuel* 31: 977–985. 10.1021/acs.energyfuels.6b01829.

103 Han, X., Jiang, H., Zhou, Y. et al. (2018). A high performance nitrogen-doped porous activated carbon for supercapacitor derived from pueraria. *J. Alloys Compd.* 744: 544–551. 10.1016/j.jallcom.2018.02.078.

104 Liu, Z., Zhu, Z., Dai, J., and Yan, Y. (2018). Waste biomass based-activated carbons derived from soybean pods as electrode materials for high-performance supercapacitors. *ChemistrySelect* 3: 5726–5732. 10.1002/slct.201800609.

105 Farma, R., Fadilah, R., Awitdrus, N.K. et al. (2018). Corn cob based activated carbon preparation using microwave assisted potassium hydroxide activation for sea water purification. *J. Phys. Conf. Ser.* 1120: 10.1088/1742-6596/1120/1/012017.

106 Girgis, B.S., Temerk, Y.M., Gadelrab, M.M., and Abdullah, I.D. (2007). X-ray diffraction patterns of activated carbons prepared under various conditions. *Carbon Lett.* 8: 95–100. 10.5714/cl.2007.8.2.095.

107 Surya, K. and Michael, M.S. (2021). Hierarchical porous activated carbon prepared from biowaste of lemon peel for electrochemical double layer capacitors. *Biomass Bioenergy* 152: 106175. 10.1016/j.biombioe.2021.106175.

108 Karnan, M., Subramani, K., Sudhan, N. et al. (2016). Aloe vera derived activated high-surface-area carbon for flexible and high-energy supercapacitors. *ACS Appl. Mater. Interfaces* 8: 35191–35202. 10.1021/acsami.6b10704.

109 Taer, E., Taslim, R., Aini, Z. et al. (2017). Activated carbon electrode from banana-peel waste for supercapacitor applications. *AIP Conf. Proc.* 1801: 10.1063/1.4973093.

110 Liu, J., Liu, B., Wang, C. et al. (2017). Walnut shell – derived activated carbon: synthesis and its application in the sulfur cathode for lithium–sulfur batteries. *J. Alloys Compd.* 718: 373–378. 10.1016/j.jallcom.2017.05.206.

111 Ma, G., Yang, Q., Sun, K. et al. (2015). Nitrogen-doped porous carbon derived from biomass waste for high-performance supercapacitor. *Bioresour. Technol.* 197: 137–142. 10.1016/j.biortech.2015.07.100.

112 Yu, J., Gao, L.Z., Li, X.L. et al. (2016). Porous carbons produced by the pyrolysis of green onion leaves and their capacitive behavior. *Xinxing Tan Cailiao/New Carbon Mater.* 31: 475–484. 10.1016/S1872-5805(16)60026-4.

113 Senthilkumar, S.T., Selvan, R.K., Melo, J.S., and Sanjeeviraja, C. (2013). High performance solid-state electric double layer capacitor from redox mediated gel polymer electrolyte and renewable tamarind fruit shell derived porous carbon. *ACS Appl. Mater. Interfaces* 5: 10541–10550. 10.1021/am402162b.

114 Feng, H., Hu, H., Dong, H. et al. (2016). Hierarchical structured carbon derived from bagasse wastes: a simple and efficient synthesis route and its improved electrochemical properties for high-performance supercapacitors. *J. Power Sources* 302: 164–173. 10.1016/j.jpowsour.2015.10.063.

115 Ranaweera, C.K., Kahol, P.K., Ghimire, M. et al. (2017). Orange-peel-derived carbon: designing sustainable and high-performance supercapacitor electrodes. *C.* 3: 25. 10.3390/c3030025.

116 Zhang, Q., Han, K., Li, S. et al. (2018). Synthesis of garlic skin-derived 3D hierarchical porous carbon for high-performance supercapacitors. *Nanoscale* 10: 2427–2437. 10.1039/c7nr07158b.

117 Wang, X., Li, Y., Lou, F. et al. (2017). Enhancing capacitance of supercapacitor with both organic electrolyte and ionic liquid electrolyte on a biomass-derived carbon. *RSC Adv.* 7: 23859–23865. 10.1039/c7ra01630a.

118 Ahmed, S., Ahmed, A., and Rafat, M. (2018). Supercapacitor performance of activated carbon derived from rotten carrot in aqueous, organic and ionic liquid based electrolytes. *J. Saudi Chem. Soc.* 22: 993–1002. 10.1016/j.jscs.2018.03.002.

119 Govindan, B., Alhseinat, E., Darawsheh, I.F.F. et al. (2020). Activated carbon derived from phoenix dactylifera (palm tree) and decorated with MnO_2 nanoparticles for enhanced hybrid capacitive deionization electrodes. *ChemistrySelect* 5: 3248–3256. 10.1002/slct.201901358.

120 Ali, U.F.M., Azmi, N.H., Isa, K.M. et al. (2018). Optimization study on preparation of amine functionalized sea mango (*Cerbera odollam*) activated carbon for carbon dioxide (CO_2) adsorption. *Combust. Sci. Technol.* 190: 1259–1282. 10.1080/00102202.2018.1448393.

121 Li, R., Zhang, L., and Wang, P. (2015). Rational design of nanomaterials for water treatment. *Nanoscale* 7: 17167–17194. 10.1039/c5nr04870b.

122 Abdullah, M.O., Tan, I.A.W., and Lim, L.S. (2011). Automobile adsorption air-conditioning system using oil palm biomass-based activated carbon: a review. *Renew. Sust. Energ. Rev.* 15: 2061–2072. 10.1016/j.rser.2011.01.012.

123 Danish, M. and Ahmad, T. (2018). A review on utilization of wood biomass as a sustainable precursor for activated carbon production and application. *Renew. Sust. Energ. Rev.* 87: 1–21. 10.1016/j.rser.2018.02.003.

124 Yalçin, N. and Sevinç, V. (2000). Studies of the surface area and porosity of activated carbons prepared from rice husks. *Carbon N. Y.* 38: 1943–1945. 10.1016/S0008-6223(00)00029-4.

125 Liu, Q.S., Zheng, T., Wang, P., and Guo, L. (2010). Preparation and characterization of activated carbon from bamboo by microwave-induced phosphoric acid activation. *Ind. Crop. Prod.* 31: 233–238. 10.1016/j.indcrop.2009.10.011.

126 Singh, G., Lakhi, K.S., Sil, S. et al. (2019). Biomass derived porous carbon for CO_2 capture. *Carbon N. Y.* 148: 164–186. 10.1016/j.carbon.2019.03.050.

127 Sharma, P.P., Wu, J., Yadav, R.M. et al. (2015). Nitrogen-doped carbon nanotube arrays for high-efficiency electrochemical reduction of CO_2: on the understanding of defects, defect density, and selectivity. *Angew. Chem. Int. Ed.* 54: 13701–13705. 10.1002/anie.201506062.

128 Amiinu, I.S., Liu, X., Pu, Z. et al. (2018). From 3D ZIF nanocrystals to $Co–N_x/C$ nanorod array electrocatalysts for ORR, OER, and Zn–air batteries. *Adv. Funct. Mater.* 28: 1–9. 10.1002/adfm.201704638.

129 Jia, Y., Chen, J., and Yao, X. (2018). Defect electrocatalytic mechanism: concept, topological structure and perspective. *Mater. Chem. Front.* 2: 1250–1268. 10 .1039/c8qm00070k.

130 Wang, S., Jiang, H., and Song, L. (2019). Recent progress in defective carbon-based oxygen electrode materials for rechargeable Zink-air batteries. *Batteries Supercaps* 2: 509–523. 10.1002/batt.201900001.

131 Tang, C. and Zhang, Q. (2017). Nanocarbon for oxygen reduction electrocatalysis: dopants, edges, and defects. *Adv. Mater.* 29: 10.1002/adma.201604103.

132 Jiang, Y., Yang, L., Sun, T. et al. (2015). Significant contribution of intrinsic carbon defects to oxygen reduction activity. *ACS Catal.* 5: 6707–6712. 10.1021/ acscatal.5b01835.

133 Hu, Y.H. and Ruckenstein, E. (2020). Comment on "dry reforming of methane by stable ni–mo nanocatalysts on single-crystalline MgO". *Science (80-.)* 368: 777–781. 10.1126/science.abb5459.

134 Dong, Y., Zhang, S., Du, X. et al. (2019). Boosting the electrical double-layer capacitance of graphene by self-doped defects through ball-milling. *Adv. Funct. Mater.* 29: 1–10. 10.1002/adfm.201901127.

135 Xue, L., Li, Y., Liu, X. et al. (2018). Zigzag carbon as efficient and stable oxygen reduction electrocatalyst for proton exchange membrane fuel cells. *Nat. Commun.* 9: 2–9. 10.1038/s41467-018-06279-x.

136 Dou, S., Tao, L., Wang, R. et al. (2018). Plasma-assisted synthesis and surface modification of electrode materials for renewable energy. *Adv. Mater.* 30: 1–24. 10.1002/adma.201705850.

137 Tao, L., Wang, Q., Dou, S. et al. (2016). Edge-rich and dopant-free graphene as a highly efficient metal-free electrocatalyst for the oxygen reduction reaction. *Chem. Commun.* 52: 2764–2767. 10.1039/c5cc09173j.

138 Zhu, J., Huang, Y., Mei, W. et al. (2019). Effects of intrinsic pentagon defects on electrochemical reactivity of carbon nanomaterials. *Angew. Chem. Int. Ed.* 58: 3859–3864. 10.1002/anie.201813805.

139 Wu, J., Sharifi, T., Gao, Y. et al. (2019). Emerging carbon-based heterogeneous catalysts for electrochemical reduction of carbon dioxide into value-added chemicals. *Adv. Mater.* 31: 1–24. 10.1002/adma.201804257.

140 Liu, S., Yang, H., Huang, X. et al. (2018). Identifying active sites of nitrogen-doped carbon materials for the CO_2 reduction reaction. *Adv. Funct. Mater.* 28: 1–7. 10.1002/adfm.201800499.

141 Li, L., Huang, Y., and Li, Y. (2020). Carbonaceous materials for electrochemical CO_2 reduction. *EnergyChem* 2: 100024. 10.1016/j.enchem.2019.100024.

142 Feng, W., Long, P., Feng, Y.Y., and Li, Y. (2016). Two-dimensional fluorinated graphene: synthesis, structures, properties and applications. *Adv. Sci.* 3: 1–22. 10.1002/advs.201500413.

143 Kim, J., Zhou, R., Murakoshi, K., and Yasuda, S. (2018). Advantage of semi-ionic bonding in fluorine-doped carbon materials for the oxygen evolution reaction in alkaline media. *RSC Adv.* 8: 14152–14156. 10.1039/c8ra01636d.

144 Wu, J., Liu, M., Sharma, P.P. et al. (2016). Incorporation of nitrogen defects for efficient reduction of CO_2 via two-electron pathway on three-dimensional graphene foam. *Nano Lett.* 16: 466–470. 10.1021/acs.nanolett.5b04123.

145 Nakata, K., Ozaki, T., Terashima, C. et al. (2014). High-yield electrochemical production of formaldehyde from CO_2 and seawater. *Angew. Chem. Int. Ed.* 53: 871–874. 10.1002/anie.201308657.

146 Su, D.S., Zhang, B., and Schlögl, R. (2015). Electron microscopy of solid catalysts – transforming from a challenge to a toolbox. *Chem. Rev.* 115: 2818–2882. 10.1021/cr500084c.

147 Jia, Y., Zhang, L., Du, A. et al. (2016). Defect graphene as a trifunctional catalyst for electrochemical reactions. *Adv. Mater.* 28: 9532–9538. 10.1002/adma.201602912.

148 Zhang, J., Zhou, H., Zhu, J. et al. (2017). Facile synthesis of defect-rich and s/n co-doped graphene-like carbon nanosheets as an efficient electrocatalyst for primary and all-solid-state Zn-air batteries. *ACS Appl. Mater. Interfaces* 9: 24545–24554. 10.1021/acsami.7b04665.

149 Wang, W., Shang, L., Chang, G. et al. (2019). Intrinsic carbon-defect-driven electrocatalytic reduction of carbon dioxide. *Adv. Mater.* 31: 1–7. 10.1002/adma.201808276.

150 Xue, X., Yang, H., Yang, T. et al. (2019). N,P-coordinated fullerene-like carbon nanostructures with dual active centers toward highly-efficient multi-functional electrocatalysis for CO_2RR, ORR and Zn-air battery. *J. Mater. Chem. A* 7: 15271–15277. 10.1039/c9ta03828k.

151 Jiao, Y., Zheng, Y., Jaroniec, M., and Qiao, S.Z. (2014). Origin of the electrocatalytic oxygen reduction activity of graphene-based catalysts: a roadmap to achieve the best performance. *J. Am. Chem. Soc.* 136: 4394–4403. 10.1021/ja500432h.

152 Yan, D., Li, Y., Huo, J. et al. (2017). Defect chemistry of nonprecious-metal electrocatalysts for oxygen reactions. *Adv. Mater.* 29: 1–20. 10.1002/adma.201606459.

153 Wu, P., Qian, Y., Du, P. et al. (2012). Facile synthesis of nitrogen-doped graphene for measuring the releasing process of hydrogen peroxide from living cells. *J. Mater. Chem.* 22: 6402–6412. 10.1039/c2jm16929k.

154 Liu, Q., Tian, H., Dai, Z. et al. (2020). Nitrogen-doped carbon nanospheres-modified graphitic carbon nitride with outstanding photocatalytic activity. *Nano-Micro Lett.* 12: 1–15. 10.1007/s40820-019-0358-x.

155 Mrozowski, S. (1971). Electronic properties and band model of carbons. *Carbon N. Y.* 9: 97–109. 10.1016/0008-6223(71)90123-0.

156 Hastening, B. (1998). A model of the electronic properties of activated carbon. *Phys. Chem. Chem. Phys.* 102: 229–237. 10.1002/bbpc.19981020214.

3

Biomass-derived Carbon-based Materials for Microbicidal Applications

Selvamuthu Preethi[1,2], *Arunachalam Arulraj*[3], *Ramalinga Viswanathan Mangalaraja*[3], *Velayutham Ravichandran*[4], *and Natesan Subramanian*[4]

[1] *Department of Pharmaceutical Technology, University College of Engineering-BIT Campus, Anna University, Tiruchirappalli, 620024, Tamil Nadu, India*
[2] *Big Bang Boon Solutions Pvt. Ltd., Adambakkam, Chennai, 600088, Tamil Nadu, India*
[3] *Faculty of Engineering & Sciences, Universidad Adolfo Ibáñez, Peñalolén, Santiago, Chile*
[4] *National Institute of Pharmaceutical Education and Research (NIPER), Department of Pharmaceutics, Kolkata, 700032, West Bengal, India*

3.1 Introduction

Microbial diversity withstands the bionetwork and succeeds in an environmental condition involving various habitations as well as adapting the biogeochemical sequence of all living organisms in the world. When a microorganism freaks by hindering its resistive mechanism against the pathogen, then it is referred to be infectious. At this cause, antibiotic is needed to recover the microorganisms to fight against the external pathogens [1, 2]. In the past few decades, the impulsive use of antibiotics had a rapid hike in the incidence of drug resistance among the microorganisms increasing the concern toward the development of a new strategy for microbicidal inhibition. The antimicrobial resources are usually fighting against the group of the pathogenic microbes such as *Listeria monocytogenes (L. monocytogenes), Escherichia coli (E. coli), Staphylococcus aureus (S. aureus), Salmonella typhimurium (S. Typhimurium), Pseudomonas aeruginosa (P. aeruginosa), Enterococcus faecalis (E. faecalis), Streptococcus mutans (S. mutans), Vibrio harveyi (V. harveyi),* and *Staphylococcus epidermidis (S. epidermidis),* which can cause infections and are severe threats to human health. In this regard, researchers have suggested different routes of strategies such as bacteriophages, antimicrobial peptides, immunomodulators, quorum sensing inhibitors, and even predatory microorganisms. Each strategy has its own merits over the drug-resisting microorganisms as well as demerits like complexity, high cost, and random risk involved in human health [3, 4].

On the other hand, the nanotechnological methods for microbicidal disinfection using metal/metal oxide nanoparticles and carbon-based derivatives are comparatively simple, inexpensive, have high adaptability, and possess great biocidal

Biomass-Derived Carbon Materials: Production and Applications, First Edition.
Edited by Alagarsamy Pandikumar, Perumal Rameshkumar, and Pitchaimani Veerakumar.

properties. Even though, the uses of antimicrobial nanomedicine in clinical trials are very much limited [5–9]. As for now, silver (Ag) nanoparticles have succeeded in clinical trials among all other antimicrobial nanoagents. Numerous silver nanoparticles comprising antimicrobial nanoagents are existing in the market, and some combinations related to Ag nanoparticles are under clinical trials [6, 9, 10]. Yet, the silver-based antimicrobial agents are available as a topical ointment formulation; it is not a flawless answer for the replacement of antibiotic drugs. Initially, Ag nanoparticles will get accumulated in the tissues over a time, which may result in systemic side effects such as allergy, skin discoloration, and toxicity. Moreover, microorganisms are also developed as resistant against Ag nanoparticles because of its metal ions acting via the interruption of pathways similar to other commonly available antibiotics [11–14]. On the other hand, carbonaceous materials, such as graphene, carbon dots (CDs), and carbon nanotubes, act through the cell envelopes by various physicochemical routes with its unique properties such as electrical, optical, magnetic, catalytic, and microbicidal properties. Even though the biological effect of carbonaceous nanomaterials is not clinically confined, it still finds hope in the development of the microbial resistance, which is an utmost need in the current situation [15–18].

Carbonaceous materials can be prepared from the abundant resources of carbon includes biomass such as coconut shells [19], wood [20], and rice husk [21, 22] via different synthesis approaches such as pyrolysis, microwave hydrothermal, chemical vapor deposition, thermal combustion, hydrothermal/solvothermal, arc discharge, and other biological methods [23–26]. The synthesis of carbonaceous materials from biomass is very much considered in recent years because of the advantages such as abundant source, environmentally friendly, inexpensive, and much more. In the following section, it is emphasized about the overview of the biomass-derived carbon-based materials, its beneficiary microbicidal property against specific resistant bacteria, and the corresponding future perspectives.

3.2 Biomass Materials

Biomass refers to the organic materials that are derived from living organisms (plants or animals). Biomass can be divided into two major groups: virgin biomass (crops, trees, fruits, and vegetables) and aquatic biomass (water plants and algae). It comprises the waste from major resources of agricultural, municipal, food, and industrial sectors. Waste management is one of the important factors toward sustainable growth of the nation (particularly of the developing and underdeveloped nations). It can be achieved by the maximum utilization of wastes into commodities apart from evading production waste [27]. Currently, plenty of biomass (rice husk, woods, walnut peel, orange peel, lychee seeds, coconut shell, corn, etc.) has been used as sources for synthesizing carbonaceous materials (active carbon, CDs, graphite, graphene, etc.). The synthesized carbonaceous materials are applied in diverse fields such as water treatment [28], biosensing [29], electrochemical sensing [30], solar cells [31], catalyst [32], supercapacitors [33], antimicrobial products [34], imaging [24], drug delivery [35], and other biomedical applications [36].

3.2.1 Carbon and Its Derivatives

Carbon dots or carbon quantum dots are groups of carbon atoms with sizes usually less than 10 nm, which have been discovered in 2004 as a by-product from single-walled carbon nanotubes synthesis [37]. It exhibits physical and optical properties similar to the known semiconductor quantum dots with higher biocompatibility, lesser toxicity, and exhibits greater fluorescence comprising photostability resistance. This property tends to allow the CDs for development in several biomedical applications, together with their use against antimicrobial-resistant microorganisms. Currently, synthesis of CDs by hydrothermal carbonization of materials is emerging where both the carbonization and functionalization are done in a single step using the various biomass as a precursor and with low reaction time and temperature [38]. Wang et al. prepared CDs by hydrothermal carbonization, annealing, and laser ablation from soybean residuals (biomass), which show strong photoluminescent blue emission with great water wettability [39]. Similarly, a report on CDs from oil palm biomass without any synthetic chemical shows excellent optical properties and pH sensitivity. These as-synthesized CDs can be used in a broad area of applications such as antioxidants, bioimaging, and pH sensors [40]. Cheng et al. adopted an inexpensive method to form a controlled synthesis of triple color emissive fluorescent CDs from cellulose (willow catkin) based biomass. The prepared CDs show good photoluminescence (PL) stability, lower cytotoxicity, and excitation independent emission performances. This indicates an advantage of a long-wavelength emission and bulk production over other reported biomass-based CDs [41]. Prasannan et al. developed a simple and one-pot synthesized fluorescent CDs (2–7 nm) from orange peels by hydrothermal carbonization method (Figure 3.1a), which consists of sp^2 and sp^3 bonded carbon atoms with attractive luminescence property. This prepared CDs have their advantages due to its high solubility, possible applications as fluorescent markers, and effective catalysts in the field of bio and energy sciences [42].

Graphene is a 2D hexagonal patterned honeycomb lattice structured, single atomic thick, nanosized, and affords large surface area with adaptable surface chemistry to form complexes. It contains sp^2-bonded carbon atoms and represented to be the world's thinnest material, which will be stable even in its free form [44]. Graphene-based materials have been categorized based on the oxygen binding, number of layers in the sheet, and their chemical composition. Graphene oxide (GO) and reduced graphene oxide (rGO) are the derivatives of the graphene with slight differences in bonding corresponding to their carbon atom.

Recent reports and research are existing on the preparation of GO and rGO using natural, agro, and industrial biomass like vegetation and fruit wastes, animal waste as a source of a precursor. Akhavan et al. have synthesized GO and rGO from seven different natural biomass materials such as black mulberry wood, plane tree leaves, bagasse from sugarcane factory waste, the rind of an orange, chicken bone, cow dung, and newspapers. The surface morphology, structure, chemical state, and electrical properties are the same for all the precursor sources, and it also comparable to the graphene sheet obtained by hierarchical porous graphene (HPG).

Figure 3.1 Schematic representation of (a) carbon dots from orange peel wastes (Source: From Ref. [42] / with permission of American Chemical Society) and (b) carbonaceous materials derived from sugarcane baggage. Source: From Ref. [43] /MDPI /CC BY 4.0.

These reports show that many kinds of solid biomass can be used for the production of high-quality graphene sheets, which will be cost-effective, successful recycling of low value, or even hazardous wastes into valuable materials [20]. Somanathan et al. synthesized GO through a single-step conversion of sugarcane bagasse agricultural waste (Figure 3.1b), which shows a well-graphitized structure and the method used to prepare will be more eco-friendly compared to other methods as it evades toxic gas during synthesis procedure [43].

3.3 Microbicidal

Microbicidal also referred to as antimicrobials agents that eradicate or suppress the spread of microorganisms including viruses, bacteria, fungi (mold), parasites, protozoans, and mildew. The antimicrobial products can be regulated in two categories: (i) drug/antiseptics (used to treat or avoid diseases on peoples, animals, and other living beings) and (ii) pesticides (used on objects such as toys, countertops, grocery carts, and hospital equipment). The first antimicrobial agent was prepared for syphilis called salvarsan by Ehrlich in 1910. Later, Domagk and other researchers synthesized sulfonamides in 1935. Yet, these antimicrobial agents are prepared using synthetic compounds and had its disadvantages over safety issues and efficacy [45].

In 1928, Penicillin was discovered by Alexander Fleming from the growth of *S. aureus*, which was inhibited in a zone of contaminated blue mold (penicillium genus) in the petriplates. This leads to finding that microorganisms will be able to produce constituents that could inhibit other microorganisms growth and in the late 1940s it came into clinical use with the name "penicillin." The antibiotic (penicillin) supports healing the soldiers during World War II owing to its outstanding safety and efficacy [45]. Later, for the following two eras, various classes of antimicrobial agents were established in sequence. In 1944, streptomycin developed from a soil bacterium (*Streptomyces griseus*) followed by tetracycline, chloramphenicol, glycol-peptide, and macrolide were found from soil bacteria. Enhancements in every class of antimicrobial agents sustained to attain a wide range of antimicrobial spectrum and greater antimicrobial activity. This leads to developing β-lactams like penicillin (for gram-positive organisms: *S. aureus*), methicillin, ampicillin (for both gram-positive and gram-negative: *Enterobacteriaceae*), Piperacillin (*P. aeruginosa*), cephems (*E. coli, Enterobacteriaceae, P. aeruginosa*), carbapenems, and monobactams [45].

Continuous developments have been made for antimicrobial agents in numerous aspects to attain improved pharmacodynamics with better oral absorption drugs, distribution in inflammation, and antimicrobial chemotherapy also recognized including drug safety. Lately, the development of new classes of antimicrobial agents has been decreased, even with the presence of a lot number of antimicrobial agent producing companies. On the contrary, contagious diseases have been continuing to dose living beings as evolving, adaptable infectious diseases, and more importantly infections with the drug-resistive microorganisms.

3.3.1 Mechanism of Action

The innovation of penicillin for treating infections made the antimicrobial field to enlarge and leads to develop more antimicrobial therapeutic drugs. But some of the microorganisms tend to resist these antibiotics causing major health issues across the globe [46]. The reason behind this may be the microbe efflux pumps that eject the antibiotics from the cell before reaching its corresponding sites [47]. In general, these antibiotics exhibit their efficacy through five mechanisms in contrast to bacterial cells as inhibition of cell wall synthesis, protein synthesis, nucleic acid replication, and transcription, metabolites, and damage to the plasma membrane. Similar to the bacterial cell, fungus cell also deals with similar mechanisms along with DNA synthesis function, inhibition of microbial cell wall synthesis, folic acid metabolism [48, 49].

3.3.2 Microbicidal Resistance

Microbicidal-resistant threats became questionable for the cure and hindrance of a broad range of infections caused by microorganisms. Antimicrobial resistance will be developed when the microbes become resistive against the drugs. The main reason for this resistance is due to the use of therapeutic drugs unreasonably. The recombination capacity of microorganisms, DNA/RNA mutations, and obstruction of metabolic processes lead to develop the resistance for antibiotics. It can affect people at any period of life, along with agricultural, veterinary, and healthcare industries, which makes the issue the most serious public health problem worldwide. Due to this reality, it is essential to take action to prevent the emerging global crisis in health care [50].

Over the history of antimicrobial resistance, *S. aureus* is the most familiar resistive bacteria in clinical practice among all other resistant bacteria due to its prompt resistance. Penicillin was used against this bacterium once, but over the period, the resistant strains were increased. Though the *methicillin-resistant S. aureus* (MRSA) was isolated in the subsequent year [51]. The World Health Organization (WHO) reported that the few fatal bacterial infections also develop resistance that comprises diarrhea, syphilis, meningitis, respiratory tract diseases, and gonorrhea [52]. Tuberculosis caused by *Mycobacterium tuberculosis* bacteria managed to be resistant toward rifampicin and isoniazid drugs [53]. An influenza virus causes FLU that gets resistive against Flumadine and amantadine [54]. *Candida albicans* causes a fungal disease called candidiasis, which became resistant to fluconazole [55]. Malaria caused by protozoans with genus *plasmodium genus* developed resistance for quinine sulfate, mefloquine, and other quinine-based drugs [56].

3.3.3 Factors Affecting Microbicidal Resistance

The main factors that lead to microbicidal resistance is shown in Figure 3.2. There are four major sectors tangled in the enlargement of microbicidal resistance such as medicine in public and hospital (for humans), agriculture and food-producing animal, and the presence of resistant microbes in the environment. Nowadays, people are taking antibiotics with inappropriate prescriptions or self-medicating themselves. As of hospitals, there are a huge number of invasive medical devices and surgeries and also widespread and persistent use of antibiotics for therapies. In the agricultural sector, animal productions are increasing in our day-to-day life. These animals are treated with antibiotics regularly for their growth elevation and bulk prophylaxis. Similar to the food-producing animals, crops are also exposed to these antimicrobials to prevent fire blight. It is noted clearly that a large number of antimicrobials usage in the majority of the food-based industrial and agricultural sectors also leads to another factor environmentally available resistant microorganisms. This inappropriate usage of antimicrobials will have resulted in polluted irrigation waters in soils, organic fertilizers, and the release of a huge amount of antimicrobial agents in wastewater, which all leads to the increase of microbicidal-resistant microbes [57].

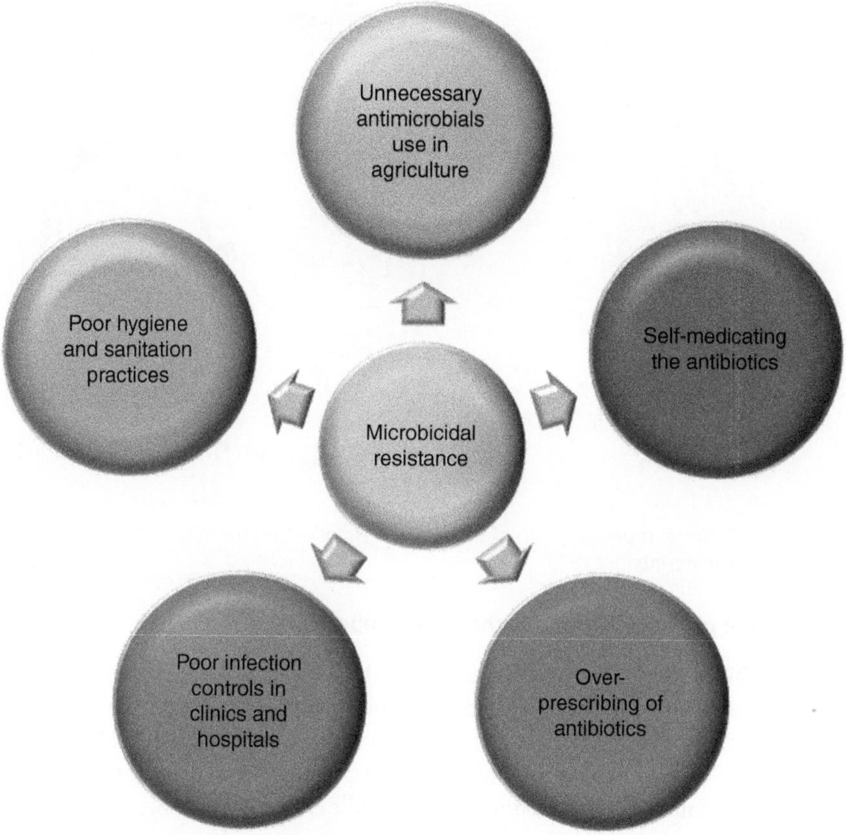

Figure 3.2 Factors affecting microbicidal resistance.

3.4 Microbicidal Performance of Biomass-Derived Carbonaceous Materials

Due to the evolving crisis of microbicidal resistance in global health, it is an urgent need for developing new classes of antibiotics, which leads the scientists and researchers to develop various types of alternatives. In this regard, the nanotechnological approaches have been giving positive results for the development of antimicrobials for resistant microorganisms with simple, greater efficacy, less cytotoxicity, and inexpensive production cost. Aforementioned effective usage of waste (carbonaceous materials) for biomedical applications (antimicrobial, antibacterial, and antifungal activities) is grabbing huge attention in recent years. The carbon-based materials are primarily engaged due to their physicochemical properties for the inhibition of microbes. This resulted to find more basic and systematic visions regarding the mechanisms of carbonaceous materials for microbial inhibition. Figure 3.3 represents the mechanism of action for the inhibition of microorganisms.

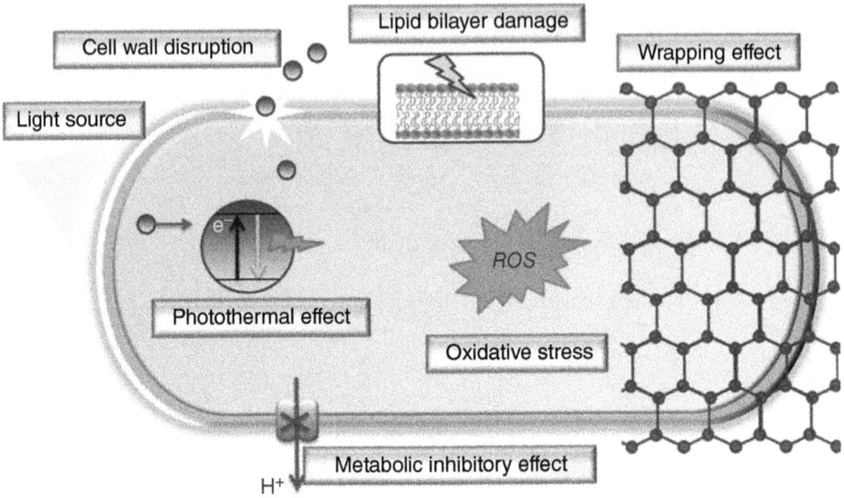

Figure 3.3 Pictorial representation of the mechanism of action for microbial inhibition by carbonaceous materials.

3.4.1 Role of Material Physicochemical Properties

The microbicidal activity of nanomaterials leads to the effects of nanotoxicology on microbial cells. Because of that, the antimicrobial activity of carbonaceous materials mainly depends on their nanosized structures and their physicochemical properties. Distinctively, carbon-based materials have a large family based upon its structure such as carbon nanotubes, fullerene, diamond-like carbon, graphene, activated carbon, and CDs. Accordingly, the antimicrobial effect varies considerably concerning their efficacy and its function. In this regard, Yallappa et al. synthesized nanocarbons from natural biomass of groundnut shells for evaluating its antimicrobial activity and to find the mechanisms behind it. The prepared nanocarbons were tested against both gram-positive and gram-negative bacteria such as *S. aureus, Bacillus subtilis, E. coli,* and *Chromobacterium violaceum* at different concentrations. By assessing the results, it is evidenced that the nanocarbons show moderate activity because of its high surface area to volume ratio, to be more specific, there will be an increase in the surface area while decreasing the size of the nanoparticles that leads to enhance the antimicrobial activity [61, 64].

3.4.1.1 Structural Destruction

In microbial cells, the cell wall is used for maintaining the cell structure, protection from mechanical stress, osmotic regulation, and fighting infections. The structural damage of the microbial cell wall will lead to its cytoplasmic components leakage, which results in the microbicidal effects. This is the most common mechanism of antimicrobial effect noted in carbon-based materials. Almost all types of carbon-based nanostructures can initiate the mechanical destruction to the microbial cell wall. For instance, GO/rGO walls show like a knife or blade-like action for microbial wall damage. Numerous accumulative studies have been reported

Table 3.1 Summarized biomass-derived carbon-based materials for antimicrobial activity and its efficacy.

Carbonaceous materials	Biomass precursor	Microorganisms	MIC ($\mu g\ mL^{-1}$)/ zone of inhibition (mm)	Antimicrobial protocol	Reference
Ag/AC	Coconut	*E. coli*	$10^7\ \mu g\ mL^{-1}$	Plate counting method	[19]
Graphene	Rice husk	*Bacillus* sp.	$3.2\times10^7\ \mu g\ mL^{-1}$ cells mL^{-1}	MIC	[21]
CNPs	Rapeseed oil cake	*S. aureus* *P. diminuta* *Y. enterocolitica* *S. enterica typhimurium* *Bacillus. cereus* *L. monocytogens* *M. smegmatis* *K. pneumonia* *C. albicans* *E. coli*	31 mm 15 mm 15 mm 15 mm 31 mm 31 mm 62 mm 7 mm 31 mm 7 mm	Well diffusion method	[58]
ACPB/AgNPs	Sugarcane bagasse	*E. coli*	$10^7\ \mu g\ mL^{-1}$	Agar well diffusion method	[59]
CNPs	Kitchen soot	*S. aureus* *S. haemolyticus* *P. refrigere* *P. aeruginosa*	19–20 mm 20–21 mm 28–29 mm 20–22 mm	Zone of inhibition, disk diffusion method	[60]
Nano carbons	Groundnut shell	*E. coli* *B. cereus* *C. vileceum*	10–17 $\mu g\ mL^{-1}$ 8–15 $\mu g\ mL^{-1}$ 7–13 $\mu g\ mL^{-1}$	Agar well diffusion method	[61]
GQDs/SnO_2	Sugarcane bagasse	*P. aeruginosa*	23 ± 0.4 mm	Disk diffusion assay	[62]
CDs	Apple juice	*M. tuberculosis* *P. aeruginosa* *Magnaporthe oryzae*	—	Zone of inhibition	[63]

that the act of sharp edges is one of the main factors for antimicrobial properties of graphene. This mechanism is known as penetration mode or insertion mode, which leads to membrane stress, resulted in both theoretical simulations and experimental studies. This has been proven by Akhavan et al. in 2010 by quantifying the release of microbial intracellular components like RNA in phosphate-buffered saline (PBS) with the exposure of graphene/reduced rGO against *S. aureus* (Figure 3.4) [65]. Das et al. showed that the CDs from the residues of rapeseed oilcake (after oil extraction) could also be used effectively for the antimicrobial applications. The report describes

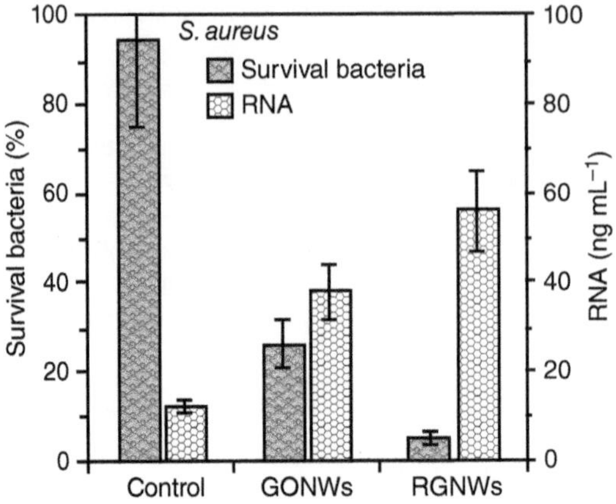

Figure 3.4 Cytotoxicity of graphene oxide nanowalls (GONWs) and reduced graphene nanowalls (RGNWs) along with the concentration of RNA in PBS to *S. aureus*. Source: Ref. [65] / with permission of American Chemical Society.

the antimicrobial ability of prepared CDs on 10 different foodborne pathogens. Among those, *S. aureus, Pseudomonas diminuta, Yersinia enterocolitica, S. enterica typhimurium, C. albicans,* and *E. coli* show high activity, *B. cereus, L. monocytogenes,* and *Klebsiella pneumonia* show moderate result, while *Mycobacterium smegmatis* shows low activity. These variations in results could be attributed to the differences in the rate of CDs permeation into the cells. TEM studies revealed that the cause of cell death is cell membrane rupture, owing to rod-like morphological of untreated *E. coli* (well-defined cell wall), whereas the CD-treated *E. coli* shows damaged cell wall thereby collapsing its network resulting in the irregularity of shape [58]. In the latest study, it exposes that the cell wall damage of microbial cell gets improved while changing the alignment of GO nanosheet from horizontal to vertical angle as it has increased density on the nanoedges. Besides, the gram-positive microbes like *S. aureus* have absence of outer membrane, which leads to being more responsive to get intact with the rGO edges [66].

Another way to find out the outer membrane damage is by using flow cytometry and confocal microscope. In one of the recent research, flow cytometry and confocal microscope were engaged to evaluate the membrane integrity of *Pseudomonas putida* cells for identifying the cell death or damage after treating it with GO. To recognize the damaged cells or cells with a ruptured membrane, a membrane-impermeable dye called propidium iodide (PI) is used for staining the cells, where the dye will be binding with DNA and releases red fluorescence indicating the cell death. The confocal microscopy revealed the rise in PI intensity after treating it with GO, and it is noted that the longer the exposure to GO, the more cell death is occurring till it reaches the threshold point (Figure 3.5) [67]. Pham et al. developed the pristine graphene nanosheets that show adaptable bactericidal efficacy for *P. aeruginosa* and *S. aureus*. The article demonstrates interaction existing between the graphene surfaces with the bacterial cell membrane and confirmed that the density of the edges of graphene plays an important factor that

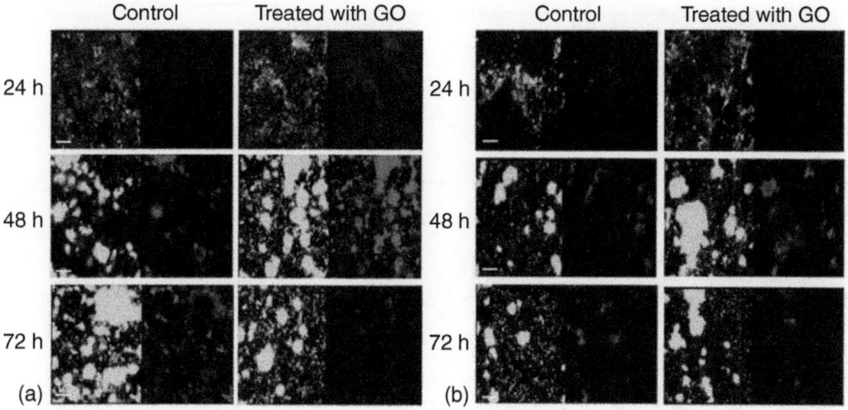

Figure 3.5 Confocal images of *P. putida* biofilm with or without GO-120 (20 μm scale bar) at 2, 48, and 72 hours: (a) 85 $\mu g\,mL^{-1}$ and (b) 8.5 $\mu g\,mL^{-1}$. Source: Ref. [67] / with permission of Springer Nature.

elevates the antimicrobial property of graphene [68]. By thoroughly analyzing the antibacterial effect of GO, rGO, graphite oxide (GtO), and graphite (Gt) for *E. coli*, Liu et al. revealed that rGO has not seemed to have antimicrobial property if the direct contact was absent indicating membrane stress leading to oxidative stress that contributed the antimicrobial activity [69].

In contrast, some researchers have proposed the presence of basal planes that regulates the antimicrobial activity of graphene rather than the sharp edges. To prove this, Mangadlao et al. developed GO sheets covered entirely with the polyethylene terephthalate (PET) substrate to restrain itself by a recognized Langmuir–Blodgett (LB) technique. In this phenomenon, the edge effect will be neglected, and remarkably, GO was able to disarm the *E. coli*, and the progressive connection was found between the basal planes and antimicrobial activity [70].

3.4.1.2 Oxidative Stress

ROS-Dependent Oxidative Stress Carbonaceous materials competent in intermediating the reactive oxygen species (ROS) generation over light-dependent or independent reactions by the absorbing O_2 from the edges and defect sites of the materials and its consequent reduction by cellular reflects with enzymes such as glutathione (GSH, antioxidant compound) and upon oxidization, it forms glutathione disulfide (GSSG). Similarly, various other antioxidants like α-tocopherol and *N*-acetylcysteine (NAC) or oxidant-sensitive dyes such as dichloro-dihydro-fluorescein diacetate (DCFH-DA) can also reveal the ROS generation. Furthermore, it is showed that pristine graphene can induce mitochondrial membrane reduction of murine RAW 264.7 macrophages concerning the dose where it is confirmed that the antimicrobial activities of graphene are correlated to mitochondrial membrane depolarization, which is a key factor related to intracellular ROS accumulation (Figure 3.6) [71].

From the various report, it is witnessed that GO and rGO together kill microbes by ROS-dependent oxidative stress when *P. aeruginosa* and *Xanthomonas oryzae* were used as models microbes. Also, it is observed that the level of ROS in GO- and

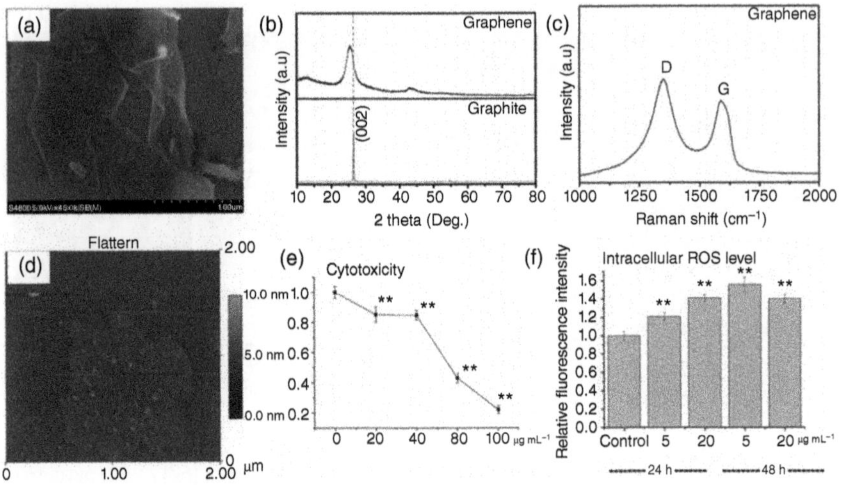

Figure 3.6 Characteristics of pristine graphene and its toxicity in RAW 264.7 macrophage cells: (a) SEM image; (b) XRD patterns; (c) Raman spectrum; (d) AFM image (1% pluronic F108); and (e and f) cell viability of RAW 264.7 cells and intracellular ROS levels of cells containing pristine graphene at 5 or 20 mg mL^{-1} for 24 or 48 hours. Source: Ref. [71] / with permission of Elsevier.

rGO-treated cells is higher compared to untreated control cells [72, 73]. Along with GO and rGO, carbon quantum dots (CQDs) also help to inhibit *P. aeruginosa*. Mohan et al. prepared sugarcane bagasse-derived CQDs with SnO_2 nanoparticles for the use of disinfectant against *P. aeruginosa*. By comparing to the previously reported studies on *P. aeruginosa*, the CQDS-SnO_2 seems to inhibit more microbes even without the light irradiation. This is because of the three stages of antimicrobial action. At first, the nanostructures adhere to the cell wall rearranging the cellular components, followed by the generation of ROS, and the final stage is the membrane damage and leakage. Here, the detrimental effects of SiO_2 might also have helped to generate ROS due to the metal ions discharge into the cell. Henceforth, the addition of numerous oxygen functionalities on to the carbon surfaces along with the SnO_2 enhances the antimicrobial properties of prepared biomass-derived CQDs-SnO_2 [62].

The aforementioned CDs also have their influence on the inhibition of microorganisms with the production of ROS. In recent years, several studies are done based upon CDs because of their high efficacy toward antimicrobial activities. For instance, Li et al. report the synthesis of CDs using vitamin C (as source) via the one-step electrochemical method and witnessed (as shown in Figure 3.7) broad spectrum over the antimicrobial activity against bacteria (like *B. subtilis, S. aureus, E. coli, Bacillus* sp. WL-6) and also for the ampicillin-resistant *E. coli* and some of the fungi like *Rhizoctonia solani* and *Pyricularia Grisea* [74].

Likewise, Jhonsi et al. demonstrate the improved antifungal activities of CDs against *C. albicans*. When the microbes were exposed to blue light with the presence of graphene quantum dots or CDs, an evident fluorescent signal from ROS-sensitive dye was noticed that indicates the ROS production from GQDs/CDs. Furthermore,

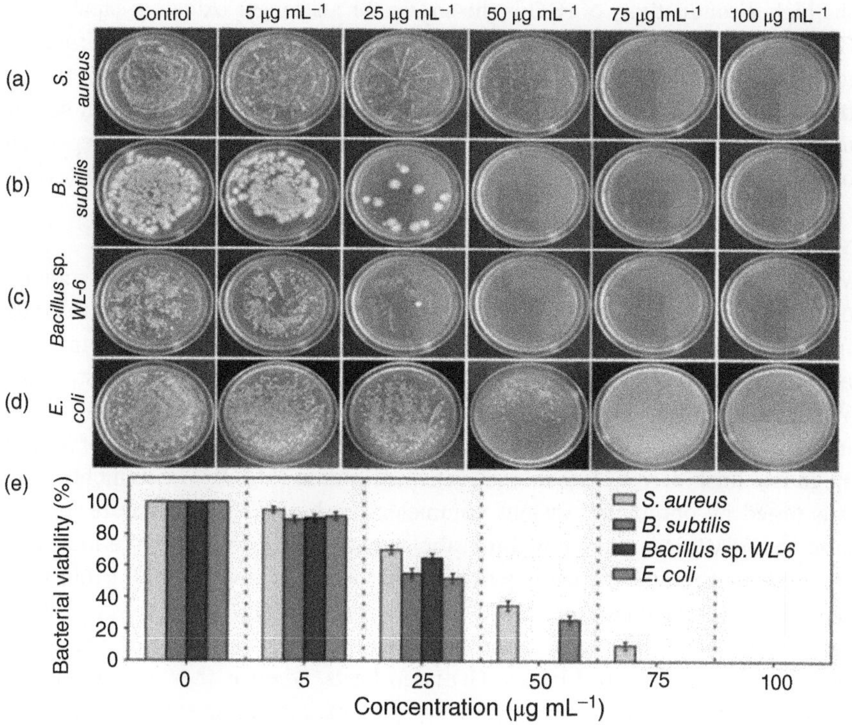

Figure 3.7 (a–d) Typical photographs of *S. aureus, B. subtilis, Bacillus* sp. WL-6, and *E. coli* after the treatment with different concentrations CDs (0, 5, 25, 50, 75, and 100 μg mL^{-1}) for 24 hours. Source: From Ref. [74], American Chemical Society. (e) Bacterial viability of *S. aureus, B. subtilis, Bacillus* sp. WL-6, and *E. coli* evaluated with different concentrations of CDs by UV–vis spectroscopy method. Source: Ref. [74] / with permission of American Chemical Society.

this was advanced by functionalizing the CDs with electron-donating groups like $-NH_2$, which influences decreasing the highest occupied molecular orbital (HOMO)-lowest unoccupied molecular orbital (LUMO) gap [75].

ROS-independent Oxidative Stress Eventhough there are many numbers of experimental proofs emphasizing the leading part of ROS-dependent oxidative stress about the antimicrobial action of carbonaceous materials, certain researchers have demanded that the carbon-based materials can be able to induce antimicrobial effect by ROS-independent oxidative stress also. There are various reports that have been given for supporting ROS-independent oxidative stress based antimicrobial mechanism for graphene. Liu et al. worked in the XTT (2,3-bis (2-methoxy-4-nitro-5-sulfophenyl)-2Htetrazolium-5-carboxamide) method used for measuring the $O_2^{\cdot -}$ levels in GO and rGO dispersals and used Ellman's assay to estimate the GSH oxidation. By this, it has been found that graphene has various oxidation capacities to GSH, which confirmed that graphene can able to facilitate O_2^{-}-independent oxidative stress. Therefore, it was found that by comparing GO, the higher conductivity of rGO is the source for a stronger oxidative capability to GSH, and rGO can act as a conductive link for lipid bilayer and external environment to transfer an electron from microbial intracellular components. In the meantime, the GSH strong oxidation of rGO proposes that conductive graphene can be able to oxidize thiols or cellular components, which together explain ROS-independent oxidative stress, also provides an increase in graphene antimicrobial properties [76].

3.4.1.3 Wrapping Effect

Wrapping effect is the third possible mechanism used for the antimicrobial activities of graphene because of its layered structure (flexible), which insulate the microbes from their surrounding environment caused by its 2D lateral structure, made up of sole layered sp^2 carbon atoms organized in the hexagonal crystal structure. Wrapping the microbial cells and isolating them from getting their nutrients result in microbial cell death. Sekaran et al. investigated about immobilization of *Bacillus* sp. in rice husk-derived mesoporous activated carbon (MAC). The immobilization was tested by optimizing various parameters such as pH, temperature, particle size, and MAC. It is to be noted that the activity gets decreased while increasing the temperature and increases when the particle size decreases [21]. Liu et al. reported that GO nanosheet could also isolate microbial cells; however, it has higher permeability than graphene nanosheets. It was exposed that the widely held *E. coli* were wrapped by less GO nanosheets, whereas they were trapped in aggregates of huge rGO sheets (Figure 3.8) [76]. Mejias Carpio et al. described that both GO and poly-*N*-vinyl carbazole-graphene oxide (PVK-GO) nanocomposite can

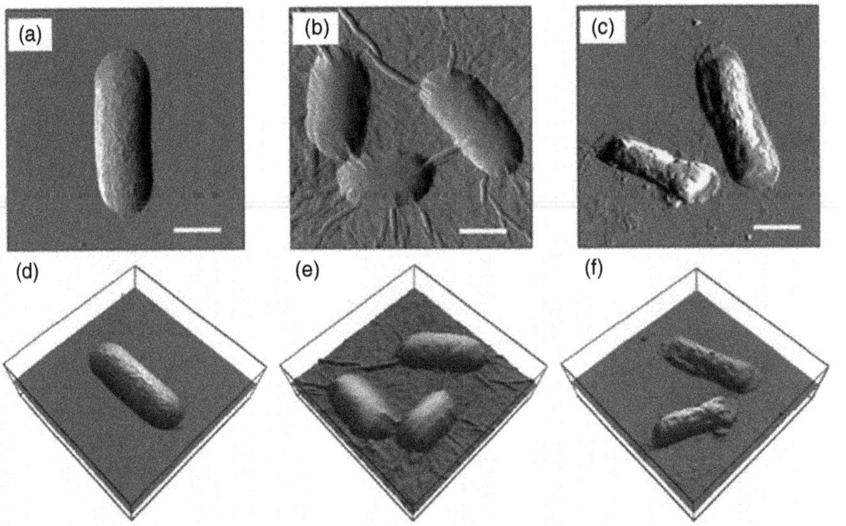

Figure 3.8 AFM amplitude and 3D images of *E. coli* cells after incubation with GO sheets (scale bar 1 μm): (a and d) *E. coli* incubation with deionized water for 2 hours; (b and e) *E. coli* incubation with the 40 μg mL^{-1} GO-0 suspension for 2 hours, and (c and f) *E. coli* after incubation with the 40 μg mL^{-1} GO-240 suspension for 2 hours. Source: Ref. [76] / with permission of American Chemical Society.

provide a strong antimicrobial activity for gram-positive and gram-negative bacteria through wrapping effect that leads to the inhibition of bacterial production [77]. In contrast to the other mechanisms such as the GONWs or rGONWs penetrating the outer membrane and inhibiting cell damage, it was found that the microbial structure and reliability endured [65]. As an alternative, the wrapping effect in the antimicrobial properties verifies that it reduces the metabolic activity of microbes and cell viability by quantifying the nicotinamide adenine dinucleotide hydrogen/phosphate through cellular metabolic assay.

Numerous studies observed that the wrapping and isolation of microbes can also inhibit microbial cell membrane perturbation to some extent with minor variations. For instance, Chen et al. established that the GO sheets entangled the pathogens once it gets contact with it. The SEM indicates the membrane integrity perturbation on *Pseudomonas syringae* and *Xanthomonas compestris pvundulosa*, but there is no wound has been showed. Further, it exposed the membrane depolarization to exhibit severe structural damage for wrapped microbial cells by membrane possible experiments [78]. The wrapping theory has been sustained by noticing the size-dependent antimicrobial activity by both experimental study and theoretical simulations [79].

3.4.1.4 Photothermal Effect

Carbon-based materials can work as photosensitizers to kill microbial cells by photo- dependent ROS production with the presence of photoexcitation. Otherwise, it can also emit fluorescence or create a photothermal effect through nonradiative relaxation pathways when it decays to the ground state. So, the carbonaceous materials with low fluorescence quantum yield and high absorption efficiency tend to have a great photothermal effect, which can be applied to the antimicrobial applications [80]. The photothermal effect for killing microbial cells was first studied in carbon nanotubes because of high quantum efficiency and strong NIR absorption which can be converted into thermal energy [81]. Here, the microbial sustainability was decreased by the heat produced when the CNT was exposed to NIR laser irradiation. Over some time, the antimicrobial effect of GO and rGO has also been identified. But, rGO showed higher photothermal efficacy than GO because of its higher NIR absorbance and intrinsic thermal conductivity. Generally, these study shows that carbon-based materials can have greater antimicrobial properties with the enhanced NIR absorption that improves the thermal conductivity and photothermal conversion [82–84].

Furthermore, Xiang et al. prepared an injectable hydrogel consists of CDs/ZnO fabricated with folic acid-conjugated polydopamine (PDA) for rapid inhibition of microbes by dual-light triggered ROS generation and membrane permeability. The *in vitro* antibacterial activity of fabricated hydrogel was treated against gram-negative bacteria (*E. coli*) and gram-positive bacteria (*S. aureus*) with and without the presence of 660 and 808 nm light irradiation for 15 minutes. After the irradiation, the inhibition of bacteria was far higher than compared to the group of bacteria without light. Under irradiation, the heat and ROS generation causes the cellular membrane destruction, and the Zn^{2+} ions pass through the membrane to disturb the metabolic activities and damaging the intracellular proteins, which enhance the antibacterial

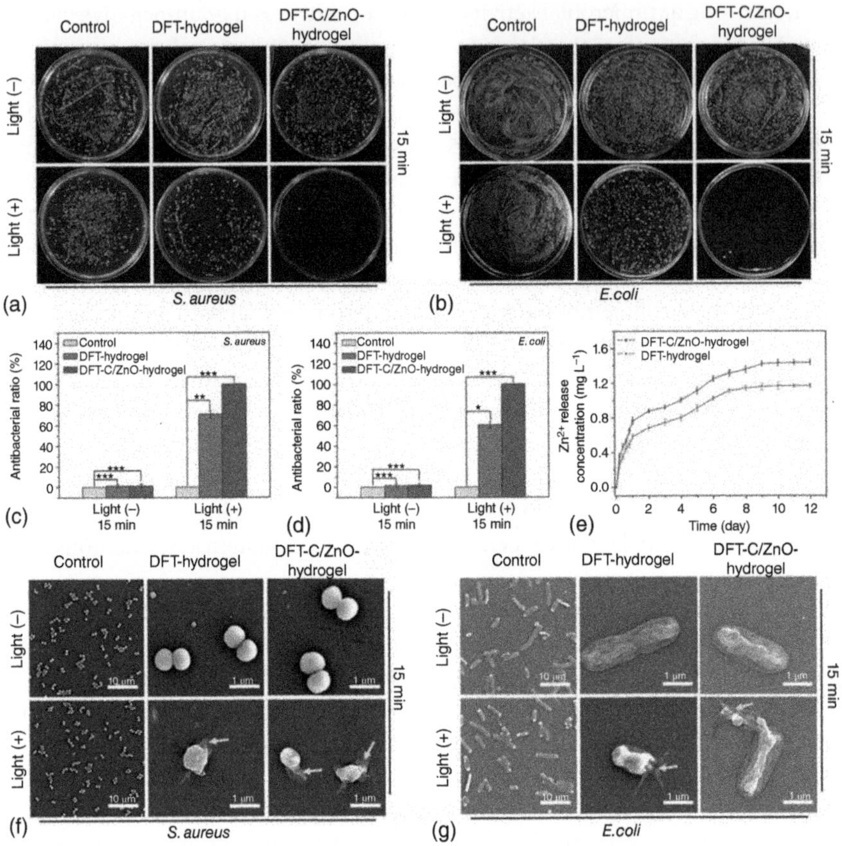

Figure 3.9 Representative images of viable (a) *S. aureus* and (b) *E. coli* grown on different samples with or without 15 minutes mixed-light irradiation; Antibacterial ratio of different samples against (c and d) *S. aureus* and *E. coli* via agar plating method; (e) cumulative amounts of Zn ions released from the hydrogels; FESEM morphology of (f) *S. aureus* and (g) *E. coli* treated with different samples. Source: Ref. [85] / with permission of John Wiley & Sons.

effect in 15 minutes. The field emission scanning electron microscope (FESEM) images show normal structures of both types of bacteria in control groups, whereas under irradiation, the membranes of two types of bacteria got ruptured or broken and the leakage of cytoplasm inside the bacteria has been marked (Figure 3.9). This study clearly documented that the CDs/ZnO hydrogels can provide highly efficient photothermal and photodynamic properties under dual light irradiation [85].

3.4.1.5 Extraction of Lipid

Microbial cells also have lipid membranes alike mammalian cells, which is necessary to uptake nutrients and neglect a toxic molecule. So, disturbing the lipid membrane will also cause microbial cell death. The graphene and microbial cell interactions are more complicated than originally predicted. Killing microorganisms through lipid extraction is a new mechanism compared to the others. Theoretical simulations and experimental studies have revealed the 2D graphene structure with entire sp^2 carbons plays a wide role in the interaction between the graphene and lipid membranes, which exceeds the lipid molecules attraction in the membrane. Accordingly, a lot of phospholipids can be vigorously removed from the lipid bilayer to the surface of graphene, which finally leads to cell wall damage and microbial cell death [86].

Researchers have found the strong hydrophobic interaction of graphene and lipid molecules known as "nanoscale dewetting" other than the van der Walls interaction, to give a driving force for cell wall damage. Furthermore, lipid extraction is a critical process in the CNT toxicity effect for mammalian cells. According to this, both the hydrophilic edges and hydrophobic plan are present in GO, and a wide range of surface areas in carbon materials regulate the binding efficiency with lipid molecules, which are crucial for their antimicrobial efficacy by lipid extraction [87–89].

3.4.1.6 Metabolic Inhibitory Effect

Microbial metabolism involves the process of nutrient to energy conversion and waste molecules elimination, which allows the microbes to be alive, develop, and replicate. Unlike eukaryotic cells, microbial cells are absent with mitochondria and this lead to energy metabolism and ATPase that are processed inside the cell membrane. Conductive carbon-based materials such as CNTs and graphenecould employ antimicrobial activity by removing electrons from membrane respiratory chain [90]. Peptidoglycan (PG) is a microbial building block that helps to inhibit the green synthesis, which was considered as a potential active pathway for emerging antimicrobial agents. Here, graphene quantum dots or CDs with and without functionalization of D/L-GLu by PG synthesis were developed by one-pot pyrolysis reaction with the use of citric acid precursor. All the synthesized GQDs show low toxicity to humans cells with good fluorescent properties, though the antimicrobial activity was high only in D-Glu-GQDs. Molecular and cellular results disclose that it can be able to enter the microbial cell wall and bind with MurD ligase and inhibit the catalyzing activity for PG biosynthesis, which leads to the rupture in the cell membrane and causes cell death. On the contrary, the GODs and L-Glu-GQDs could also enter into the cell wall, but they were comparatively weak for their binding interaction with MurD. And also, the molecular dynamics simulation data show the same as experimental data that hydrogen bonding and van der Waals interaction with MurD and D-Glu-GQDs are stronger than in the other two, which confirms the greater inhibitory effect of D-Glu-GQDs [91].

3.5 Bioengineering Prospective Toward Carbonaceous Materials

As a matter of fact, there is no record on graphene-based materials toward antimicrobial activity in the clinical trial and even in preclinical studies, and the cases are rare, except for the recent case of nanodiamonds in antibacterial dental fillings fabrication [92]. Hence, we are going to see about the proof of concept in the various usages of carbon-based materials in antimicrobial applications such as fabrication of antimicrobial agents in wound dressings, a coating on the medical device surfaces, and nanoantibiotic formulations.

3.5.1 Wound Dressing

In recent years, AgNP- based wound bandages/dressings exposed some of the clinical success, specifically by controlling the infections and treating superficial wounds. Similarly, carbon-based materials might be used in wound dressing applications to recognize its functions such as reducing the infections, keeping the wounded area moist, speeding up to close the wound, and stimulating healing without any scar. Several approaches have been reported for the application of carbon-based materials in wound healing such as antimicrobial composites, hybrids with hydrogen peroxide and AgNPs, structural stability for wound dressing, and improving the microbial photothermal inhibitions. Fan et al. prepared a series of cross-linked Ag-graphene composite hydrogels using acrylic acid and *N*,*N*′-methylene bisacrylamide and glucose as a reducing agent, which reduces the toxicity on tissues. Ag-graphene composite hydrogel with a high ratio of 5 : 1 shows desired antimicrobial activities against both gram-positive *S. aures* and gram-negative *E. coli*, and the wound healing time process tends to be less compared to the other ratios. Additionally, with the help of histological analysis, it proved that Ag-graphene composite hydrogel with higher ratio has its potential application toward wound treatment. It is necessary to get a proper dressing over the wound to protect it from external mechanical stress and keep the place moist because people may even die due to severe infections. In this regard, a group of researchers prepared a novel wound dressing membranes using polyurethane/siloxane membranes with GO nanoplatelets as an antimicrobial and reinforcing agent. Using GO on the dressing tends to have good tensile strength and flexibility in both dried and hydrated states, which confirms the dressing to keep the damaged tissues from external stress. In addition to this, the GO also kills the microbial strains without presenting any adverse effects on growth and proliferation of fibroblast cells, which is confirmed by the designed dressing used in *in vivo* assay on rat model [93].

Also, GO-based wound dressing, several studies are done using CDs. In this regard, Shaikh et al. developed CQDs from *Citrus limetta* fruit extract and formed as an anti-biofilm agent. These synthesized CQDs have been shown as effective against *C. albicans* [94]. Xiang et al. fabricated the ZnO/CDs composite hydrogel with folic acid-conjugated PDA and evaluated its efficacy in wound models in rats. The circular wound of 5 mm diameters on the back of the rats and infested with *S. aureus* and fabricated hydrogel was applied to cure as a wound dressing. It shows an obvious result on the fifth day with the epidermal regeneration occurrences of the CDs/ZnO hydrogels treated rats compared to the others, and it also shows more collagen fibers in alignment like normal skin, which indicates its excellent performance toward wound healing [85]. Omidi et al. developed a novel pH-sensitive CDs/chitosan nanocomposite and used against *S. aureus* both in *in vitro* and *in vivo*. The result exposed that the developed composites could enhance the wound healing process and has great potential applications toward tissue engineering due to its excellent antibacterial properties and nontoxic effects [95].

Sun et al. reported the efficiency of GQD/CDs bandages in wound disinfection by *in vivo* studies using mice. Fascinatingly, CDs that are derived from GO do not have any antimicrobial property *in vitro*, but the intrinsic peroxidase-like activity of CDs facilitates the H_2O_2 breakdown for OH radicals generation with strong antimicrobial activity. This allows wound healing using CDs bandages with a low concentration of H_2O_2 [96]. Subsequently, the occurrence of wounds (on the skin) leads to microbial permeations that cause more damages to the muscle fibers too when it was not treated properly. Regarding this, developing wound dressing based on hydrogels or patches combined with light irradiation may lead to effective prevention of infections [97, 98].

3.5.2 Surface Modifications (Coating) on Medical Devices

Microbial contamination of transcutaneous medical devices such as dental/bone implants, and catheters before the medical procedures is usually unavoidable. Always, there will be an absence of integration between the tissues and the device crossing points, which leads to swelling, inflammation, accumulation of bursal fluids, and finally infections will occur. This case happens in almost all the biomaterials based on the medical device [99, 100]. Hence, as a prevention against infections and early failure, it is critical to fabricate a medical device surface with antimicrobial activity [101]. Even though there are no clinical trials still now, numerous basic studies have been proposed on the antimicrobial coatings by carbonaceous materials. Choudhary et al. prepared protein conjugated water-soluble rGO by a simple one-pot biosynthesis method and coated on a glass substrate. As the rGO stick with glass substrates (covalently), there will be an increase in the hydrophobicity on the surface. This caused a color change of glass substrate to dark brown and exhibits adequate durability by conducting repeated washings. The fabricated protein-rGO on glass substrate has been used to treat against *E. coli*, which shows a clear zone of inhibition and again it has been confirmed by fluorescence microscopy by staining the cells with SYTO 9 (stains both live and dead cells in green) and PI (membrane-permeable dye that can only go through the ruptured membrane that results in the emission of red fluorescence) [102]. Recently, Loczechin et al. investigated the seven differently functionalized CQDs for the treatment of highly pathogenic human coronavirus (HCoV) infections, which are urgently needed in the current scenario. These CQDs were shown interference significantly in a dose-dependent manner, and CQDs-4 shows moderate activity. Optimized CQDs lead to the development of first-generation anti-HCoV nanomaterials with effective dose-dependency to achieve 50% of inhibition. The mechanism of action for the inhibition of HCoV-229E entry is based on the interactions of the boronic acid functionalization with S protein through pseudolectin interactions that give enormous inhibition activity at viral replication steps. These results are tremendously promising for the replacement of commercial antiviral agents, and at the same time, it is also necessary to be tested in clinical applicable MERS-CoV to evaluate the potential as an alternative to anti-MERS therapeutics [103]. Similarly, researchers also created PDA-functionalized titanium surfaces with AgNPs-decorated GO. Due to the release of Ag^+ and generation of light-induced ROS from prepared AgNPs-GO particles, the titanium surfaces exposed higher antimicrobial activity. It was observed that the mice treated with nanocomposite coatings could fight microbial infections in a few days, but the bare implants take two weeks. Also, the collagen coating with AgNPs-GO shows improved biocompatibility on the Ti surfaces [104].

The antimicrobial properties of carbonaceous materials are favorable for dental applications. An early study has established a substantial reduction of cariogenic species like *S. mutans* in ZnO-GM-coated biofilm for acrylic teeth implants without any toxic effects. Shih et al. fabricated the nitrogen-doped GO bioglass that shows enhanced antimicrobial activity against *E. coli* [105, 106]. Thampi et al. described a common route for reducing the microbial attachment on polycarbonate-urethane membranes catheter fabrication using GO. The research group investigated its antimicrobial activity on *S. aureus* and *P. aeruginosa* that shows reduced microbial adhesion of 64 and 85%, respectively [107]. Lu et al. developed a novel CD-doped chitosan/hydroxyapatite scaffold for bone rejuvenation, and it is noted that CDs can enhance osteogenesis within the scaffolds and also promotes the vascularized bone tissue formation. Furthermore, photothermal therapy based on the prepared scaffold under NIR laser could efficiently reduce tumors, and it promises the antimicrobial properties of the scaffold in clinical infections. So, the CD-doped chitosan/hydroxyapatite scaffolds can be used for bone defects restoration and for treating microbial infections [108].

3.5.3 Nanoantibiotic Formulations

Ironic chemistry and its tunability of carbon-based materials enable the conjugation of capping agents for improved bio circulation. Significantly, they let the attachments like antibodies that will be useful for controlled and targeted delivery. Even though their usage is not anticipated in the nearby years, but we trust that carbonaceous formulations may come into use as smart nanoantibiotic formulations. Most favorable developments are based on light-activated strategies so far. For instance, Tian et al. synthesized a multifunctional, synergistic, and recyclable iron oxide and silver-coated GO nanocomposites (GO-IONP-Ag), which show enhanced antimicrobial efficacy against both gram-negative *E. coli* and Gram-positive *S. aureus* compared to the plain AgNPs. Taking advantage of absorbing light irradiation using GO in nanocomposites, effective photothermal activity is achieved for cellular inhibition toward the microbes [109].

With further improvement in the fabrication, Halouane et al. established a light-activated antimicrobial nanocomposite using rGO and surface-functionalized IONPs for targeting urinary tract pathogens. Since pyrene-PEG is used as an XPS marker, it can be easily modified with the other functional ligands onto the magnetic nanoparticles. Also, the photothermal activity of the *E. coli* captures nanocomposites that will lead to the complete ablation of the removed pathogens. This novel pathogen-capturing matrix gives an attractive strategy for decontaminating the pathogens and is useful for other biomedical purposes [110].

3.6 Biosafety

The growing productions and applications of carbon-based materials have boosted the risk of their exposure involuntarily. Though there are some contradictory results on their biosafety, and biocompatibilities partly are shown because of the variations in their physicochemical properties. Despite having various advantages toward antimicrobial applications, such as high surface area, determined antimicrobial effect, and exceptional dispersions in aqueous, its biocompatibility still rests as a challenge. To hustle the commercial or clinical use of carbonaceous materials, it is necessary to evaluate its biosafety in mammals and have to check their interactions with biological macromolecules such as protein and nucleic acids. Tan et al. exposed the PEGylated graphene generated nanointerface, which is used to reduce serum protein binding and accompanied C3 activations [111]. Jung et al. established that the antibody will link to carboxylic acid edges and the graphene-folded structures [112]. Lately, there have been some of the reports relating to the potential toxicity of carbonaceous materials in mammalian cells and animals. It is found that the carbon-based materials tend to induce ineligible toxic effects in mammalian systems through damaging the plasma membrane, lysosome dysfunction, or mitochondrial disruption, etc. The mammalian cells' toxicity of carbonaceous materials was generally influenced by its size and shape, concentration, and surface functionalization similar to the antimicrobial activity [113, 114].

The size of the carbonaceous materials may influence its cellular uptake and bio-distributions in animal organs. A group of researchers found out that large GO was the utmost profibrogenic material with the help of prompt kinetics of pulmonary injury. It is confirmed by these results that intravenous delivery of GO is generally accumulated in the lungs. As comparing the mammalian cells and microorganisms, the microbes cannot take GO inside cytoplasm due to the absence of endocytosis pathways so there will be a similar distribution for both large and small GO [115].

A significant number of reports show that carbonaceous materials inhibit dose-dependent cytotoxicity in cells and animal organs such as apoptosis, inflammation, kidney and liver injury, and lung fibrosis. Wang et al. observed the GO effect on human fibroblast cells, the results exposed that there will be no toxicity to human fibroblast cells for less than $20\,\mu g\,mL^{-1}$ dose, but doses over $50\,\mu g\,mL^{-1}$ exhibit cytotoxicity such as inhibiting cell apoptosis, reducing the cell adhesion, and ingoing to endoplasm, lysosomes, mitochondria, and the nucleus. Though, in mice, the GO did not show any toxicity with the low dose of 0.1 mg and the middle dose of 0.25 mg but shown chronic toxic effects in a high dose of 0.4 mg (4/9 mice) and the formation of granuloma in lung, spleen, liver, and kidney, which could not be removed via kidney [116]. Zhang et al. demonstrated the biosafety of CDs by routine blood tests, chemical assay, and histological examination for three different groups of CD-based materials such as CDs, nitrogen-doped CDs, and folic acid-modified CDs along with one control group. The researchers also carried out the time- and dose-dependent administration, which reveals that the lower concentration of 10 and 15 $mg\,kg^{-1}$ shows a similar count of WBCs and hepatorenal values relatable to the normal range, but in the higher dose of 20 $mg\,kg^{-1}$ the values get higher than other groups, which may cause liver toxicity to the body. However, all the results are near to the normal range and do not show any severe pathological changes in organs such as heart, liver, lung, spleen, or kidney, which gives an idea of using CDs materials in biomedical applications with the proper dosages [117]. Subsequently, Lammel et al. report that the contact of HepG2 cells to both carboxylated graphene and GO shows time- and dose-dependent increase in ROS generations [118]. Recently, Yang et al. investigated the cytotoxicity and *in vivo* biodistribution of GO and surface-functionalized GO with PEG in various coating and sizes. There is no cellular uptake that was noted for PEG-GO in oral administration, but in the intraperitoneal injection, a huge amount of PEG-GO was accumulated in RES (reticuloendothelial system). Furthermore, histological examination and hematological analysis of organ slices show that both GO and PEG-GO were retained in mice for a long time, which leads to an insignificant assessment of the cytotoxicity [119]. The influence of surface charge has also been emphasized in cytotoxicity of GO due to its electrostatic repulsion between nonphagocytes and GO, which tends to be an important part of particle internalization [113].

Meanwhile, the biocompatibility of carbonaceous materials can be enhanced by fabricating the surface functionalities; it is evident to determine the role of specific functional groups in microbes and mammalian systems. From the perspective of the considerable toxicity of carbon-based materials in animals through the various routes of administration, the direct exposure of carbonaceous material-based antimicrobial agents to human cells may inhibit ineligible toxic effects.

3.7 Conclusion and Future Perspectives

Preventing the multidrug-resistant microbial infections has become a global need in the current scenario because of its treatment toward high morbidity and mortality rates. In this regard, several novel nanotechnological strategies using carbonaceous materials for antimicrobial agents are in development due to its various advantages such as simple, higher efficiency, and low-cost development. In addition to this, another major crisis that we are dealing with all over the world is rapidly increasing various types of biomass. Nowadays, preparing carbonaceous materials from biomass has been trending in the research field as it can solve both of our problems in a single way. The antimicrobial activity of various biomass-derived carbonaceous materials against several microorganisms has been reviewed in this chapter. The biomass-derived carbon-based materials displaying good antimicrobial activities that could be used in various applications. Since the antimicrobial efficacy of the carbonaceous materials varied according to its physicochemical properties, the mechanism of action, and also functionalizing with other nanomaterials like a metal ion/oxide nanoparticles, the detailed summary of the same has been reviewed with respective sections to provide an overview for best understanding. Furthermore, the biomedical perspective toward the prepared carbonaceous materials and its biosafety in clinical trials has also been incorporated. Few studies based on biomass-derived carbonaceous materials have been reported based even on its cytotoxicity, and their effects in mammalian cells are still questionable. The third challenge is the antimicrobial activity of carbonaceous materials was majorly tested only against two or three selected microbes like *E. coli* and *S. aureus*. It is must be tested these novel materials against other viral, fungal, and pathogens too to overcome the global crisis toward public health.

Acknowledgment

The author Preethi S. gratefully acknowledges the support of Indian Council of Medical Research (ICMR), New Delhi, for granting Senior Research Fellowship (no. 45/03/2020-Nan/BMS, dt: 11 January 2021). A. Arulraj and R.V. Mangalaraja duly acknowledge ANID-FONDECYT Postdoctoral Fellowship (Project No.: 3200076), Government of Chile and Universidad Adolfo Ibáñez, Peñalolén, Santiago, Chile for financial support.

References

1 Rai, R.V. and Bai, J.A. (2011). Nanoparticles and their potential application as antimicrobials. *Formatex* 1: 197–209.

2 Lakshmi, S.D., Avti, P.K., and Hegde, G. (2018). Activated carbon nanoparticles from biowaste as new generation antimicrobial agents: a review. *Nano-Struct. Nano-Objects* 16: 306–321.

3 Allen, H.K., Trachsel, J., Looft, T., and Casey, T.A. (2014). Finding alternatives to antibiotics. *Ann. N. Y. Acad. Sci.* 1323: 91–100.

4 Kumar Parveen, L.B., Peipei, H., and Rongzhoa, Z. (2019). Antibacterial properties of graphene based nanomaterials. *Nanomaterials.* 9: 1–32.

5 Tran, N. and Tran, P.A. (2012). Nanomaterial-based treatments for medical device-associated infections. *ChemPhysChem* 13: 2481–2494.

6 Srividya, N., Ghoora, M.D., and Padmanabh, P.R. (2017). *Antimicrobial Nanotechnology: Research Implications and Prospects in Food Safety*. Elsevier Inc. https://doi.org/10.1016/b978-0-12-804303-5.00004-3.

7 Zhu, X., Radovic-Moreno, A.F., Wu, J. et al. (2014). Nanomedicine in the management of microbial infection – overview and perspectives. *Nano Today* 9: 478–498.

8 Rizzello, L., Cingolani, R., and Pompa, P.P. (2013). Nanotechnology tools for antibacterial materials. *Nanomedicine* 8: 807–821.

9 Huh, A.J. and Kwon, Y.J. (2011). "Nanoantibiotics": a new paradigm for treating infectious diseases using nanomaterials in the antibiotics resistant era. *J. Control. Release* 156: 128–145.

10 Allahverdiyev, A.M., Kon, K.V., Abamor, E.S. et al. (2011). Coping with antibiotic resistance: combining nanoparticles with antibiotics and other antimicrobial agents. *Expert Rev. Anti-Infect. Ther.* 9: 1035–1052.

11 Hobman, J.L. and Crossman, L.C. (2015). Bacterial antimicrobial metal ion resistance. *J. Med. Microbiol.* 64: 471–497.

12 Marx, D.E. and Barillo, D.J. (2014). Silver in medicine: the basic science. *Burns* 40: S9–S18.

13 Lemire, J.A., Harrison, J.J., and Turner, R.J. (2013). Antimicrobial activity of metals: mechanisms, molecular targets and applications. *Nat. Rev. Microbiol.* 11: 371–384.

14 Rizzello, L. and Pompa, P.P. (2014). Nanosilver-based antibacterial drugs and devices: mechanisms, methodological drawbacks, and guidelines. *Chem. Soc. Rev.* 43: 1501–1518.

15 Zou, X., Zhang, L., Wang, Z., and Luo, Y. (2016). Mechanisms of the antimicrobial activities of graphene materials. *J. Am. Chem. Soc.* 138: 2064–2077.

16 Szunerits, S. and Boukherroub, R. (2016). Antibacterial activity of graphene-based materials. *J. Mater. Chem. B* 4: 6892–6912.

17 Al-Jumaili, A., Alancherry, S., Bazaka, K., and Jacob, M.V. (2017). Review on the antimicrobial properties of carbon nanostructures. *Materials (Basel)* 10: 1–26.

18 Wang, L., Hu, C., and Shao, L. (2017). The-antimicrobial-activity-of-nanoparticles - present-situation. *Int. J. Nanomedicine* 12: 1227–1249.

19 Zhao, Y., Wang, Z.Q., Zhao, X. et al. (2013). Antibacterial action of silver-doped activated carbon prepared by vacuum impregnation. *Appl. Surf. Sci.* 266: 67–72.

20 Akhavan, O., Bijanzad, K., and Mirsepah, A. (2014). Synthesis of graphene from natural and industrial carbonaceous wastes. *RSC Adv.* 4: 20441–20448.

21 Sekaran, G., Karthikeyan, S., Gupta, V.K. et al. (2013). Immobilization of *Bacillus* sp. in mesoporous activated carbon for degradation of sulphonated phenolic compound in wastewater. *Mater. Sci. Eng. C* 33: 735–745.

22 Muramatsu, H., Kim, Y.A., Yang, K.S. et al. (2014). Rice husk-derived graphene with nano-sized domains and clean edges. *Small* 10: 2766–2770.

23 Bhatia, S.K., Joo, H.S., and Yang, Y.H. (2018). Biowaste-to-bioenergy using biological methods - a mini-review. *Energy Convers. Manag.* 177: 640–660.

24 De Yro, P.A.N., Quaichon, G.M.O., Cruz, R.A.T. et al. (2019). Hydrothermal synthesis of carbon quantum dots from biowaste for bio-imaging. *AIP Conf. Proc.* 2083: 1–6.

25 Kukreja, D., Mathew, J., Lakshmipathy, R., and Sarada, N.C. (2015). Synthesis of fluorescent carbon dots from mango peels. *Int. J. ChemTech Res.* 8: 61–64.

26 Meng, W., Bai, X., Wang, B. et al. (2019). Biomass-derived carbon dots and their applications. *Energy Environ. Mater.* 2: 172–192.

27 Nzihou, A. (2020). *Handbook on Characterization of Biomass, Biowaste and Related by-Products*. Switzerland: Springer https://doi.org/10.1007/978-3-030-35020-8.

28 Kim, J. and Van Der Bruggen, B. (2010). The use of nanoparticles in polymeric and ceramic membrane structures: review of manufacturing procedures and performance improvement for water treatment. *Environ. Pollut.* 158: 2335–2349.

29 Shi, H., Wei, J., Qiang, L. et al. (2014). Fluorescent carbon dots for bioimaging and biosensing applications. *J. Biomed. Nanotechnol.* 10: 2677–2699.

30 Bhat, V.S., Supriya, S., and Hegde, G. (2020). Review – biomass derived carbon materials for electrochemical sensors. *J. Electrochem. Soc.* 167: 037526.

31 Xu, S., Liu, C., and Wiezorek, J. (2018). 20 Renewable biowastes derived carbon materials as green counter electrodes for dye-sensitized solar cells. *Mater. Chem. Phys.* 204: 294–304.

32 Kumar, J., Mallampati, R., Adin, A., and Valiyaveettil, S. (2014). Functionalized carbon spheres for extraction of nanoparticles and catalyst support in water. *ACS Sustain. Chem. Eng.* 2: 2675–2682.

33 Phiri, J., Dou, J., Vuorinen, T. et al. (2019). Highly porous willow wood-derived activated carbon for high-performance supercapacitor electrodes. *ACS Omega* 4: 18108–18117.

34 Ganewatta, M.S. (2017). Antimicrobial Biomaterials and Sustainable Polymers from Renewable Biomass. Doctoral dissertation. https://scholarcommons.sc.edu/etd/4109.

35 Chung, H.K., Wongso, V., Sambudi, N.S., and Isnaeni (2020). Biowaste-derived carbon dots/hydroxyapatite nanocomposite as drug delivery vehicle for acetaminophen. *J. Sol-Gel Sci. Technol.* 93: 214–223.

36 Iravani, S. and Varma, R.S. (2020). Green synthesis, biomedical and biotechnological applications of carbon and graphene quantum dots. A review. *Environ. Chem. Lett.* https://doi.org/10.1007/s10311-020-00984-0.

37 Xu, X., Ray, R., Gu, Y. et al. (2004). Electrophoretic analysis and purification of fluorescent single-walled carbon nanotube fragments. *J. Am. Chem. Soc.* 126: 12736–12737.

38 Tuerhong, M., Xu, Y., and Yin, X.B. (2017). Review on carbon dots and their applications. *Chinese J. Anal. Chem.* 45: 139–150.

39 Wang, S., Sun, W., Yang, D., and Yang, F. (2020). Soybean-derived blue photoluminescent carbon dots. *Beilstein J. Nanotechnol.* 11: 606–619.

40 Mahat, N.A. and Shamsudin, S.A. (2020). Transformation of oil palm biomass to optical carbon quantum dots by carbonisation-activation and low temperature hydrothermal processes. *Diam. Relat. Mater.* 102: 107660.

41 Cheng, C., Xing, M., and Wu, Q. (2019). Preparation of carbon dots with long-wavelength and photoluminescence-tunable emission to achieve multicolor imaging in cells. *Opt. Mater. (Amst).* 88: 353–358.

42 Prasannan, A. and Imae, T. (2013). One-pot synthesis of fluorescent carbon dots from orange waste peels. *Ind. Eng. Chem. Res.* 52: 15673–15678.

43 Somanathan, T., Prasad, K., Ostrikov, K.K. et al. (2015). Graphene oxide synthesis from agro waste. *Nanomaterials* 5: 826–834.

44 Novoselov, K.S., Geim, A.K., Morozov, S.V. et al. (2016). Electric field effect in atomically thin carbon films. *Science* 306: 666–669.

45 Saga, T. and Yamaguchi, K. (2009). History of antimicrobial agents and resistant bacteria. *Jpn Med. Assoc. J.* 52: 103–108.

46 Rizzo, L., Manaia, C., Merlin, C. et al. (2013). Urban wastewater treatment plants as hotspots for antibiotic resistant bacteria and genes spread into the environment: a review. *Sci. Total Environ.* 447: 345–360.

47 Tenover, F.C. (2006). Mechanisms of antimicrobial resistance in bacteria. *Am. J. Med.* 119: https://doi.org/10.1016/j.amjmed.2006.03.011.

48 Neu, H.C. (1992). The crisis in antibiotic resistance. *Science* 257: 1064–1073.

49 Chakraborty, S.P., Pramanik, P., and Roy, S. (2012). A review on emerggence of antibiotic resistant *Staphylococcus aureus* and role of chitosan nanoparticle in drug delivery. *Int. J. Life Sci. Pharma Res.* 2: 96–115.

50 Ahmad, I., Mehmood, Z., and Mohammad, F. (1998). Screening of some Indian medicinal plants for their antimicrobial properties. *J. Ethnopharmacol.* 62: 183–193.

51 Bush, K. (2004). Antibacterial drug discovery in the 21st century. *Clin. Microbiol. Infect.* 10: 10–17.

52 (2017). Who publishes list of bacteria for which new antibiotics are urgently needed. *Saudi Med. J.* 38: 444.

53 Ojha, A.K., Baughn, A.D., Sambandan, D. et al. (2008). Growth of *Mycobacterium tuberculosis* biofilms containing free mycolic acids and harbouring drug-tolerant bacteria. *Mol. Microbiol.* 69: 164–174.

54 Borders-Hemphill, V. and Mosholder, A. (2012). U.S. utilization patterns of influenza antiviral medications during the 2009 H1N1 influenza pandemic. *Influenza Other Respir. Viruses* 6: 129–133.

55 Zaidi, S.T.R. and Roberts, J.A. (2016). *Drug Dosing in Obesity: Volume I: Antimicrobials.* New York: Springer https://doi.org/10.1007/978-3-319-44034-7.

56 Sinden, R.E., Delves, M., and Blagborough, A. (2012). Targeting the parasite in the mosquito: rationale and practicality. *Malar. J.* 11: 2012. https://doi.org/10.1186/1475-2875-11-s1-o5.

57 Prestinaci, F., Pezzotti, P., and Pantosti, A. (2015). Antimicrobial resistance: a global multifaceted phenomenon. *Pathog. Glob. Health* 109: 309–318.

58 Das Purkayastha, M., Manhar, A.K., Mandal, M., and Mahanta, C.L. (2014). Industrial waste-derived nanoparticles and microspheres can be potent antimicrobial and functional ingredients. *J. Appl. Chem.* 2014: 1–12.

59 Gonçalves, S.P.C., Strauss, M., Delite, F.S. et al. (2015). Activated carbon from pyrolysed sugarcane bagasse: silver nanoparticle modification and ecotoxicity assessment. *Sci. Total Environ.* 565: 833–840.

60 Varghese, S., Kuriakose, S., and Jose, S. (2013). Antimicrobial activity of carbon nanoparticles isolated from natural sources against pathogenic gram-negative and gram-positive bacteria. *J. Nanosci.* 2013: 1–5.

61 Yallappa, S., Deepthi, D.R., Yashaswini, S. et al. (2017). Natural biowaste of groundnut shell derived nano carbons: synthesis, characterization and itsin vitro antibacterial activity. *Nano-Struct. Nano-Objects* 12: 84–90.

62 Mohan, A.N. and Manoj, B. (2019). Biowaste derived graphene quantum dots interlaced with SnO2 nanoparticles-a dynamic disinfection agent against: *Pseudomonas aeruginosa*. *New J. Chem.* 43: 13681–13689.

63 Mehta, V.N., Jha, S., Basu, H. et al. (2015). One-step hydrothermal approach to fabricate carbon dots from apple juice for imaging of mycobacterium and fungal cells. *Sensors Actuators B Chem.* 213: 434–443.

64 Karahan, H.E., Wiraja, C., Xu, C. et al. (2018). Graphene materials in antimicrobial nanomedicine: current status and future perspectives. *Adv. Healthc. Mater.* 7: 1–18.

65 Akhavan, O. and Ghaderi, E. (2010). Toxicity of graphene and graphene oxide nanowalls against bacteria. *ACS Nano* 4: 5731–5736.

66 Lu, X., Feng, X., Werber, J.R. et al. (2017). Enhanced antibacterial activity through the controlled alignment of graphene oxide nanosheets. *Proc. Natl. Acad. Sci. U. S. A.* 114: E9793–E9801.

67 Fallatah, H., Elhaneid, M., Ali-Boucetta, H. et al. (2019). Antibacterial effect of graphene oxide (GO) nano-particles against *Pseudomonas putida* biofilm of variable age. *Environ. Sci. Pollut. Res.* 26: 25057–25070.

68 Pham, V.T.H., Truong, V.K., Quinn, M.D.J. et al. (2015). Graphene induces formation of pores that kill spherical and rod-shaped bacteria. *ACS Nano* 9: 8458–8467.

69 Liu, S., Zeng, T.H., Hofmann, M. et al. (2011). Antibacterial activity of graphite, graphite oxide, graphene oxide, and reduced graphene oxide: membrane and oxidative stress. *ACS Nano* 5: 6971–6980.

70 Mangadlao, J.D., Santos, C.M., Felipe, M.J.L. et al. (2015). On the antibacterial mechanism of graphene oxide (GO) Langmuir-Blodgett films. *Chem. Commun.* 51: 2886–2889.

71 Li, Y., Liu, Y., Fu, Y. et al. (2012). The triggering of apoptosis in macrophages by pristine graphene through the MAPK and TGF-beta signaling pathways. *Biomaterials* 33: 402–411.

72 Gurunathan, S., Han, J.W., Abdal Dayem, A. et al. (2012). Oxidative stress-mediated antibacterial activity of graphene oxide and reduced graphene oxide in *Pseudomonas aeruginosa*. *Int. J. Nanomedicine* 7: 5901–5914.

73 Chen, J., Wang, X., and Han, H. (2013). A new function of graphene oxide emerges: inactivating phytopathogenic bacterium *Xanthomonas oryzae* pv. Oryzae. *J. Nanopart. Res.* 15: https://doi.org/10.1007/s11051-013-1658-6.

74 Li, H., Huang, J., Song, Y. et al. (2018). Degradable carbon dots with broad-spectrum antibacterial activity. *ACS Appl. Mater. Interfaces* 10: 26936–26946.

75 Jhonsi, M.A., Ananth, D.A., Nambirajan, G. et al. (2018). Antimicrobial activity, cytotoxicity and DNA binding studies of carbon dots. *Spectrochim. Acta A Mol. Biomol. Spectrosc.* 196: 295–302.

76 Liu, S., Hu, M., Zeng, T.H. et al. (2012). Lateral dimension-dependent antibacterial activity of graphene oxide sheets. *Langmuir* 28: 12364–12372.

77 Mejías Carpio, I.E., Santos, C.M., Wei, X., and Rodrigues, D.F. (2012). Toxicity of a polymer-graphene oxide composite against bacterial planktonic cells, biofilms, and mammalian cells. *Nanoscale* 4: 4746–4756.

78 Chen, J., Peng, H., Wang, X. et al. (2014). Graphene oxide exhibits broad-spectrum antimicrobial activity against bacterial phytopathogens and

fungal conidia by intertwining and membrane perturbation. *Nanoscale* 6: 1879–1889.
79 Dallavalle, M., Calvaresi, M., Bottoni, A. et al. (2015). Graphene can wreak havoc with cell membranes. *ACS Appl. Mater. Interfaces* 7: 4406–4414.
80 Kotagiri, N., Lee, J.S., and Kim, J.W. (2013). Selective pathogen targeting and macrophage evading carbon nanotubes through dextran sulfate coating and PEGylation for photothermal theranostics. *J. Biomed. Nanotechnol.* 9: 1008–1016.
81 Meng, D., Yang, S., Guo, L. et al. (2014). The enhanced photothermal effect of graphene/conjugated polymer composites: photoinduced energy transfer and applications in photocontrolled switches. *Chem. Commun.* 50: 14345–14348.
82 Xiao, L., Sun, J., Liu, L. et al. (2017). Enhanced photothermal bactericidal activity of the reduced graphene oxide modified by cationic water-soluble conjugated polymer. *ACS Appl. Mater. Interfaces* 9: 5382–5391.
83 Xu, Y., Feng, T., Yang, T. et al. (2018). Utilizing intramolecular photoinduced electron transfer to enhance photothermal tumor treatment of Aza-BODIPY-based near-infrared nanoparticles. *ACS Appl. Mater. Interfaces* 10: 16299–16307.
84 Kim, J.W., Shashkov, E.V., Galanzha, E.I. et al. (2007). Photothermal antimicrobial nanotherapy and nanodiagnostics with self-assembling carbon nanotube clusters. *Lasers Surg. Med.* 39: 622–634.
85 Xiang, Y., Mao, C., Liu, X. et al. (2019). Rapid and superior bacteria killing of carbon quantum dots/ZnO decorated injectable folic acid-conjugated PDA hydrogel through dual-light triggered ROS and membrane permeability. *Small* 15: 1–15.
86 Tu, Y., Lv, M., Xiu, P. et al. (2013). Destructive extraction of phospholipids from *Escherichia coli* membranes by graphene nanosheets. *Nat. Nanotechnol.* 8: 594–601.
87 Liu, P., Huang, X., Zhou, R., and Berne, B.J. (2005). Observation of a dewetting transition in the collapse of the melittin tetramer. *Nature* 437: 159–162.
88 Zhou, R., Huang, X., Margulis, C.J., and Berne, B.J. (2004). Hydrophobic collapse in multidomain protein folding. *Science* 305: 1605–1609.
89 Berne, B.J., Weeks, J.D., and Zhou, R. (2009). Dewetting and hydrophobic interaction in physical and biological systems. *Annu. Rev. Phys. Chem.* 60: 85–103.
90 Mashino, T., Usui, N., Okuda, K. et al. (2003). Respiratory chain inhibition by fullerene derivatives: hydrogen peroxide production caused by fullerene derivatives and a respiratory chain system. *Bioorg. Med. Chem.* 11: 1433–1438.
91 Xin, Q., Liu, Q., Geng, L. et al. (2017). Chiral nanoparticle as a new efficient antimicrobial nanoagent. *Adv. Healthc. Mater.* 6: 1–6.
92 Lee, D.K., Kee, T., Liang, Z. et al. (2017). Clinical validation of a nanodiamond-embedded thermoplastic biomaterial. *Proc. Natl. Acad. Sci. U. S. A.* 114: E9445–E9454.
93 Fan, Z., Liu, B., Wang, J. et al. (2014). A novel wound dressing based on Ag/graphene polymer hydrogel: effectively kill bacteria and accelerate wound healing. *Adv. Funct. Mater.* 24: 3933–3943.

94 Shaikh, A.F., Tamboli, M.S., Patil, R.H. et al. (2018). Bioinspired carbon quantum dots: an antibiofilm agents. *J. Nanosci. Nanotechnol.* 19: 2339–2345.

95 Omidi, M., Yadegari, A., and Tayebi, L. (2017). Wound dressing application of pH-sensitive carbon dots/chitosan hydrogel. *RSC Adv.* 7: 10638–10649.

96 Sun, H., Gao, N., Dong, K. et al. (2014). Graphene quantum dots-band-aids used for wound disinfection. *ACS Nano* 8: 6202–6210.

97 Schreml, S., Landthaler, M., Schaferling, M., and Babilas, P. (2011). A new star on the H2O2 rizon of wound healing? *Exp. Dermatol.* 20: 229–231.

98 Loo, A.E.K., Wong, Y.T., Ho, R. et al. (2012). Effects of hydrogen peroxide on wound healing in mice in relation to oxidative damage. *PLoS One* 7: https://doi.org/10.1371/journal.pone.0049215.

99 Shah, S.R., Tatara, A.M., D'Souza, R.N. et al. (2013). Evolving strategies for preventing biofilm on implantable materials. *Mater. Today* 16: 177–182.

100 Koidou, V.P., Argyris, P.P., Skoe, E.P. et al. (2018). Peptide coatings enhance keratinocyte attachment towards improving the peri-implant mucosal seal. *Biomater. Sci.* 6: 1936–1945. https://doi.org/10.1039/c8bm00300a.

101 Campoccia, D., Montanaro, L., and Arciola, C.R. (2006). The significance of infection related to orthopedic devices and issues of antibiotic resistance. *Biomaterials* 27: 2331–2339.

102 Choudhary, P. and Das, S.K. (2019). Bio-reduced graphene oxide as a nanoscale antimicrobial coating for medical devices. *ACS Omega* 4: 387–397.

103 Łoczechin, A., Séron, K., Barras, A. et al. (2019). Functional carbon quantum dots as medical countermeasures to human coronavirus. *ACS Appl. Mater. Interfaces* 11: 42964–42974.

104 Xie, X., Mao, C., Liu, X. et al. (2017). Synergistic bacteria killing through photodynamic and physical actions of graphene oxide/Ag/collagen coating. *ACS Appl. Mater. Interfaces* 9: 26417–26428.

105 Shih, S.J., Chen, C.Y., Lin, Y.C. et al. (2016). Investigation of bioactive and antibacterial effects of graphene oxide-doped bioactive glass. *Adv. Powder Technol.* 27: 1013–1020.

106 Shih, S.J., Hong, B.J., and Lin, Y.C. (2017). Novel graphene oxide-containing antibacterial mesoporous bioactive glass. *Ceram. Int.* 43: S784–S788.

107 Thampi, S., Nandkumar, A.M., Muthuvijayan, V., and Parameswaran, R. (2017). Differential adhesive and bioactive properties of the polymeric surface coated with graphene oxide thin film. *ACS Appl. Mater. Interfaces* 9: 4498–4508.

108 Lu, Y., Li, L., Li, M. et al. (2018). Zero-dimensional carbon dots enhance bone regeneration, osteosarcoma ablation, and clinical bacterial eradication. *Bioconjug. Chem.* 29: 2982–2993.

109 Tian, T., Shi, X., Cheng, L. et al. (2014). Graphene-based nanocomposite as an effective, multifunctional, and recyclable antibacterial agent. *ACS Appl. Mater. Interfaces* 6: 8542–8548.

110 Halouane, F., Jijie, R., Meziane, D. et al. (2017). Selective isolation and eradication of: *E. coli* associated with urinary tract infections using anti-fimbrial modified magnetic reduced graphene oxide nanoheaters. *J. Mater. Chem. B* 5: 8133–8142.

111 Tan, X., Feng, L., Zhang, J. et al. (2013). Functionalization of graphene oxide generates a unique interface for selective serum protein interactions. *ACS Appl. Mater. Interfaces* 5: 1370–1377.

112 Jung, J.H., Cheon, D.S., Liu, F. et al. (2010). A graphene oxide based immuno-biosensor for pathogen detection. *Angew. Chem.* 122: 5844–5847.

113 Ou, L., Song, B., Liang, H. et al. (2016). Toxicity of graphene-family nanoparticles: a general review of the origins and mechanisms. *Part. Fibre Toxicol.* 13: https://doi.org/10.1186/s12989-016-0168-y.

114 Cacaci, M., Martini, C., Cinzia, G. et al. (2019). Graphene oxide coatings as tools to prevent microbial biofilm formation on medical device. *Adv. Exp. Med. Biol.* https://doi.org/10.1007/5584_2019_434.

115 Wang, X., Duch, M.C., Mansukhani, N. et al. (2015). Use of a pro-fibrogenic mechanism-based predictive toxicological approach for tiered testing and decision analysis of carbonaceous nanomaterials. *ACS Nano* 9: 3032–3043.

116 Wang, K., Ruan, J., Song, H. et al. (2011). Biocompatibility of graphene oxide. *Nanoscale Res. Lett.* 6: 1–8.

117 Zhang, S., Pei, X., Xue, Y. et al. (2019). Bio-safety assessment of carbon quantum dots, N-doped and folic acid modified carbon quantum dots: a systemic comparison. *Chin. Chem. Lett.* https://doi.org/10.1016/j.cclet.2019.09.018.

118 Lammel, T., Boisseaux, P., Fernández-Cruz, M.L., and Navas, J.M. (2013). Internalization and cytotoxicity of graphene oxide and carboxyl graphene nanoplatelets in the human hepatocellular carcinoma cell line Hep G2. *Part. Fibre Toxicol.* 10: 1–21.

119 Cheng, L., Yang, K., Shao, M. et al. (2011). In vivo pharmacokinetics, long-term biodistribution and toxicology study of functionalized upconversion nanoparticles in mice. *Nanomedicine* 6: 1327–1340.

4

Carbon-Based Nanomaterials Prepared from Biomass for Catalysis

A. Rajeswari, E. Jackcina Stobel Christy, and Anitha Pius

The Gandhigram Rural Institute – Deemed to be University, Gandhigram, Dindigul, 624302, Tamil Nadu, India

4.1 Introduction

The rapid progression of human development has acquired incredible worldwide difficulties in the form of energy crisis and water shortage. It is evaluated that the world will require twice over its energy supply by 2050 [1]. Nowadays, water issues and energy problems are interrelated because water is widely utilized in the energy creation process, while energy is required to extract, transport, and decontaminate water resources [2]. These energy demands have stimulated voices clamoring for other energy alternatives with high efficiency, low cost, and environmental benignity [3–5].

Nanotechnology is expected to contribute significantly to environmental issues by saving raw materials, energy, and water as well as by reducing greenhouse gases and hazardous wastes. Nanomaterials have revolutionized all major industrial areas, from drug delivery to agriculture, food industry, etc. [6]. It is highly desirable to reduce loads of chemical contamination on the environment by developing new and convenient methods to overcome the drawbacks of chemical techniques while at the same time improving the yield and reducing cost. Mostly, carbon-based nanomaterials have produced stimulation due to their unique physico-chemical properties such as enhanced magnetic, electrical, and optical catalytic properties [7–10]. They could be used in various fields of applications such as water treatment, antimicrobial products, catalyst support, biosensing, and other biomedical applications [11–15].

Carbon is a major component of living beings. Carbon is one of the basic building blocks for organic life together with nitrogen and oxygen. Carbon is an old and new catalytic material [16]. Activated carbons (ACs) are used commercially in many catalytic formulations, particularly for hydrogenation catalysts, excellent properties of dispersion of metal particles (particularly based on noble metals), and absence (or limited) presence of sites, which may catalyze side reactions [17]. The different adsorption properties of carbon materials with respect to other type of supports, metal oxides for example, are another relevant element determining they large

Biomass-Derived Carbon Materials: Production and Applications, First Edition.
Edited by Alagarsamy Pandikumar, Perumal Rameshkumar, and Pitchaimani Veerakumar.

industrial use, particularly in selective hydrogenation reactions. However, carbon can also be considered as a new catalytic material. There are various nanostructures of carbon materials among which fullerenes, graphene, carbon nanofibers, and carbon nanotubes (CNTs) have most important applications [18, 19].

4.2 Preparation of Biomass-Derived Carbon-Based Nanomaterials

With the increase of agricultural and industrial production and consumption, the generation of biomass has been greatly accelerated. Waste materials ranging from household wastes to industrial residues cause detrimental effects on environment and human health [20]. However, wastes can be exploited as utility resources to prepare value-added products [21]. Figure 4.1 shows the schematic representation of preparation of biomass-derived carbon-based nanomaterials.

However, wastes can be exploited as utility resources to prepare value-added products [21]. One of the important techniques to achieve this objective is pyrolysis. Pyrolysis belongs to thermal decomposition, which is operated in air-free condition. Besides generating gases, various liquid phases and solid char can be obtained by manipulating pyrolysis parameters, such as operating temperature and heating rate [22]. Recycling is the conversion of biomass wastes into various valuable chemicals such as biofuels and alcohols [23].

Recycling waste materials into high-valued products is markedly feasible that have inspired researchers to synthesize carbon-based nanomaterials from wastes resources for various potential applications. Also being small confers advantages in terms of negotiating biological barriers, which may be desirable, but nanoscale size per se is not sufficient to qualify as a nanotechnology. Carbon-based nanomaterials, however, possess intrinsic physicochemical properties that can potentially be exploited. Carbon-based nanomaterials, including CNTs, graphene oxide (GO), fullerenes, and nanodiamonds are potential candidates for various applications, and they have been materials of priority for research in nanotechnology.

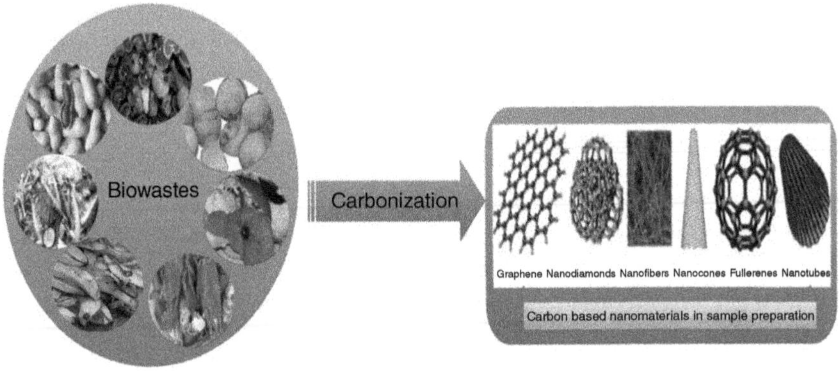

Figure 4.1 Schematic representation of preparation of biomass-derived carbon-based nanomaterials. Source: ChiccoDodiFC/Adobe Stock.

4.3 Graphene

Graphene, which is considered the basic building block of all graphitic forms, possesses a single layer of carbon atoms in a closely packed honeycomb two-dimensional lattice. Graphene has a large specific surface area (theoretical value 2630 $m^2\ g^{-1}$) [24], and both sides of its planar sheets are available for molecule adsorption. Furthermore, the large delocalized π-electron system of graphene can form strong π-stacking interaction with the benzene ring, which might make graphene a good choice for the extraction of benzenoid form compounds. Finally, graphene can be easily modified with functional groups, especially via graphene oxide. The exceptional properties of graphene make it a superior candidate as a good adsorbent in different sample preparation methods. This material has driven research in nanoscience and nanotechnology because of graphene exceptional electrical, mechanical, and chemical properties. Graphene can be used in sensors, batteries, supercapacitors, solar and fuel cells, and in biotechnology. The recognition of the importance of graphene resulted in its discoverers being awarded the Nobel Prize in Physics in 2010 [25–29].

4.3.1 Preparation of Graphene

Generally, there are two fundamental sources for the preparation of graphene: graphite and organic molecules. Conversely, there are several approaches to prepare graphene that are as follows, CVD on metallic films (chemical vapor deposition, bottom-up approach) [30] and top-down approach include liquid exfoliation of graphite crystal [31], epitaxial growth on silicon carbide [32], mechanical cleavage [33], and chemical reduction of graphene oxide [34]. W. Ren et al. [35] have demonstrated the detailed strategy of each method that is shown in Figure 4.2. These are usual approaches to prepare graphene. Every method has its own merits and demerits.

4.3.2 Graphene from Different Sources

The requirement of the sustainable resource is inevitable for any application, so renewable biomass is a sensible entrant for graphene precursor. In addition, coal has been used as a starting material, though it is not a renewable resource yet we included due to the innovative and sensible approach involved in the production of graphene quantum dots. This chapter illuminates the use of various precursors such as rice husk (RH), glucose, hemp, and disposable paper cups. Table 4.1 demonstrates the different methods and precursors used for the preparation of graphene.

Wang et al. [37] created unique interconnected partially graphitic carbon nanosheets with high specific surface area, significant volume fraction of mesoporosity, and good electrical conductivity from hemp bast fiber, which is shown in Figure 4.3. The nanosheets are ideally suited for low- through high-temperature ionic liquid-based supercapacitor applications. This novel precursor synthesis route

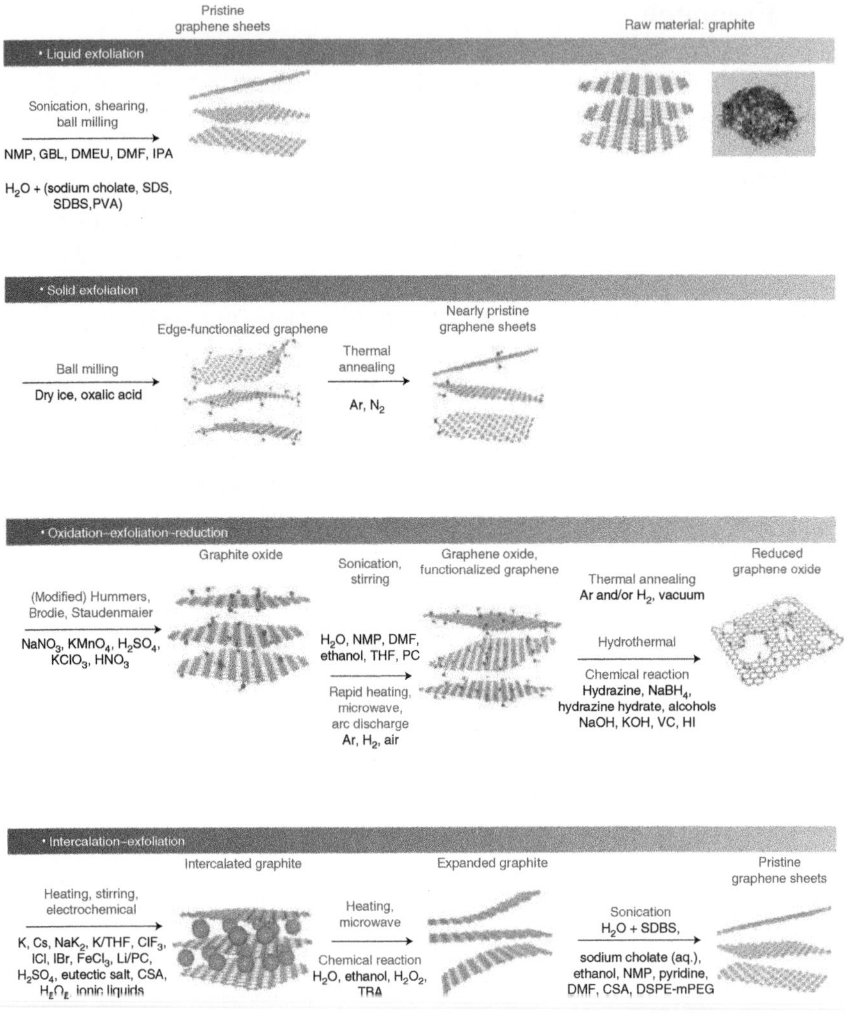

Figure 4.2 Four typical methods for the mass production of small graphene sheets by exfoliation of bulk graphite. Source: Ref. [35] / with permission of Springer Nature.

presents a great potential for facile large-scale production of high-performance carbons for a variety of diverse applications including energy storage.

Likewise, Hong Zhao et al. [38] used disposable paper cups for the formation of graphene sheets with Fe^{2+} as a catalyst, which is shown in Figure 4.4. The proposed synthesis strategy not only enables graphene sheets to be produced in high yield and high quality but also results in two added bonus products: Fe/graphene and Pt/graphene.

G. Ruan et al. [39] have developed chemical vapor and solid deposition methods to grow graphene from organic gases or solid carbon sources, which is shown in Figure 4.5. Most of the carbon sources used were purified chemicals that could be

Table 4.1 Different methods and precursors used for the preparation of graphene.

S. no.	Precursor and treatment	Method	Graphene	Reference
1	Rice husk/chemical activation with KOH	Combustion	Graphene sheets	[36]
2	Hemp/activation with KOH	Hydrothermal	Interconnected carbon nanosheets	[37]
3	Paper cups/with the aid of Fe^{2+} catalyst	Chemical treatment	N-doped and undoped graphene sheets	[38]
4	Food, insect/grown at Cu foil under inert atmosphere	CVD	Monolayer graphene	[39]
5	Chitosan/chemical treatment under inert atmosphere	Pyrolysis	N-doped graphene	[40]

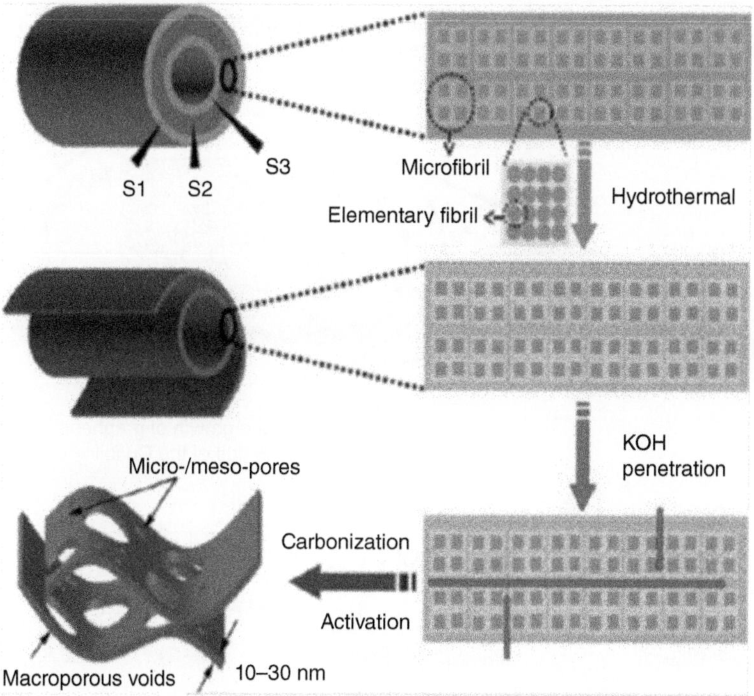

Figure 4.3 Schematic of the synthesis process for the hemp-derived carbon nanosheets, with the three different structural layers S1, S2, and S3. Source: Ref. [37] / with permission of American Chemical Society.

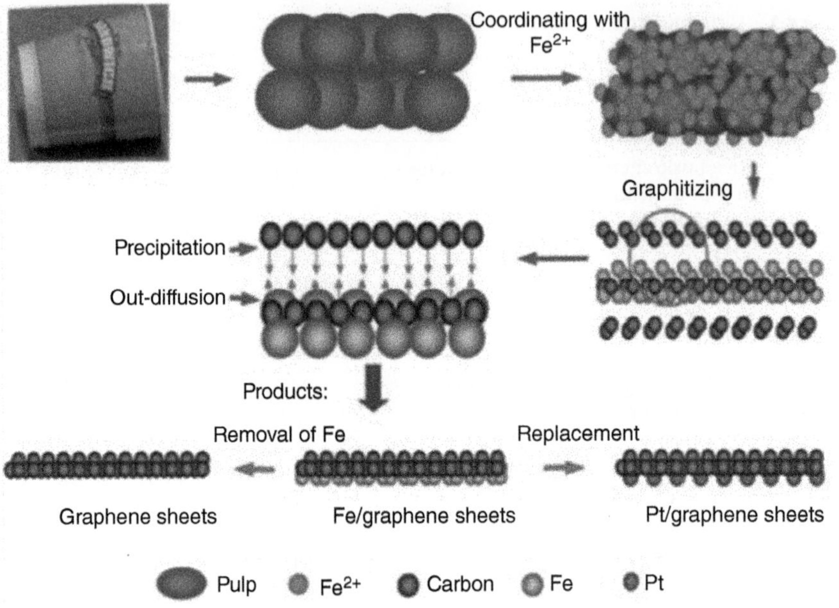

Figure 4.4 The supposed formation mechanism of the Fe/graphene sheets, graphene sheets, and Pt/graphene sheets from disposable paper cups. Source: Ref. [38] / with permission of Royal Society of Chemistry.

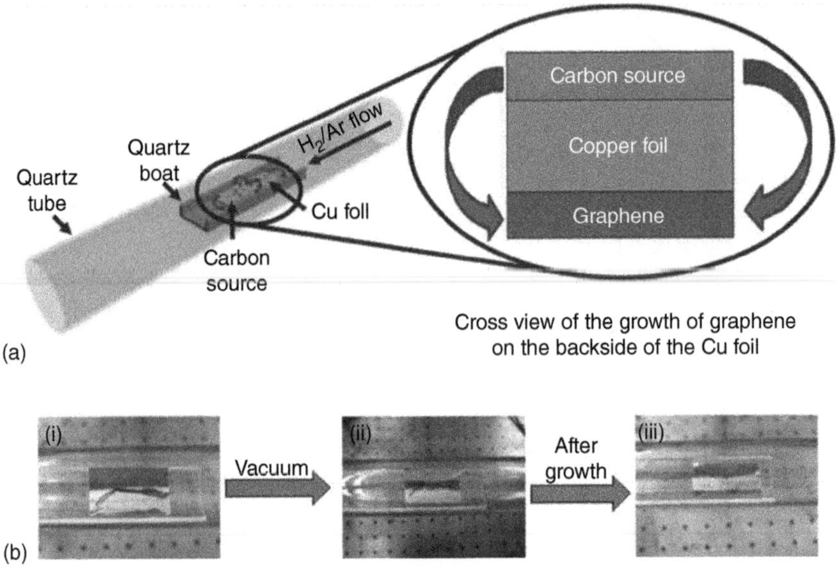

Figure 4.5 (a) Diagram of the experimental apparatus for the growth of graphene from food, insects, or waste in a tube furnace. On the left, the Cu foil with the carbon source contained in a quartz boat was placed at the hot zone of a tube furnace. The growth was performed at 1050 °C under low pressure with a H_2/Ar gas flow. On the right is across view which represents the formation of pristine graphene on the backside of the Cu substrate. (b) Growth of graphene from a cockroach leg. (i) One roach leg on top of the Cu foil. (ii) Roach leg under vacuum. (iii) Residue from the roach leg after annealing at 1050 °C for 15 minutes. Source: Ref. [39] / with permission of American Chemical Society.

expensive for mass production. Also they have developed a less expensive approach using six easily obtained, low, or negatively valued raw carbon-containing materials used without prepurification (cookies, chocolate, grass, plastics, roaches, and dog feces) to grow graphene directly on the backside of a Cu foil at 1050 °C under H_2/Ar flow. The nonvolatile pyrolyzed species were easily removed by etching away the front side of the Cu.

4.4 Carbon Nanotubes (CNTs)

CNTs are among the most studied nanostructures in recent decades [41]. They are a few nanometers in diameter and form one-dimensional structures because the length is orders of magnitude is larger than the diameter. They have different properties from other allotropes of carbon, such as graphite and diamond, with unique mechanical and electrical characteristics. The discovery of CNTs has not been fully elucidated, but they have been known in the scientific community since 1991 because of the publication by Sumio Iijima [42]. The importance of CNTs for nanotechnology is reflected in their numerous application possibilities. In engineering, CNTs have been studied to create new composites for the aeronautics industry [43]. They have also been used in nano devices and electronic nanocircuits to manufacture new chips for computers [44].

CNTs are formed only by carbon and have a cylindrical shape. The structure can be compared to a graphene sheet rolled into a tube shape in which the ratio between the length and diameter makes it almost a one-dimensional structure [43, 45]. CNT are classified as single-walled carbon nanotubes (SWCNTs) and multiwalled carbon nanotubes (MWCNTs). The structure of SWNTs is like a rolled-up graphene sheet, the unit of which is in hexagonal shape constituted by sp^2 hybrid carbon atoms with σ bonding (in plane) and π bonding (out of plane) [46–49].

4.4.1 Synthesis of CNTs

There are many methods for synthesizing CNTs that produce different types with different properties. The three main methods are electric arc, pulsed laser, and chemical vapor deposition (CVD) [50].

4.4.2 Synthesis of CNTs Using Biomass Materials

As products of plants photosynthesis, biomass can be the by-products or residues from agricultural refuse, ligneous plants, and waste water deposits. The biomass is not only valuable for sustainable energy resources but also have the potential to be used as carbon sources for fabrication of CNTs.

A.B. Suriani et al. [51] have been synthesized vertically aligned carbon nanotubes (VACNTs) using waste chicken fat as the starting material. Chicken fat oil, which was obtained through a rendering process, was directly mixed with 5.33 wt% ferrocene as a catalyst to form the synthesis stock. A mixture of single- and multiwalled

VACNTs was synthesized at a fixed temperature of 750 °C in a thermal chemical vapor deposition furnace. Field emission scanning electron microscopy (FESEM), micro-Raman spectroscopy, and thermogravimetric analyses (TGA) showed that the produced VACNTs were of excellent quality, comparable to those obtained using conventional carbon sources, with a purity of 88.2% and tube diameters ranging from 18 to 78 nm. Based on our study, waste chicken fat is a promising carbon source for the synthesis of high-quality and high-purity VACNTs. TGA and derivative thermogravimetric analysis (DTGA) curves for chicken fat oil, FESEM images of VACNTs grown from waste chicken fat on a Si substrate with increasing magnification, and high-resolution transmission electron microscopy (HRTEM) image of multiwalled CNTs are shown in Figure 4.6.

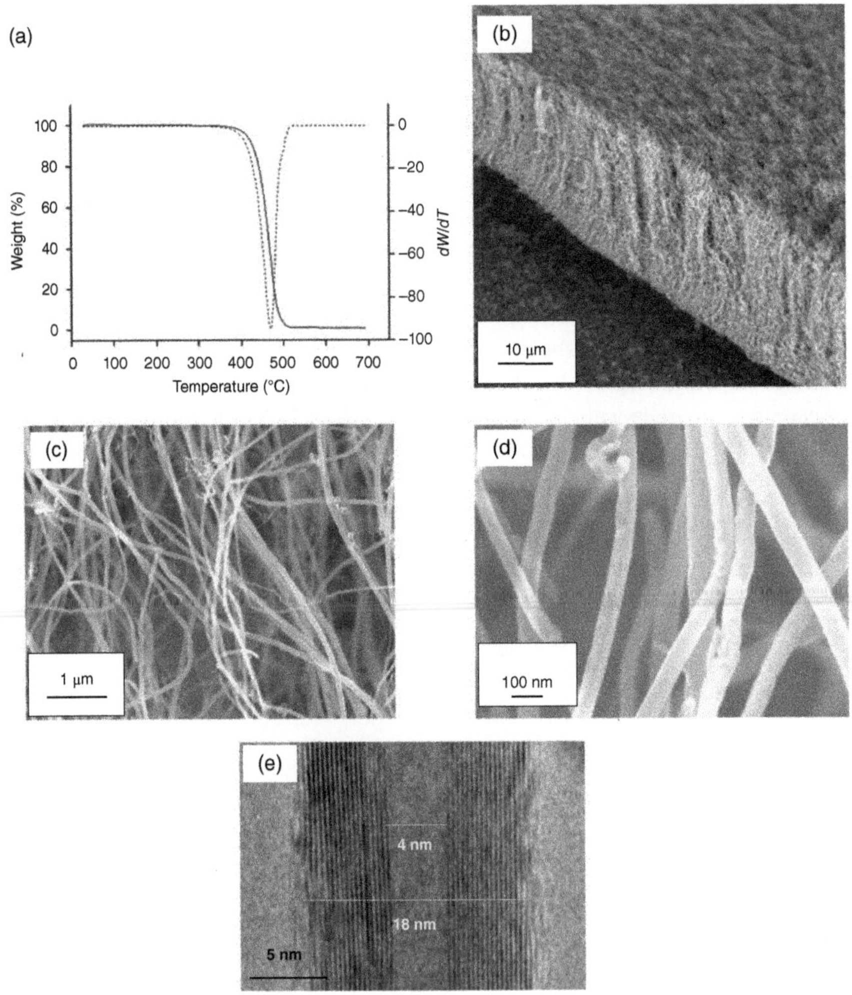

Figure 4.6 (a) TGA and DTGA curves for chicken fat oil, (b)–(d) FESEM images of VACNTs grown from waste chicken fat on a Si substrate within creasing magnification, (e) HRTEM image of multiwalled CNTs. Source: Ref. [51] / with permission of Elsevier.

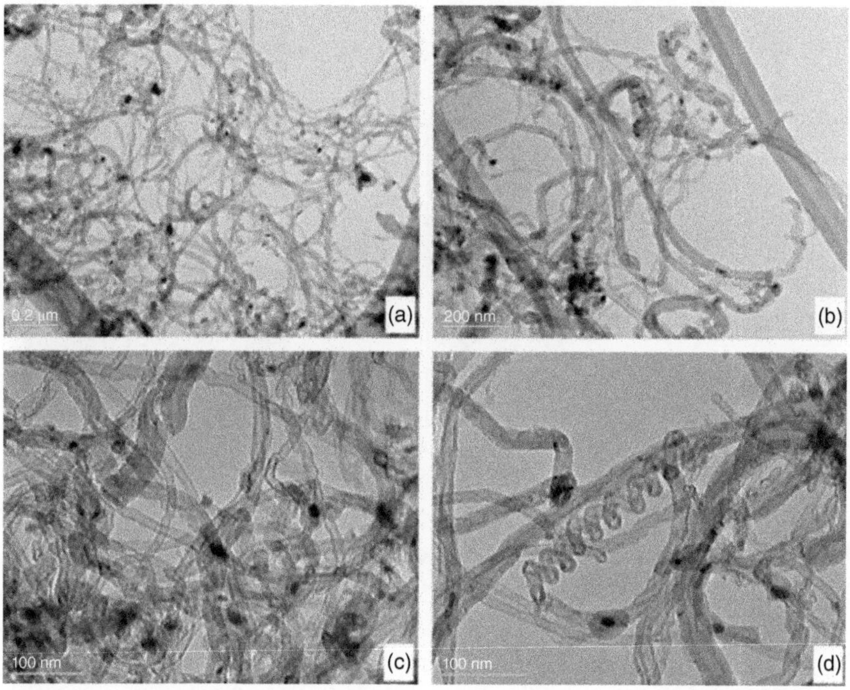

Figure 4.7 TEM images of (a) synthesized MWCNTs produced by fluidized bed chemical vapor deposition at 650 °C using a calcined red mud catalyst, (b) and (c), MWCNTs 6–35 nm in diameter, many with encapsulated catalyst particles (d) a coiled MWCNT with diameter of 16 nm. Source: Ref. [52] / with permission of Elsevier.

Oscar M. Dunens et al. [52] used red mud, a toxic waste product from bauxite processing as a catalyst for the synthesis of MWCNTs by fluidised bed chemical vapor deposition. The products were analyzed using thermogravimetric analysis, Raman spectroscopy, and transmission electron microscopy. Using ethylene at 650 °C a MWCNT yield of 375% (with respect to Fe loading) was obtained. Carbon products were approximately 75% MWCNTs from Raman spectroscopy of 1.43. The production technique and reaction conditions used are conducive to large-scale CNT production, offering a potential value-added commercial use for red mud. Figure 4.7 shows TEM images of (a) synthesized MWCNTs produced by fluidised bed chemical vapor deposition at 650 °C using a calcined red mud catalyst, (b) and (c), MWCNTs 6–35 nm in diameter, many with encapsulated catalyst particles (d) a coiled MWCNT with diameter of 16 nm.

Muhammad Asnawi et al. [53] utilized microwave oven to fabricate carbon nanostructure, specifically CNTs, from waste RH powders. It has been shown that the use of carbon source, catalyst, and commercial microwave oven to induce plasma is necessary to carry on this synthesis. The plasma enhances and speeds up the catalytic decomposition of RH in the presence of ferrocene. FESEM, TGA, and Raman spectroscopy were utilized to confirm the presence and quality of produced carbon nanomaterials. In addition, these results suggest the conversion of ferrocene to iron

Figure 4.8 The conversion process of RHs soot-like material on the top of aluminum case surface including spherical and tubular structures. Source: Ref. [53] / Hindawi Publishing Corporation/ CC BY 4.0.

(II, III) oxide with notable conversion rate. Figure 4.8 shows the schematic diagram of MO (Modus Operandi) technique for derivation of carbon nanomaterials from RH.

4.5 Carbon Quantum Dots (CQDs)

Carbon quantum dots (CQDs) have attracted remarkable attention as a novel and promising fluorescent carbon material with their unique optical properties, good water solubility, high stability, low toxicity, excellent biocompatibility, and low environmental impact. The CQDs are emerging nanostructures that consist of carbon atoms in core and the surface, which is functionalized with organics or biomolecules. The CQDs are typically below 10 nm in size. The CQDs have an amorphous to a nanocrystalline core which the carbon atoms predominantly have sp^2 hybridization. The functional groups such as carbonyl, carboxyl, hydroxyl, and epoxy on the CQDs result in their water solubility and make platforms for functionalization. The optical properties of the CQDs and in particular their fluorescence emission have attracted great attention in recent years. The CQDs have superior properties such as water solubility, cytotoxicity, easy synthesis, easy functionalization, chemical inertness, good fluorescence emission, and photo bleaching resistance [54].

4.5.1 CQDs from Biomass

In the past few years, there have been numerous reports of natural biomass being used as a carbon source for CQD synthesis as an alternative to chemical carbon sources. Natural biomasses include orange juice, hair, sugarcane bagasse pulp, walnut shells, petroleum coke, and egg, etc. These carbon sources present to be cheap, green, and sustainable. However, the quantum yield (QY) of natural biomass is low and has to be improved. The QY is a vital property of CQDs and dictates the possible applications. Heteroatom doping of CQDs not only improves the fluorescence efficiency but also provides active sites in the CQDs to broaden their potential applications in the analysis and sensing. Due to the atomic size and chemical valence of nitrogen atoms, the nitrogen-doped CQDs (N-CQDs) are of great interests.

For example, Ankit Tyagi et al. [55] synthesized water-soluble carbon quantum dots (wsCQDs) from lemon peel waste using a facile and cost-effective hydrothermal process. Schematic illustration for the synthetic procedure of wsCQDs by hydrothermal treatment of lemon peel waste precursor is shown in Figure 4.9.

As synthesized wsCQDs were 1–3 nm in size with spherical morphology and oxygen-rich surface functionalities. These wsCQDs manifest excellent photoluminescent (PL) properties and exhibited (QY ~14% with high aqueous stability. wsCQDs were further used to design the economic, green, and highly sensitive fluorescent probe for the detection of Cr^{6+} ions with detection limit of ~73 nM. This wsCQD-based fluorescent probe could provide a simple, rapid, convenient technique for the sensitive and selective detection of Cr^{6+} in water purification processes. Further, wsCQDs were immobilized over electrospun TiO_2 nanofibers, and photocatalytic activity for such TiO_2-wsCQDs composite was demonstrated using methylene blue (MB) dye as a model pollutant. Photocatalytic activity for TiO_2-wsCQDs composite was found ~2.5 times more than that TiO_2 nanofibers. The synthesis method for wsCQDs could be easily scaled up for gram scale synthesis of CQDs. Morphological characterization of wsCQDs was analyzed with TEM.

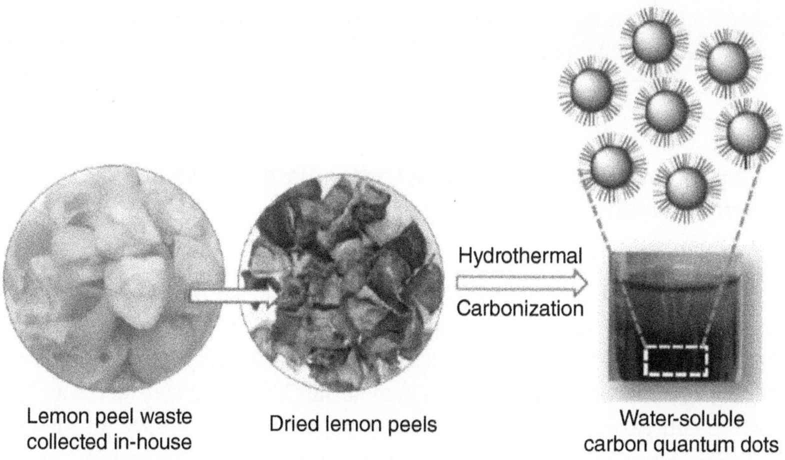

Figure 4.9 Schematic illustration for the synthetic procedure of wsCQDs by hydrothermal treatment of lemon peel waste precursor. Source: Ref. [55] / with permission of Royal Society of Chemistry.

Likewise, Madrakian et al. [56] proposed an isotretinoin detection system with a 0.03 mM limit of detection (LOD) using an N-CQD-based fluorescence sensor, which had been applied as a bio-probe for isotretinoin detection in pharmaceutical preparations and human serum samples.

Niu et al. [57] used a controllable electrochemical/electro analytical approach to generate aspartic acid (Asp) based NCQDs (Asp-N-CQDs), which exhibited the highest QY and were applied as probes for the Fe^{3+} detection in cells. The Asp-N-CQDs as nano sensor probes for the Fe^{3+} detection showed a low LOD of 0.5 mM.

4.6 Catalytic Applications of Carbon-Based Nanomaterials

In recent years, in order to improve the efficiency of the fuel cell, a lot of studies are carried out in different ways. In order to improve the efficiency of the fuel cell, the catalyst is needed. The noble metal Pt has the good catalytic efficiency. However, Pt is expensive, which is a huge obstacle for fuel cell to be commercialized. So we need to reduce the use of Pt and develop new catalysts improving the oxidation–reduction reaction (ORR) of the electrode. The support of catalyst can help improve the capability of catalyst.

Common catalysts support includes carbon, graphene, carbon nanotube (CNT), and other forms of carbon. Low content of platinum catalyst being thin film layer has been reported to be effective and provides a higher quality of Pt utilization and activity. Graphene nanosheet is an ideal alternative compared to the traditional carbon support materials has a high electrical activity of the catalyst and superior durability than the commercial Pt/C catalyst. Furthermore, the support of iron-based catalyst is porous carbon used in polymer electrolyte fuel cell, which can improve the fuel cell oxygen reducing ability. So the supports are really important for fuel cells. CNT as a support can effectively improve the catalyst performance and utilization. Nowadays, the CNTs adopted in fuel cells are receiving wide attention.

4.6.1 Potential Advantages in Using Carbon-Based Nanomaterials for Advanced Catalysts

There is a number of potential advantages given by the use of nanocarbon materials to develop advanced catalysts, in comparison to the use of catalysts, other types of carbon materials such as carbon gels and aerogels, or different nanostructured catalytic materials such as mixed oxides. There are thus many motivations for their use. It may be remarked that often carbon materials have been considered an "inert" support, in comparison to mixed oxides, for example. As discussed also later, it is evident that carbon materials, and particularly nanocarbons, are instead showing a rich functional surface activity, with different type of surface groups (depending on the preparation and functionalization of nanocarbon) showing itself a rich catalytic chemistry (metal-free carbon catalysts).

Although knowledge is growing, it is still missing a complete picture of how to put in relation the presence of these surface functional groups to the characteristic nano-dimension and nano-architecture of the nanocarbon. Stability of many of these functional groups is also still an issue in comparison with "conventional" catalysts, as well as the possibility to maximize their concentration to improve the catalyst productivity.

4.6.2 Photocatalysts

A photocatalyst is a material that absorbs the energy of the light leading to a higher energy level and releases such energy to a one or more substrates (reagents) catalyzing a specific reaction. The use of CNPs as photosensitizers displays several advantages, such as abundance, nontoxicity, stability, and sustainability.

These nanoparticles show a broad absorption band extended to 600 nm, with an average particle size of ca. 25 nm, with the presence of an external shell of 0.26 nm due to the presence of ZnO. These nanoparticles have been used as catalysts for the degradation of three different industrial dyes by using the daylight irradiation: malachite green, methylene blue, and fluoresce in. For example, CNPs doped with ZnO show higher degradation efficiency of malachite green dye (100%), if compared to simple Zn—O doped with nitrogen (60%), demonstrating the positive role of the nanocatalyst in the dye degradation.

In the case of methylene blue, nanocatalyst degraded completely the dye in 45 minutes, while the control experiments show 82% of degradation in 60 minutes. With fluoresce in, a 95% of degradation occurs in 15 minutes (100% in 30 minutes) with the synthesized nanocatalyst, while in the control experiments, the 92% of degradation requires 60 minutes. Furthermore, the nanocatalysts have been reused four times, preserving the efficiency.

Benjamin C. M. Martindale et al. [58] established CQDs as excellent photosensitizers in combination with a molecular catalyst for solar light driven hydrogen production in aqueous solution. The inexpensive CQDs can be prepared by straight forward thermolysis of citric acid in a simple one-pot, multigram synthesis and are therefore scalable. The CQDs produced reducing equivalents under solar irradiation in a homogeneous photocatalytic system with a Ni-bis-(diphosphine) catalyst, giving an activity of 398 μmol_{H_2} $(gCQD)^{-1}$ h^{-1} and a "per Ni catalyst" turnover frequency of 41 h^{-1}. The CQDs displayed activity in the visible region beyond $\lambda > 455$ nm and maintained their full photocatalytic activity for at least one day under full solar spectrum irradiation. A high quantum efficiency of 1.4% was recorded for the noble- and toxic-metal-free photocatalytic system. Thus, CQDs are shown to be a highly sustainable light-absorbing material for photocatalytic schemes, which are not limited by cost, toxicity, or lack of scalability. The photocatalytic hybrid system was limited by the lifetime of the molecular catalyst, and intriguingly, no photocatalytic activity was observed using the CQDs and 3D transition metal salts or platinum precursors. This observation highlights the advantage of using a molecular catalyst over commonly used heterogeneous catalysts in this photocatalytic system. Figure 4.10 describes solar H_2 production

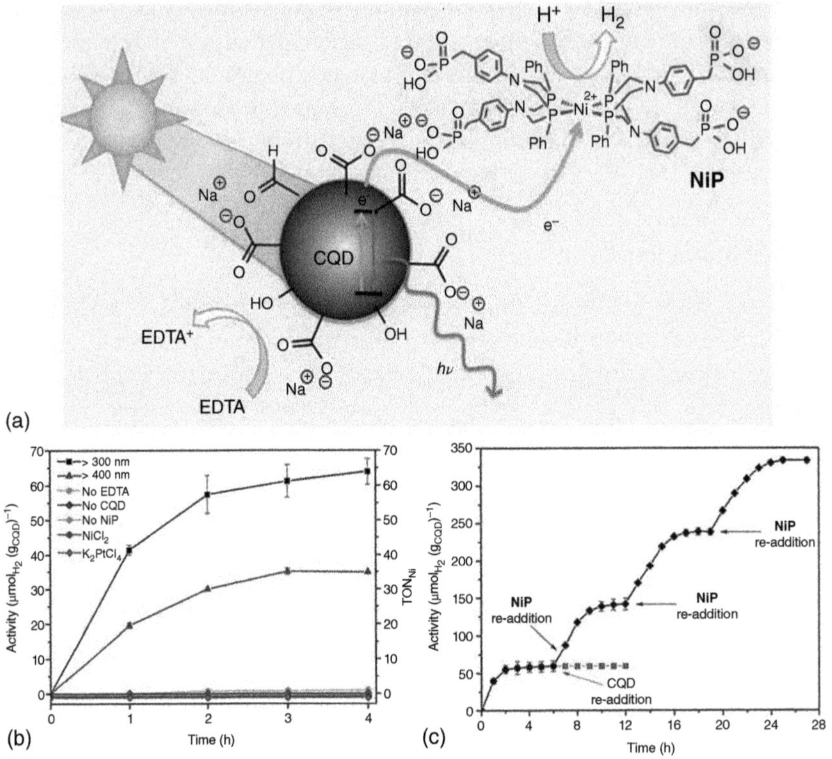

Figure 4.10 (a) Schematic illustration of solar H_2 production using the hybrid CQDs–NiP system. (b) Optimized H_2 generation using CQDs (10 mg) and NiP (10 nmol) in aqueous EDTA solution (0.1 M, pH 6) under sun irradiation in the absence ($l > 300$ nm) and presence of a 400 nm UV filter. Control experiments without EDTA, CQDs, and NiP, as well as with heterogeneous catalyst precursors, $NiCl_2$ (30 nmol) and K_2PtCl_4 (128 nmol, 0.5 wt.%), are also shown. (c) Photocatalytic H_2 generation using CQDs (10 mg) and NiP (10 nmol) in aqueous EDTA solution under 1 sun full solar spectrum irradiation. Activity is recovered upon each readdition of the catalyst, NiP (10 nmol), but not upon readdition of CQDs (10 mg). Source: Ref. [59] / with permission of Royal Society of Chemistry.

using the hybrid CQD–NiP system. Optimized H_2 generation using CQDs (10 mg) and NiP (10 nmol) in aqueous EDTA solution (0.1 M, pH 6) under sun irradiation in the absence (l > 300 nm) and presence of a 400 nm UV filter. Control experiments without EDTA, CQDs, NiP, as well as with heterogeneous catalyst precursors, $NiCl_2$ (30 nmol) and K_2PtCl_4 (128 nmol, 0.5 wt%), are also shown. Photocatalytic H_2 generation using CQDs (10 mg) and NiP (10 nmol) in aqueous EDTA solution under 1 sun full solar spectrum irradiation. Activity is recovered upon each readdition of the catalyst, NiP (10 nmol), but not upon readdition of CQDs (10 mg).

Dario Mosconi et al. [60] proposed convenient routes to produce hybrid polymers that covalently enclosed or confined N-doped CQDs. We focus our attention on polyamide, polyurea urethane, polyester, and polymethylmetacrylate polymers, some of the most common resources used to create everyday materials. These hybrid materials can be easily prepared and processed to obtain macroscopic objects of different shapes, i.e. fibers, transparent sheets, and bulky forms, where the characteristic luminescence properties of the native N-doped CQDs are preserved. More importantly, we explore the potential use of these hybrid composites to achieve photochemical reactions as those of photoreduction of silver ions to silver nanoparticles (under UV light), the selective photo-oxidation of benzylalcohol to the benzaldehyde (under vis-light), and the photocatalytic generation of H_2 (under UV light).

4.6.3 Electro Catalysts

Nanocomposites based on transition metal oxides and carbon nanostructures with low-cost, high activity, and good stability are promising catalysts toward electrochemical water oxidation, which is desirable but remains challenging. Here, Shunyan Zhao et al. [61] reported the design and synthesis of nanocomposite based on CQDs, SnO_2, and CO_3O_4 (CQDs/SnO_2–CO_3O_4) as the electrocatalyst for highly efficient oxygen evolution reaction (OER). In alkaline media, the complex exhibited high electrocatalytic activity and long-term stability, which was better than either pristine CO_3O_4 or SnO_2–CO_3O_4 composite. Moreover, the CQDs/SnO_2–CO_3O_4 with a molar ratio of Sn:Co at 1 : 3 revealed the highest catalytic activity toward OER among different molar ratios of CQDs/SnO_2–CO_3O_4 composites. Experimental results indicated that the Co atoms were considered as the active centre, the nano-sized SnO_2-enhanced electronic conductivity, and the insoluble CQDs layer on the surface effectively protected the catalyst, thus the composite structure resulted in the excellent electrocatalytic activity and high stability.

4.7 Conclusions, Future Outlook, and Challenges

Until today, the nanocatalysis has been dominated by metal nanoparticles. Only recently, the wide diffusion of carbon nanoparticles allows applying these new nanomaterials also in the catalytic field. This chapter explains different carbon nanoparticles derived by biomass. The different catalytic applications presented earlier confirm the potential of the carbon nanomaterials to be used as catalyst. Although carbon-based nanomaterials are used from a long time in catalysis as support and for catalytic applications. It offers unconventional ways for their utilization and to address some of the new challenges derived from moving to a more sustainable future. Versatile and multifunctional carbon catalyst derived from biomass should be established. The sustainable chemistry and advanced green processes in the industry pursue the development of versatile and multifunctional catalysts.

Acknowledgments

The authors would like to thank the University Grants Commission, Government of India for the financial support (ref. no. – F.No.25-1/2014-15/(BSR)/7-225/2008/(BSR), dt: October 2015) and also thank the authorities of Gandhigram Rural Institute – Deemed to be University for the encouragement.

References

1 Dai, L., Chang, D.W., Baek, J.B., and Lu, W. (2012). Carbon nanomaterials for advanced energy conversion and storage. *Small* 8: 1130–1166.

2 Schnoor, J.L. (2011). Water–energy nexus. *Environ. Sci. Technol.* 45: 5065.

3 Zhai, Y., Zhu, Z., Zhu, C. et al. (2014). Multifunctional water-soluble luminescent carbon dots for imaging and Hg^{2+} sensing. *J. Mater. Chem. B* 2: 6995–6999.

4 Su, C. and Loh, K.P. (2013). Carbocatalysts: graphene oxide and its derivatives. *Acc. Chem. Res.* 46: 2275–2285.

5 Mestl, G., Maksimova, N.I., Keller, N. et al. (2001). Carbon nanofilaments in heterogeneous catalysis: an industrial application for new carbon materials? *Angew. Chem. Int. Ed.* 40: 2066–2068.

6 Adams, W.W., Baughman, R.H., and Smalley, R.E. (2007). Synthesis approaches of zinc oxide nanoparticles: the dilemma of ecotoxicity. *J. Nanomater.* 2007: 1–15.

7 Avti, P.K., Talukdar, Y., Sirotkin, V.M. et al. (2013). Toward single-walled carbon nanotube-gadolinium complex as advanced MRI contrast agents: pharmacodynamics and global genomic response in small animals. *J. Biomed. Mater. Res. Part B* 101: 1039–1049.

8 Mashal, A., Sitharaman, B., Li, X. et al. (2010). Toward carbon-nanotube-based theragnostic agents for microwave detection and treatment of breast cancer: enhanced dielectric and heating response of tissue-mimicking materials. *IEEE Trans. Biomed. Eng.* 57: 1831–1834.

9 Sitharaman, B., Rajamani, S., and Avti, P.K. (2011). Time-resolved red luminescence from europium-catalyzed single walled carbon nanotubes. *Chem. Commun. (Camb.)* 47: 1607–1609.

10 Talukdar, Y., Avti, P.K., Sun, J., and Sitharaman, B. (2014). Multimodal ultrasoundphotoacoustic imaging of tissue engineering scaffolds and blood oxygen saturation in and around the scaffolds. *Tissue Eng. Part C Methods* 20: 1–65.

11 Kim, J. and Bruggen, D.V.B.J. (2010). The use of nanoparticles in polymeric and ceramic membrane structures: review of manufacturing procedures and performance improvement for water treatment. *Environ. Pollut.* 158: 2335–2349.

12 Rai, M., Yadav, A., and Gade, A. (2009). Silver nanoparticles as a new generation of antimicrobials. *Biotechnol. Adv.* 27: 76–83.

13 Kumar, J., Mallampati, R., Adin, A., and Valiyaveettil, S. (2014). Functionalized carbon spheres for extraction of nanoparticles and catalyst support in water. *ACS Sustain. Chem. Eng.* 2: 2675–2682.

14 Xu, J.Y., Weinberg, G., Liu, X. et al. (2008). Nanoarchitecturing of activated carbon: facile strategy for chemical functionalization of the surface of activated carbon. *Adv. Funct. Mater.* 18: 3613–3619.

15 Shivakumar, M., Nagashree, K.L., Yallappa, S. et al. (2016). Biosynthesis of silver nanoparticles using pre-hydrolysis liquor of eucalyptus wood and its effective antimicrobial activity. *Enzym. Microb. Technol.* 97: 9755–9762.

16 Su, D.S., Perathoner, S., and Centi, G. (2013). Nanocarbons for the development of advanced catalysts. *Chem. Rev.* 113: 5782–5816.

17 Cafer Saka, B.E.T. (2012). TG–DTG, FT-IR, SEM, iodine number analysis and preparation of activated carbon from acorn shell by chemical activation with $ZnCl_2$. *J. Anal. Appl. Pyrolysis* 95: 21–24.

18 Wang, D.W. and Su, D.S. (2014). Heterogeneous nanocarbon materials for oxygen reduction reaction. *Energy Environ. Sci.* 7: 576–591.

19 Sun, H.Q. and Wang, S.B. (2014). Research advances in the synthesis of nanocarbon-based photocatalysts and their applications for photocatalytic conversion of carbon dioxide to hydrocarbon fuels. *Energy Fuel* 28: 22–36.

20 Ghiban, A., Negoita, O.I., and Negoita, O.D. (2012). Management of the waste materials. *JEEE* 5: 73–78.

21 Zhuo, C. and Levendis, Y.A. (2014). Upcycling waste plastics into carbon nanomaterials: a review. *J. Appl. Polym. Sci.* 131: 39931–39945.

22 Miskolczi, N., Angyal, A., Bartha, L., and Valkai, I. (2009). Fuels by pyrolysis of waste plastics from agricultural and packaging sectors in a pilot scale reactor. *Fuel Process. Technol.* 90: 1032–1040.

23 Donate, P.M. (2014). Green synthesis from biomass. *Chem. Biol. Technol. Agric.* 1: 1–8.

24 Katz, E. and Willner, I. (2004). Biomolecule-functionalized carbon nanotubes: applications in nanobioelectronics. *ChemPhysChem* 5: 1085–1104.

25 Geim, K. (2009). Graphene: status and prospects. *Science* 324: 1530–1534.

26 Allen, M.J., Tung, V.C., and Kaner, R.B. (2010). Honeycomb carbon: a review of graphene. *Chem. Rev.* 110: 132–145.

27 Novoselov, K.S., Faĺko, V.I., Colombo, L. et al. (2012). Aroadmap for graphene. *Nature* 490: 192–200.

28 Sambasivudu, K. and Yashwant, M. (2012). Challenges and opportunities for the mass production of high quality graphene: an analysis of worldwide patents. *Nanotech Insights* 3: 6–19.

29 Dhand, V., Rhee, K.Y., Kim, H.J., and Jung, D.H. (2013). A comprehensive review of graphene nanocomposites: research status and trends. *J. Nanomater.* 2013: 1–14.

30 Strudwick, A.J., Weber, N.E., Schwab, M.G. et al. (2015). Chemical vapor deposition of high quality graphenefilms from carbon dioxide atmospheres. *ACS Nano* 9: 31–42.

31 Coleman, J.N. (2013). Liquid exfoliation of defect-free graphene. *Acc. Chem. Res.* 46: 14–22.

32 Mishra, N., Boeckl, J., Motta, N., and Iacopi, F. (2016). Graphene growth on silicon carbide: a review. *Phys. Status Solidi* 213: 2277–2289.

33 Bonaccorso, F., Lombardo, A., Hasan, T. et al. (2012). Production and processing of graphene and 2d crystals. *Mater. Today* 15: 564–589.

34 Abdolhosseinzadeh, S., Asgharzadeh, H., Seop Kim, H., and Novoselov, K.S. (2015). Fast andfully-scalable synthesis of reduced graphene oxide. *Sci. Rep.* 5: 10160.

35 Ren, W. and Cheng, H.-M. (2014). The global growth of graphene. *Nat. Nanotechnol.* 9: 726–730.

36 Muramatsu, H., Kim, Y.A., Yang, K.-S. et al. (2014). Rice husk-derived graphene withnano-sized domains and clean edges. *Small* 10: 2766–2770.
37 Wang, H., Xu, Z., Kohandehghan, A. et al. (2013). Interconnected carbon nanosheets derived from hemp for ultrafastsupercapacitors with high energy. *ACS Nano* 7: 5131–5141.
38 Zhao, H., Zhao, T.S., Lee, C. et al. (2013). Graphene sheets fabricated from disposable paper cups as a catalyst support material for fuel cells. *J. Mater. Chem. A* 1: 183–187.
39 Ruan, G., Sun, Z., Peng, Z., and Tour, J.M. (2011). Growth of graphene from food, insects, and waste. *ACS Nano* 5: 7601–7607.
40 Primo, A., Atienzar, P., Sanchez, E. et al. (2012). From biomass wastes to large-area, high-quality, N-doped graphene: catalyst-free carbonization of Chitosan coatings on arbitrary substrates. *Chem. Commun.* 48: 9254.
41 Li, Y., Wu, J., and Chopra, N. (2015). Nano-carbon-based hybrids and heterostructures: progress in growth and application for lithium-ion batteries. *J. Mater. Sci.* 50: 7843–7865.
42 Iijima, S. (1991). Helical microtubules of graphitic carbon. *Nature* 354: 56–58.
43 Ohardani, O., Elola, M.C., and Elizetxea, C. (2014). Potential and prospective implementation of carbon nanotubes on next generation aircraft and space vehicles: a review of current and expected applications in aerospace sciences. *Prog. Aerosp. Sci.* 70: 42–68.
44 Maiti, U.N., Lee, W.J., Lee, J.M. et al. (2014). 25th Anniversary article: chemically modified/doped carbon nanotubes & graphene for optimized nanostructures & nanodevices. *Adv. Mater.* 26: 40–67.
45 Thostenson, E.T., Ren, Z.F., and Chou, T.W. (2001). Advances in the science and technology of carbon nanotubes and their composites: a review. *Compos. Sci. Technol.* 61: 1899–1912.
46 Jie, H. (2004). Structures and properties of carbon nanotubes. In: *Carbon Nanotubes*, vol. 1 (ed. M. Meyyappan), 1–24. CRC Press.
47 Yanagi, K. (2014). Differentiation of carbon nanotubes with different chirality. In: *Carbon Nanotubes and Graphene*, Seconde (ed. K.T. Iijima), 19–38. Oxford: Elsevier.
48 Mc Euen, P.L., Fuhrer, M.S., and Park, H. (2002). Single-walled carbon nanotube electronics. *IEEE Trans. Nanotechnol.* 1: 78–85.
49 Corral, E.L., Wang, H., Garay, J. et al. (2011). Effect of single-walled carbon nanotubes on thermal and electrical properties of silicon nitride processed using spark plasma sintering. *J. Eur. Ceram. Soc.* 31: 391–400.
50 Koziol, K., Boskovic, B.O., and Yahya, N. (2010). Synthesis of carbon nanostructures by CVD method. In: *Carbon and Oxide Nanostructures, Advanced Structured Materials*, 5e (ed. N. Yahya). Berlin, Heidelberg: Springer-Verlag.
51 Suriani, A.B., Dalila, A.R., Mohamed, A. et al. (2013). Vertically aligned carbon nanotubes synthesized from waste chicken fat. *Mater. Lett.* 101: 61–64.
52 Dunens, O.M., Mac Kenzie, K.J., and Harris, A.T. (2010). Synthesis of multi-walled carbon nanotubes on 'red mud' catalysts. *Carbon* 48: 2375–2377.

53 Asnawi, M., Azhari, S., Hamidon, M.N. et al. (2018). Synthesis of carbon nanomaterials from rice husk via microwave oven. *J. Nanomater.* 2018: 1–5.

54 Amjadi, M., Hallaj, T., Asadollahi, H. et al. (2017). Facile synthesis of carbon quantum dot/silver nanocomposite and its application for colorimetric detection of methimazole. *Sens. Actuators B Chem* 244: 425–432.

55 Tyagi, A., Tripathi, K.M., Singh, N. et al. (2016). Green synthesis of carbon quantum dots from lemon peel waste: applications in sensing and photocatalysis. *RSC Adv.* 6: 72423–72432.

56 Madrakian, T., Maleki, S., Gilak, S., and Afkhami, A. (2017). Turn-off fluorescence of amino functionalized carbon quantum dots as effective fluorescent probes for determination of isotretinoin. *Sens. Actuators B Chem* 247: 428–435.

57 Niu, F., Xu, Y., Liu, J. et al. (2017). Controllable electrochemical/electroanalytical approach to generate nitrogen-doped carbon quantum dots from varied amino acids: pinpointing the utmost quantum yield and the versatile photoluminescent and electrochemiluminescent applications. *Electrochim. Acta* 236: 239–251.

58 Martindale, B.C.M., Hutton, G.A.M., Caputo, C.A., and Reisner, E. (2015). Solar hydrogen production using carbon quantum dots and a molecular nickel catalyst. *J. Am. Chem. Soc.* 137: 6018–6025.

59 Wang, R., Kang-Qiang, L., Tang, Z.-R., and Yi-Jun, X. (2017). Recent progress in carbon quantum dots: synthesis, properties and applications in photocatalysis. *J. Mater. Chem. A* 5: 3717–3734.

60 Mosconi, D., Mazzier, D., Silvestrini, S. et al. (2015). Synthesis and photochemical applications of processable polymers enclosing photoluminescent carbon quantum dots. *ACS Nano* 9: 4156–4164.

61 Zhao, S., Li, C., Liu, J. et al. (2015). Carbon quantum dots/SnO_2-CO_3O_4 composite for highly efficient electrochemical water oxidation. *Carbon* 92: 64–73.

5

Biomass-Derived Carbon Quantum Dots for Fluorescence Sensors

Somasundaram Anbu Anjugam Vandarkuzhali[1], Jeyabalan Shanmugapriya[2], Chinna Ayya Swamy P[3], Subramanian Singaravadivel[4], and Gandhi Sivaraman[5]

[1] *Indian Institute of Technology Madras, National Centre for Catalysis Research, Chennai 600036, Tamil Nadu, India*
[2] *The Madura College, Department of Chemistry, Madurai 625011, Tamil Nadu, India*
[3] *National Institute of Technology, Department of Chemistry, Calicut 673601, Kerala, India*
[4] *SSM Institute of Engineering and Technology, Dindigul 624002, Tamil Nadu, India*
[5] *Gandhigram Rural Institute-Deemed to be University, Gandhigram 624302 Tamil Nadu, India*

5.1 Introduction

Carbon dots (CDs) are a novel sort of carbon-based zero-dimensional material and obligate drawn substantial attention due to their excellent photostability, favorable biocompatibility, low toxicity, outstanding water solubility, high sensitivity, and excellent selectivity for target analytes, tunable fluorescence emission and excitation, high quantum yield, and broad stokes shift. The fluorescent carbon nanoparticles were accidentally discovered by Xu et al. [1] in 2004 during electrophoretic purification of single-walled carbon nanotubes (SWCNTs). In 2006, Sun et al. [1] reported the synthesis of fluorescent carbon particles of size less than 10 nm and termed them as CDs. The discovery of CDs led to intensive research exploring this fascinating young material with extraordinary potentials. The popularity of CDs is due to the unique and remarkable properties like multicolor wavelength-tuned emission, up-conversion photoluminescence (PL), high quantum yield, aqueous dispersibility, and high biocompatibility.

Several individuals pay consideration to merely the nature and application of CDs, disregarding the prominence of carbon sources. Biomass is a upright carbon source for the preparation of biomass CDs (BCDs). Biomass carbon sources are eco-friendly natural products paralleled to further carbon sources and obligate many recompenses in preparing BCDs, comprising existence economical, informal to acquire, green, and abundant. Further, the production of BCDs from natural biomass can adapt low-value biomass waste into valued and expedient constituents. The biomass encompassing heteroatoms is the excellent raw material for the preparation of CDs, in distinction to the CDs of human-made carbon sources that entail the addition of external heteroatoms [2].

Biomass-Derived Carbon Materials: Production and Applications, First Edition.
Edited by Alagarsamy Pandikumar, Perumal Rameshkumar, and Pitchaimani Veerakumar.

In recent years, the occurrence of bioinspired approaches for the synthesis of multifunctional nanomaterials has fascinated ample attention from ultimate exploration to innumerable scientific applications [3–6]. In precise, scholars are enthusiastically probing for greener and renewable carbon sources to produce biocompatible carbonaceous [7]. Such exertions steered to the probe of biomolecules or other natural means through good biocompatibility and indispensible biological functions as precursors to synthesize bioinspired CDs, an evolving class of fluorophores with unified biological and/or chemical properties, which is known as "biodots" [3, 4]. A wide spectrum of natural resources extending from small molecules (e.g. glucose) and biopolymers (e.g. proteins or nucleic acids) to complex matrices such as plant- and/or animal-derived materials (e.g. leaves or eggs) has remained effectively engaged as natural carbon sources to synthesize biodots with intrinsic biological or biomimetic functions. In this perspective, the synthesis routes of biodots are considered centered on the sorts of precursors used.

Especially carbon-based nanomaterials were deliberated as a milestone in the field of sensors and their environmental solicitations remaining to their exclusive physical and chemical properties [8–13]. Biomass-derived CDs has turned out to be a new area in the green synthesis with an economic gain in full applications. In the facet of nanomaterials synthesis, the carbon sources could be eagerly obtainable from the natural bioresources for simple, commercial, and green synthesis of carbon nanomaterials. Variety of CDs can be prepared from biomasses with varying the experimental conditions. Several studies have been reported in the synthesis of CDs from biomasses, natural resources, and devastate resources [14–18]. Though lots of biomasses have been used as carbon sources, biomasses residues are hitherto obtainable to employ broadly in an effective manner. This chapter is projected toward the recent advancement of biomasses-derived fluorescent CDs and their sensing applications.

5.2 Characterization of CDs

Carbon quantum dots (CQDs) are generally referred in terms of a carbogenic core consisting of amorphous and crystalline structures with surface functional groups and sp^2/sp^3 carbon [19, 20]. The CDs surface is usually involved or adapted chemical groups, for instance, amino groups, oxygen groups, etc. TEM and AFM are used to characterize the morphology of CDs. Most BCDs are usually spherical with an average diameter of less than 10 nm. The sizes and phases of the CDs were measured via high-resolution transmission electron microscopy (HRTEM). HRTEM images spectacle the lattice spacing of BCDs in the range of 0.18–0.24 nm, consistent to diverse diffraction planes, yet some have no significant spacing. The characteristic XRD profiles of CDs indicate their crystalline natures. The XRD spectra of biomass-derived CDs generally have a broad diffraction peak in the 2θ range of 20°–25° and lattice spacing between 0.31 and 0.38 nm [21–24]. The FTIR analysis has been used to confirm the functional groups of CDs [25]. Raman spectrum can similarly deliver a substantial indication of a link between the graphene structure and the CDs. The characteristic peak at 1580 cm^{-1} (G band) is related through vibrations of sp^2-hybridized carbon atoms in a 2D hexagonal lattice and the E2g vibration mode of graphite, suggesting the ordered graphite structure, while the peak at 1300 cm^{-1} (D band) reflects the disorder of sp^3-hybridized carbon atoms. The relative intensity ratio of disordered D band and crystalline G band (I_D/I_G) leads research to estimate the degree of graphitization and crystallization of CDs. More excellent the value of I_D/I_G, the proportion of amorphous structures of CDs is higher [25].

5.3 Optical Properties

5.3.1 Absorbance

The UV–visible absorption spectra of biomass-derived CDs characteristically revealed to have peak in UV region with a tail encompassing in the visible region. The π–π^* transition originating from the sp^2-hybridized domains in the core of CDs, and the n–π^* transitions of surface groups are a joint source that strictly linked to the absorption peak of CDs in the ultraviolet region (260–410 nm) with a tail extending to the visible region. Besides, CDs produced by diverse methods also display diverse optical absorption characteristics [24, 26–28].

5.3.2 Fluorescence

The fluorescence properties diverge sensitively with the size of the CDs due to the dependence of the HOMO-LUMO gap on the size, with amassed size ensuing in a red-shifted emission. Several explanations for the same have been given which includes size dependence, surface defects, surface states, the degree of oxidation, etc. Replacement with electron-donating character advances the HOMO to sophisticated energy, instigating a constricting of the energy gap. Nitrogen-substituted aromatics induce a further energy level amid the LUMO and HUMO, distinctly constricting the bandgap [29].

5.4 Methods for the Synthesis of CDs

The diversity of carbon-based quantum elements is connected to a great number of synthetic approaches obtainable. Biomass-derived carbon nanostructure aimed at the synthetic methods is mostly categorized into two primary sorts: top-down and bottom-up. While the top-down strategy is based on breaking bulk carbon materials into small pieces by physical or chemical methods, the bottom-up approach uses selected molecular precursors, forming CDs upon dehydration and carbonization processes. The precursors generally possess —OH, —COOH, and NH_2 groups, which can be dehydrated at elevated temperatures. Here, we emphasis on the bottom-up approach accomplished by the combination of organic

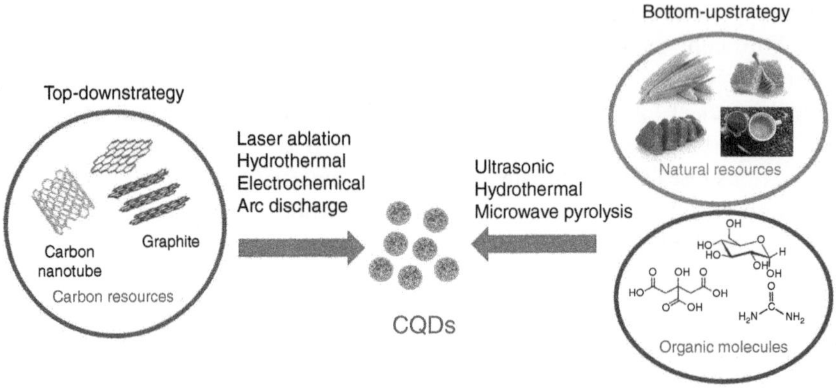

Figure 5.1 Synthetic approaches for carbon-based nanomaterials. Source: Atoss / Adobe Stock; Dionisvera / Adobe Stock; karandaev / Adobe Stock, Lucid_dream / Adobe Stock.

molecules beneath the thermal/hydrothermal/solvent thermal circumstances or the carbonation of organic precursors. Commonly discourse, the bottom-up method has a high yield, and it is expedient to familiarize heteroatom doping in the synthesis process (Figure 5.1).

5.4.1 Hydrothermal Carbonization Method

Hydrothermal carbonization is a low-cost, ecologically friendly, and nontoxic route to produce carbon-based materials from numerous precursors [20]. The concern of this practice lies in the excellent control of the nucleation practices in an ecologically friendly fashion deprived of the use of strong acids or bases [30]. It has remained extensively used due to the easy control of the retort, the high reactivity of reactants, and low energy loss. The hydrothermal dealing is the thermochemical degradation of biomass in the existence of water (water/biomass ratio may range from 5 : 1 to 75 : 1) at elevated temperature and pressure is a capable practice for adapting biomass into novel carbon materials for a varied diversity of latent applications [31–37]

5.4.2 Microwave Method

Microwave synthesis, which can accomplish great solution temperatures in minutes, has converted a vital practice in synthetic chemistry, although there is an existence of a simple and cheap approach for the fast synthesis of CDs. It is deliberated as a time-saving technique for organizing BCDs owing to its transient heating properties and ominously enlightening the yield of high-quality CDs [38–41].

5.4.3 Chemical Oxidation Method

The chemical oxidation method is a technique in which an oxidizing agent oxidizes a target by instigating the target to lose electrons, and a process of oxidizing with a potent oxidizing agent such as H_2O_2 and H_2SO_4/HNO_3 can promote the carbonization of biomass. The precursors mostly employed as should be thermally stable and able to yield useful carbon residues with subsequent high-temperature processing. The carbonization practice arises over several diverse flame methods [42]. This technique similarly has the advantage of high selectivity. Still, it has prodigious hindrances: (i) it uses expensive oxidants and causes some pollution to the environment, and (ii) it outcomes in frequently recurrent production with a low production capacity. Thus far, the use of chemical oxidation is rare in preparing CDs using biomass as a carbon source [42–44].

5.4.4 Pyrolysis

Pyrolysis is the reaction practice in which a constituent is putrefied by heat. So far, the pyrolysis technique has become one of the conventional methods to prepare CDs. With biomass as the carbon source, the particle size of CDs prepared by pyrolysis is generally between 0.4 and 6 nm, and the quantum yield is between 3 and 25% [45–52].

5.5 Application of CDs

The diverse and exclusive physicochemical properties and the prospect to synthesize a large panel of carbon-based nanostructures obligate ended their curiosity for requests in numerous various fields extending from sensors and catalysis to energy and medicine. Natural biomass is used as a raw material for the synthesis of fluorescent CDs, and its use has developed further exciting and striking, due to its sustainable and workable features. Biomass is abundant in phytochemicals, which safeguards the self-passivation of CDs with extensive applications for biosensor use [14, 53, 54]. Significantly, CDs reveal low toxicity and superior biocompatibility which enrich their mark in the biological fields. The fluorescence properties of CDs are essentially allied with surface functional groups, and these are highly sensitive to the adjoining environment subsequently to make further robust interaction with analytes. At present, CDs distinctive optical properties and their perspective in the development of efficient optoelectronic devices and sensors have widened their horizon in fluorescence sensing applications [55–58].

5.5.1 Metal Ion Sensing

In general, the CDs inherently retain oxygen-based surface functional groups such as carboxylic and hydroxyl groups to create a hydrophilic nature of CDs and more surface active sites. Thus, metal ions can be efficiently interacting with CDs over surface bonding, consequently ensuing in the tuning of CDs properties. The metal ions interact with the CDs via surface bonding with high selectivity and sensitivity, which results in the formation of new electron–hole recombination via energy transfer process and results in a change in fluorescence intensity of CDs and acts as a measurable response signal [59].

5.5.1.1 Mercury (Hg^{2+}) Sensor

Mercury (Hg^{2+}) ions are pervasive impurities that can cause environmental and health problems. The Hg^{2+} can interact with —COOH groups on the CDs surface to form new nonradiative electron–hole pairs, prominent to the fluorescence quenching of CDs. Also, Hg^{2+} ions induce conversion of the —CONH— functional group to open ring amide from bridged cyclic amide and subsidize to fluorescence quenching. Further, the stronger affinity between iodide and Hg^{2+} can remove Hg^{2+} from the CDs-Hg^{2+} system, leading to the recovery of fluorescence BCDs. The CDs consequent from each orange juice, *Coccinia indica*, and barley were developed as nanosensors for the detection of Hg^{2+} (Figure 5.2) [60–62]. The self-passivated and passivated CDs were adequately validated as fluorescence sensors, which show diverse interactions with metal ions. The self-passivated CDs show selectivity in sensing Hg^{2+}, with a detection limit of 3.3 nM. In contrast, passivated CDs, including N/SCDs, N/O-CDs, and N-CDs, are sensitive and selective toward Cu^{2+}, Pb^{2+}, and Fe^{3+} ions, revealing lower detection limits of 0.045, 0.27, and 6.2 μM, respectively. Similar to biomass-derived CDs, functionalization can effectively detect the Hg^{2+} ion to yield very low LOD [63–69] (Figure 5.3).

Feng group used strawberry as a carbon source for the synthesis of fluorescent nitrogen-doped carbon nanoparticles (FNCPs) for selective Hg^{2+} sensing in liner range 10 nM to 50 μM, and detection limit of 3 nM [71]. Yeng's group achieved a detection limit of 1 nM using tea-derived CDs for Hg^{2+} with real sample

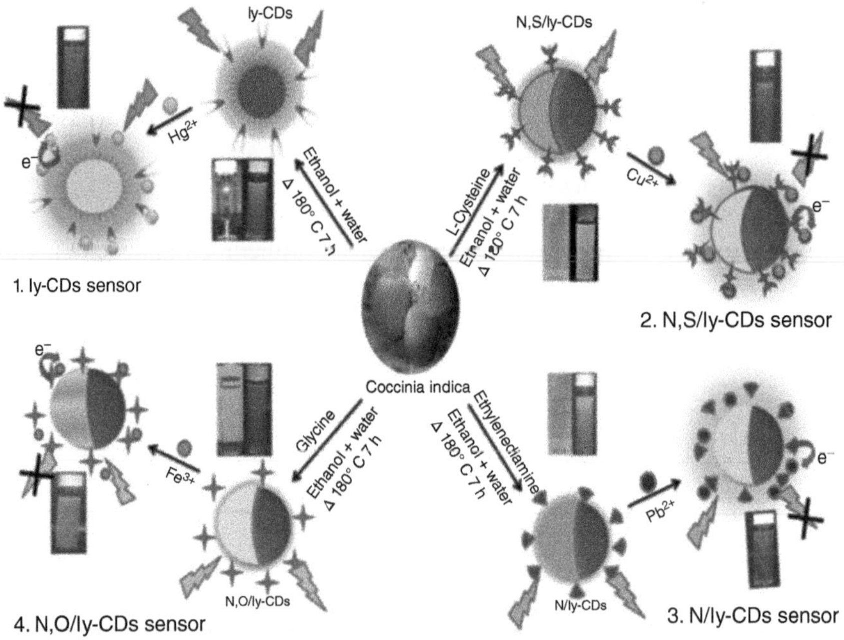

Figure 5.2 Schematic illustration for the detection process *Coccinia indica*-derived CDs toward Hg^{2+}, Cu^{2+}, Pb^{2+}, and Fe^{3+} ions. Source: Ref. [60] / with permission of Royal Society of Chemistry.

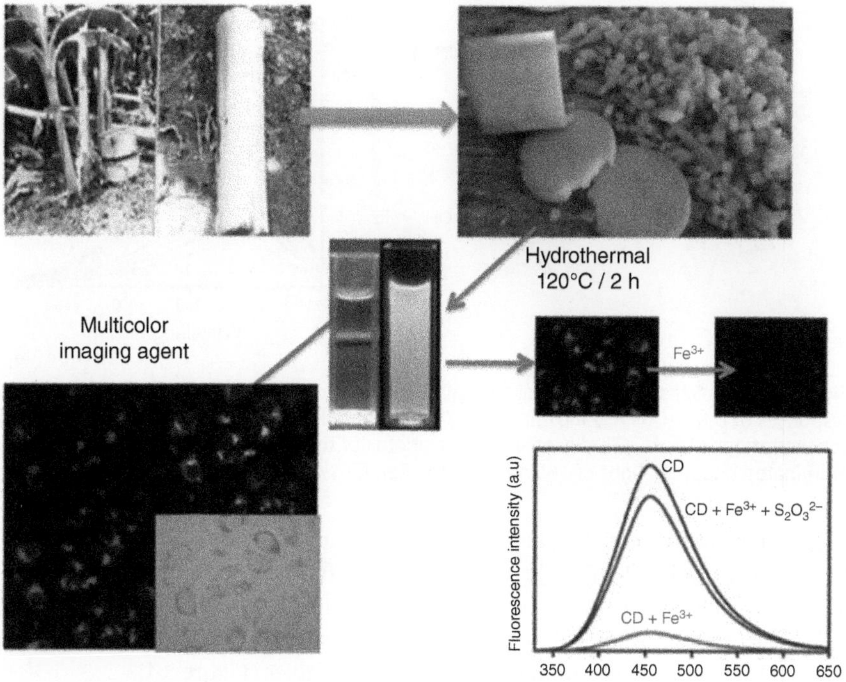

Figure 5.3 Carbon dots derived from pseudostem of banana plants as fluorescence sensor for Fe(III) ions. Source: Ref. [70] / with permission of Elsevier.

applicability [72]. Chen group made CDs using *Jinhua bergamot* which acts as a dual sensor for Hg^{2+} and Fe^{3+}ions. Hg^{2+} atoms quench the fluorescence intensity of CDs via static quenching. The detection limit was 5.5 nM for Hg^{2+} and liner range was between 0.01 and 100 μM, whereas for Fe^{3+} detection limit was 0.075 μM and liner range was 0.025–100 μM [73].

5.5.1.2 Iron (Fe^{3+}) Sensor

Recently, iron (Fe^{3+}) ions are one of the utmost collective described metal ions by fluorescent CDs detection. Due to the low solubility product constant of metal hydroxides (the Ksp of $Fe(OH)_3 = 4 \times 10^{38}$), Fe^{3+} can interact with the phenolic hydroxyl groups of CDs which lead to fluorescence quenching. Besides, the fluorescence quenching may be contributed to nonradiative electron transfer that intricates the restricted transfer of an electron in the excited state to the d-orbital of Fe^{3+}. Due to this interaction between CDs and Fe^{3+}, an electron in the excited state of CDs may transfer to the unfilled orbital of Fe^{3+} (Figure 5.2) [74]. Such an effect can interrupt the radiative transition resulting in the fluorescence quenching and consequently foremost to the high sensitivity and selectivity toward Fe^{3+} ion. The LOD in nanomolar range was accomplished for diverse kind of biomass-derived CDs such as watermelon (160 nm), grass (20 nm), banana peel (6.4 nm), radish (130 nm), lycii (21 nm), sweet potato (320 nm), black tea (250 nm), mint leaves (374 nm), chimney oil (0.18 nm), fish scale (540 nm), and mangosteen pulp (52 nm) [26, 70, 75–82].

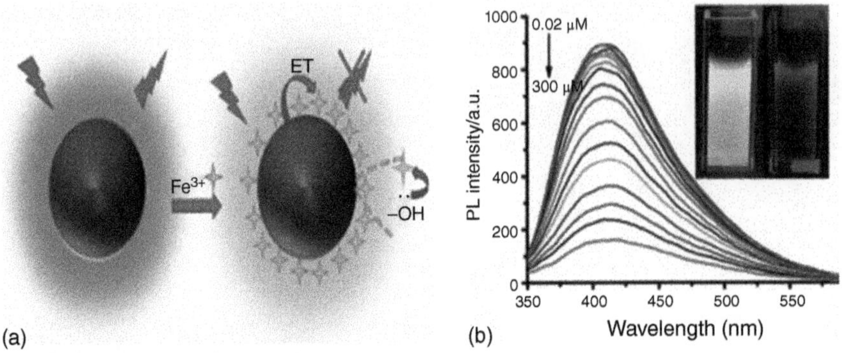

Figure 5.4 (a) Sensing principle of the N-CDs-based probe for Fe^{3+}. (b) The fluorescence responses of the N-CDs (0.2 mg mL^{-1}) in PBS solution (pH 7.4) after the addition of different concentrations of Fe^{3+}. (The inset shows the photos of the corresponding solutions illuminated under UV light of 365 nm.) Source: Ref. [74] / with permission of Elsevier.

CDs derived from corianders leaves [83] (0.4 μM), papaya [84] (0.29 μmol L^{-1}), rose heart radish [74] (0.13 μM), Prunus avium [85] (0.96 μM), and onion waste [86] (0.31 μM). The TCSPC data revealed decay in fluorescence lifetime from 5.71 to 4.89 ns as a function of Fe^{3+} concentration, revealing dynamic quenching. This sensor also worked well with tap and lake water samples (Figure 5.4).

5.5.1.3 Lead (Pb^{2+}) Sensor

Lead (Pb^{2+}) has extended amassed consideration for its extreme effects on the environment, humans, and animals. Pb^{2+} ions have a strong affinity with surface functional groups of CDs to enable the electron transfer from the excited CDs to Pb^{2+} leading to the fluorescence quenching [87]. Formerly, Kumar et al. and Bandi et al. have prepared biomass-derived CDs independently from tulsi leaves and berries, respectively; both CDs retain amine, functional groups, on the surface [88]. Both tulsi- and berries-derived CDs sensors exhibited exceptional sensitivity toward the detection of Pb^{2+} with LOD of 0.59 and 9.64 nm, respectively. *Ocimum sanctum*-derived green CDs were used as a sensor for Pb^{2+} with detection limit of 0.59 nM and linear detection range of 0.01–1.0 μM. The selective nature of this sensing system is due to the efficient binding affinity between vacant d-orbital of Pb^{2+} ions and amine group present on CDs surface. The oxygen and amino functional groups of CDs have a high binding affinity with Pb^{2+} ions that induce proximity between them which facilitates effective electron transfer [89] (Figure 5.5).

5.5.1.4 Copper (Cu^{2+}) Sensor

Copper (Cu^{2+}) ions fit an indispensable trace element that often subsists in water. Monitoring Cu^{2+} helps to apprehend their impact on human health. The existence of oxygen and nitrogen atoms in BCDs makes the probe selectively detect Cu^{2+} due to its faster chelation and higher thermodynamic affinity [59]. The bamboo leaves-derived CDs were capped with BPEI for selective detection of Cu^{2+} ions with a limit of detection 115 nM [90]. The increasing Cu^{2+} concentration caused

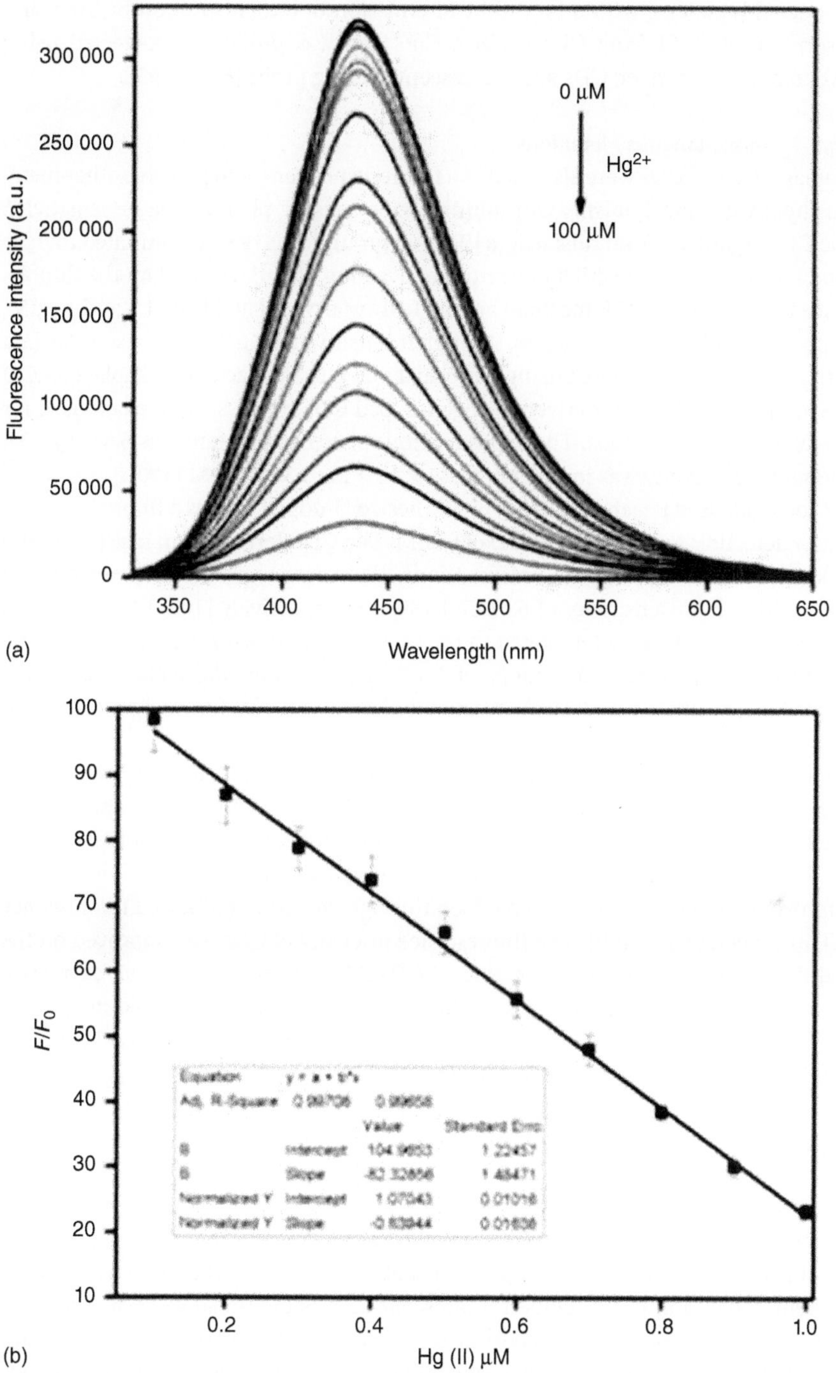

Figure 5.5 (a) Fluorescence titrations of pineapple peel-derived CD with Hg(II) in PBS buffer. (b) Stern–Volmer plot the fluorescence intensioty vs. [Hg(II)]. Source: Ref. [69] / with permission of American Chemical Society.

PL quenching along with a blue shift in emission due to adsorption of Cu^{2+} and desorption of BPEI from CDs surface. Cu^{2+} ion was proficiently detected using several biomass-derived CDs as a fluorescent sensing probe [60, 91–93].

5.5.1.5 Miscellaneous Metal Ions

Biomass-derived CDs were also used as a fluorescence sensor to perceive other metal ions by diverse mechanisms with innumerable sensing phenomena or sensitivity. The CDs organized from tulsi leaves [94] and groundnuts [95] was initiated to have good selectivity and sensitivity toward the revealing of Cr(VI) based on the fluorescence turn-off via the IFE mechanism; the LOD was calculated to be 1.6 and 1.9 μM, respectively [24, 47]. Nevertheless, the higher sensitivity of Cr(VI) was accomplished with N-doped CDs prepared using citric acid and glutamic acid and displayed LOD of 5 nm [96–99]. Pigskin-derived nitrogen-doped CDs (N-CDs) have stood used for the detection of Co^{2+} ions. The exposure limit of sensing system was 6.8×10^{-7} M. The sensing capacity was in the range of 1×10^{-6} M to 3×10^{-4} M [100].

Zhou et al. used peach gum [101] to synthesize N-doped CDs as a fluorescent sensor for detecting Au^{3+} with the LOD of 64 nM. The CDs derived from rose petals and jackfruit seed are employed as an exceptional fluorescence probe in the detection of Au^{3+} with a high sensitivity of 63.1 and 239 nm, respectively [102–105]. Likewise, N/S co-doped CDs were fabricated to the sensing of Ag^{+} with high sensitivity and selectivity [106]. In both detections of Au^{3+} and Ag^{+} ions, the metal ions have a strong binding affinity with surface functional groups of CDs initiating proximity to each other.

Ramezani et al. have described the detection of As^{3+} ion by using CDs prepared from Quince fruit [107]. Gu et al. prepared N/S co-doped CDs from scallion to detect Cd^{2+} ion based on fluorescence quenching process with LOD of 15 nm [108]. Conflicting to fluorescence quenching, Bhamore et al. have perceived fluorescence improvement in the detection of Al^{3+} ion through chelation-enhanced fluorescence (CHEF) mechanism [109]. The fluorescence intensity of CDs was improved on the addition of Al^{3+} due to the formation of CDs/Al^{3+} complex. Gao et al. developed the red-emissive fluorescent CDs for the multiple detections of noble metal ions (Pt^{2+}, Au^{3+}, and Pd^{2+}). Upon the addition of individual Pt^{2+}, Au^{3+}, and Pd^{2+} ions, the fluorescence intensity of CDs was quenched.

5.5.2 Anion Sensors

Numerous metal ions can quench the biomass-derived CDs fluorescence. Subsequently adding some explicit anions, the fluorescence was improved because of the binding between cations and anions. Yao et al. reconnoitered the sweet pepper consequent down- and up-adaptation fluorescent CDs as dual-readout assay probe for detecting hypochlorite (ClO-) [110]. It was described that several hydroxyl groups existing on the CD surface act as a reducing group and get oxidized in the existence of ClO- group. This alteration in the surface state leads to PL quenching. Zhan et al. used fish scales resultant N-CDs for the detection of hypochlorite in the range of 0–10 mM via fluorescence quenching [111]. The hypochlorite sensing was ascribed to selective photoinduced electron transfer from N atom to ClO-. Xu et al. synthesized blue-fluorescing CDs by hydrothermal of potatoes. The sensing system is sensitive to PO_4^{3-} concentration, with a detection limit of 0.8 μmol L^{-1} [112].

Based on the off–on phenomenon, our group synthesized CDs from pseudostem of the banana plant and was engaged as a turn-on sensing probe for $S_2O_3^{2-}$ detection [70]. The CDs displayed quenching in the presence of Fe^{3+}, which might be improved using $S_2O_3^{2-}$. The LOD of the CD/Fe^{3+} system for $S_2O_3^{2-}$ was 8.47×10^{-7} M.

5.5.3 Miscellaneous Molecules

Song et al. synthesized CDs over an environmental method that employs linseed as a natural precursor. The as-prepared CDs were pragmatic to the fabrication of biosensors for the sensitive detection of butyrylcholinesterase (BChE) based on a fluorescence quenching mechanism; these biosensors can be applied as nerve gases and an indicator for detecting. The limit of detection of BChE is 0.035 μ mL^{-1} [113]. Rose-derived CDs can be used to detect tetracycline (TC) based on the interactions between TC and CDs, and the proposed analytical approach tolerable for the detection of TC in a linear range of 1.0×10^{-8} to 1.0×10^{-4} mol L^{-1} with a detection limit of 3.3×10^{-9} mol L^{-1} at a signal-to-noise ratio of 3 [40].

Li et al. used grape skin to synthesize BCDs by a hydrothermal method. There is a dehydration condensation reaction between the phenolic hydroxyl groups on the picric acid and the carboxyl groups on the CDs. Furthermore, electrons are conveyed from amino groups on CDs (donor) to the nitro groups on the picric acid (acceptor). These reactions ensued in fluorescence quenching, which indorsed the nonradiative electron–hole recombination. In this research, the BCDs-based sensors were excellently used to selectively and sensitively perceive picric acid with the LOD of 10 nM [114].

5.6 Conclusion and Future Perspectives

Number of low-cost, competent, and simple preparation technique of BCDs have been considered. The physicochemical characteristics of BCDs, such as chemical stability and fluorescent properties, mend superior sensing platforms probable. Based on inimitable characters of CDs, diverse kinds of reviews are fabricated for the revealing of several metal ions, anions, small molecules, and macromolecules by monitoring quenching or recovery practice of CDs fluorescence. BCDs have revealed good retort to a diversity of chemical species with applicability in practical sample analysis. However, biomass-derived CDs with high QYs are still less informed. Further, enlightening sensitivity and selectivity is a challenge.

Correspondingly, there is always room to develop nontoxic functional materials from the organic-based environmental pollutants. Furthermore, in the aspect of synthesis, the large-scale production of CDs should need more effort for their usage in commercial purposes. Regarding the application of CDs in sensor and catalysis fields, CDs evidenced to serve as superior sensing probe and catalyst remaining to the optical properties and low toxicity with exceptional biocompatibility. Additionally, the improved efficiency of sensing or catalytic activity was accomplished for CDs-based nanohybrids/composites. By compelling improvement of surface functional groups, CDs can be easily functionalized by inorganic and organic small molecules; further, they may interact with numerous kinds of nanomaterials to form the hybrid/composite materials. The green precursors are superior to chemical carbon sources in terms of affordability and feasibility. These benefits have inspired prodigious perceptions in preparation and applications of BCDs, making them probable to accomplish green chemistry.

References

1 Xu, X.-Y., Gu, Y.-L., Ploehn, H.J. et al. (2004). *J. Am. Chem. Soc.* 126: 12736.
2 Zhao, S., Lan, M., Zhu, X. et al. (2015). *ACS Appl. Mater. Interfaces* 7: 17060.
3 Zheng, X.T., Xu, H.V., and Tan, Y.N. (2017, ch. 7, vol. 1253). *Advances in Bioinspired and Biomedical Materials*, vol. 2, 123–152. American Chemical Society.
4 Xu, H.V., Zheng, X.T., Mok, B.Y.L. et al. (2016). *J. Mol. Eng. Mater.* 4: 1640003.
5 Yu, Y., Mok, B.Y.L., Loh, X.J., and Tan, Y.N. (2016). *Adv. Healthc. Mater.* 5: 1844–1859.
6 J. Xie, Y. N. Tan and J. Y. Lee, in *Nanotechnologies for the Life Sciences*, 2012, https://doi.org/10.1002/9783527610419.ntls0209.
7 Liu, X., Pang, J., Xu, F., and Zhang, X. (2016). *Sci. Rep.* 6: 31100.
8 Mauter, M.S. and Elimelech, M. (2008). *Environ. Sci. Technol.* 42: 5843.
9 Yu, X., Cheng, H., Zhang, M. et al. (2017). *Nat. Rev. Mater.* 2: 17046.
10 Georgakilas, V., Tiwari, J.N., Kemp, K.C. et al. (2016). *Chem. Rev.* 116: 5464.
11 Zhang, Y., Wu, M., Wu, M. et al. (2018). *ACS Omega* 3: 9126.
12 Zhao, F., Wu, J., Ying, Y. et al. (2018). *Trends Anal. Chem.* 106: 62.
13 Schroeder, V., Savagatrup, S., He, M. et al. (2019). *Chem. Rev.* 119: 599.
14 Zhang, X., Jiang, M., Niu, N. et al. (2018). *ChemSusChem* 11: 11.
15 Liu, M.L., Chen, B.B., Li, C.M., and Huang, C.Z. (2019). *Green Chem.* 21: 449.
16 Das, R., Bandyopadhyay, R., and Pramanik, P. (2018). *Mater. Today Chem.* 8: 96.
17 Sharma, V., Tiwari, P., and Mobin, S.M. (2017). *J. Mater. Chem. B* 5: 8904.
18 Jeong, Y., Moon, K., Jeong, S. et al. (2018). *ACS Sustain. Chem. Eng.* 6: 4510.
19 Deng, J., You, Y., Sahajwalla, V., and Joshi, R.K. (2016). *Carbon* 96: 105.
20 Hu, B., Wang, K., Wu, L. et al. (2010). *Adv. Mater.* 22: 813.
21 Feng, J., Wang, W.J., Hai, X. et al. (2016). *J. Mater. Chem. B* 4: 387.
22 Essner, J.B., Laber, C.H., Ravula, S. et al. (2016). *Green Chem.* 18: 243.
23 Bankoti, K., Rameshbabu, A.P., Datta, S. et al. (2017). *J. Mater. Chem. B* 5: 6579.
24 Wang, W.J., Xia, J.M., Feng, J. et al. (2016). *J. Mater. Chem. B* 4: 7130.
25 Qu, S., Liu, X., Guo, X. et al. (2014). *Adv. Funct. Mater.* 24: 2689.
26 Song, P., Zhang, L., Long, H. et al. (2017). *RSC Adv.* 7: 28637.
27 Guo, Y., Zhang, L., Cao, F., and Leng, Y. (2016). *Sci. Rep.* 6: 35795.

28 Ramanan, V., Thiyagarajan, S.K., Raji, K. et al. (2016). *ACS Sustain. Chem. Eng.* 4: 4724.

29 Sun, Y.P., Zhou, B., Lin, Y. et al. (2006). *J. Am. Chem. Soc.* 128: 7756.

30 Wang, L. and Zhou, H.S. (2014). *Anal. Chem.* 86: 8902.

31 Abbas, A., Mariana, L.T., and Phan, A.N. (2018). *Carbon* 140: 77.

32 Kasibabu, B.S.B., D'Souza, S.L., Jha, S. et al. (2015). *Anal. Methods* 7: 2373.

33 Liu, X.-J., Guo, M.-L., Huang, J., and Yin, X.-Y. (2013). *Bioresources* 8: 2546.

34 Mehta, V.N., Jha, S., and Kailasa, S.K. (2014). *Mater. Sci. Eng. C* 38: 20.

35 De, B. and Karak, N. (2013). *RSC Adv.* 3: 8286.

36 Mehta, V.N., Jha, S., Basu, H. et al. (2015). *Sens. Actuator* 213: 434.

37 Zhao, X.J., Zhang, W.L., and Zhou, Z.Q. (2014). *Colloids Surf. B* 123: 493.

38 Liu, X., Li, T., Hou, Y. et al. (2016). *RSC Adv.* 6: 11711.

39 Qin, X., Lu, W., Asiri, A.M. et al. (2013). *Sensors Actuators B Chem.* 184: 156.

40 Feng, Y., Zhong, D., Miao, H., and Yang, X. (2015). *Talanta* 140: 128.

41 Si, M., Zhang, J., He, Y. et al. (2018). *Green Chem.* 20: 3414.

42 Qiao, Z.-A., Wang, Y., Gao, Y. et al. (2010). *Chem. Commun.* 46: 8812.

43 D'Angelis do, E.S.B.C., Correa, J.R., Medeiros, G.A. et al. (2015). *Chem. Eur. J.* 21: 5055.

44 Liu, H., Ye, T., and Mao, C. (2007). *Angew. Chem. Int. Ed.* 119: 6593.

45 Xue, M., Zhao, J., Zhan, Z. et al. (2018). *Nanoscale* 10: 18124.

46 Hsu, P.-C., Shih, Z.-Y., Lee, C.-H., and Chang, H.-T. (2012). *Green Chem.* 14: 917.

47 Hu, Y., Yang, J., Tian, J. et al. (2014). *Carbon* 77: 775.

48 Zhou, J., Sheng, Z., Han, H. et al. (2012). *Mater. Lett.* 66: 222.

49 Xue, M., Zhan, Z., Zou, M. et al. (2016). *New J. Chem.* 40: 1698.

50 Saxena, M. and Sarkar, S. (2012). *Diam. Relat. Mater.* 24: 11.

51 Park, S.Y., Lee, H.U., Park, E.S. et al. (2014). *ACS Appl. Mater. Interfaces* 6: 3365.

52 Dong, Y., Pang, H., Yang, H.B. et al. (2013). *Angew. Chem. Int. Ed.* 52: 7800.

53 Zhang, J. and Yu, S.-H. (2016). *Mater. Today* 19: 382.

54 Liu, R., Zhang, J., Gao, M. et al. (2015). *RSC Adv.* 5: 4428.

55 Kang, X., Wang, S.X., and Zhu, M.Z. (2018). *Chem. Sci.* 9: 3062.

56 Dong, J.Q., Li, X., Zhang, K. et al. (2018). *J. Am. Chem. Soc.* 140: 4035.

57 Zhao, Z.J., He, B.R., and Tang, B.Z. (2015). *Chem. Sci.* 6: 5347.

58 Qin, W., Zhang, P.F., Li, H. et al. (2018). *Chem. Sci.* 9: 2705.

59 Liu, S., Tian, J., Wang, L. et al. (2012). *Adv. Mater.* 24: 2037.

60 Radhakrishnan, K., Panneerselvam, P., and Marieeswaran, M. (2019). *Anal. Methods* 11: 490.

61 Li, Z., Zhang, Y., Niu, Q. et al. (2017). *J. Lumin.* 187: 274.

62 Xie, Y., Cheng, D., Liu, X., and Han, A. (2019). *Sensors* 19: 3169.

63 Wang, B.-B., Jin, J.-C., Xu, Z.-Q. et al. (2019). *J. Colloid Interface Sci.* 551: 101.

64 Bano, D., Kumar, V., Chandra, S. et al. (2019). *Opt. Mater.* 92: 311.

65 Wang, C., Wang, Y., Shi, H. et al. (2019). *Mater. Chem. Phys.* 232: 145.

66 Xu, Q., Su, R., Chen, Y. et al. (1886). *ACS Appl. Nano Mater.* 2018: 1.

67 Meng, A., Xu, Q., Zhao, K. et al. (2018). *Sensors Actuators B Chem.* 255: 657.

68 Han, Y., Shi, L., Luo, X. et al. (2019). *Carbon* 149: 355.

69 Vandarkuzhali, A.A., Sampathkumar, N., Krishnamurthy, K.R. et al. (2018). *ACS Omega* 3: 12584.

70 Vandarkuzhali, S.A.A., Jeyalakshmi, V., Sivaraman, G. et al. (2017). *Sensors Actuators B Chem.* 252: 894.

71 Huang, H., Lv, J.-J., Zhou, D.-L. et al. (2013). *RSC Adv.* 3: 21691.

72 Wei, J., Liu, B., and Yin, P. (2014). *RSC Adv.* 4: 63414.

73 Yu, J., Song, N., Zhang, Y.-K. et al. (2015). *Sensors Actuators B Chem.* 214: 29.

74 Liu, W., Diao, H., Chang, H. et al. (2017). *Sensors Actuators B Chem.* 241: 190.

75 Lu, M., Duan, Y., Song, Y. et al. (2018). *J. Mol. Liq.* 269: 766.

76 Picard, M., Thakur, S., Misra, M., and Mohanty, A.K. (2019). *RSC Adv.* 9: 8628.

77 Sun, X., He, J., Yang, S. et al. (2017). *J. Photochem. Photobiol. B* 175: 219.

78 Shen, J., Shang, S., Chen, X., and Cai, Y. (2017). *Mater. Sci. Eng. C* 76: 856.

79 Raveendran, V., Babu, A.R.S., and Renuka, N.K. (2019). *RSC Adv.* 9: 12070.

80 Das, P., Ganguly, S., Maity, P.P. et al. (2018). *J. Photochem. Photobiol. B* 180: 56.

81 Zhang, Y., Gao, Z., Yang, X. et al. (2019). *RSC Adv.* 9: 940.

82 Yang, R., Guo, X., Jia, L. et al. (2017). *Appl. Surf. Sci.* 423: 426.

83 Sachdev, A. and Gopinath, P. (2015). *Analyst* 140: 4260.

84 Wang, N., Wang, Y., Guo, T. et al. (2016). *Biosens. Bioelectron.* 85: 68.

85 Edison, T.N.J.I., Atchudan, R., Shim, J.-J. et al. (2016). *J. Photochem. Photobiol. B* 158: 235.

86 Bandi, R., Gangapuram, B.R., Dadigala, R. et al. (2016). *RSC Adv.* 6: 28633.

87 Kumar, A., Chowdhuri, A.R., Laha, D. et al. (2017). *Sensors Actuators B Chem.* 242: 679.

88 Bandi, R., Dadigala, R., Gangapuram, B.R., and Guttena, V. (2018). *J. Photochem. Photobiol. B* 178: 330.

89 Xu, J., Jie, X., Xie, F.F. et al. (2017). *Nano Res.* 11: 3648–3657.

90 Liu, Y., Zhao, Y., and Zhang, Y. (2014). *Sensors Actuators B Chem.* 196: 647.

91 Das, P., Ganguly, S., Bose, M. et al. (2017). *Mater. Sci. Eng. C* 75: 1456.

92 Srinivasan, V., Jhonsi, M.A., Kathiravan, A., and Ashokkumar, M. (2019). *Sensors Actuators B Chem.* 282: 972.

93 Murugan, N., Prakash, M., Jayakumar, M. et al. (2019). *Appl. Surf. Sci.* 476: 468.

94 Bhatt, S., Bhatt, M., Kumar, A. et al. (2018). *Colloids Surf. B* 167: 126.

95 Roshni, V., Misra, S., Santra, M.K., and Ottoor, D. (2019). *J. Photochem. Photobiol. A* 373: 28.

96 Zheng, M., Xie, Z., Qu, D. et al. (2013). *ACS Appl. Mater. Interfaces* 5: 13242.

97 Cai, F., Liu, X., Liu, S. et al. (2014). *RSC Adv.* 4: 52016.

98 Liu, Y., Hu, J., Li, Y. et al. (2015). *Talanta* 134: 16.

99 Zhang, Y., Fang, X., Zhao, H., and Li, Z. (2018). *Talanta* 181: 318.

100 Wen, X., Shi, L., Wen, G. et al. (2016). *Sensors Actuators B Chem.* 235: 179.

101 Liao, J., Cheng, Z., and Zhou, L. (2016). *ACS Sustain. Chem. Eng.* 4: 3053.

102 Yang, Y., Yin, C., Huo, F., and Chao, J. (2013). *RSC Adv.* 3: 9637.

103 Niamsa, N., Kaewtong, C., Srinonmuang, W. et al. (2013). *Polym. Chem.* 4: 3039.

104 Sharma, V., Kaur, N., Tiwari, P. et al. (2018). *Carbon* 139: 393.

105 Raji, K., Ramanan, V., and Ramamurthy, P. (2019). *New J. Chem.* 43: 11710.

106 Dang, D.K., Chandrasekaran, S., Ngo, Y.-L.T. et al. (2018). *Sensors Actuators B Chem.* 255: 3284.

107 Ramezani, Z., Qorbanpour, M., and Rahbar, N. (2018). *Colloids Surf. A Physicochem. Eng. Asp.* 549: 58.

108 Gu, D., Hong, L., Zhang, L. et al. (2018). *J. Photochem. Photobiol. B* 186: 144.

109 Bhamore, J.R., Jha, S., Singhal, R.K. et al. (2018). *J. Mol. Liq.* 264: 9.

110 Yin, B., Deng, J., Peng, X. et al. (2013). *Analyst* 138: 6551.

111 Wu, G., Feng, M., and Zhan, H. (2015). *RSC Adv.* 5: 44636.

112 Xu, J., Zhou, Y., Cheng, G. et al. (2015). *Luminescence* 30: 411.

113 Song, Y., Yan, X., Li, Z. et al. (2018). *J. Mater. Chem. B* 6: 3181.

114 Li, J., Zhang, L., Li, P. et al. (2018). *Sensors Actuators B Chem.* 258: 580.

6

Biomass-Derived Mesoporous Carbon Nanomaterials for Drug Delivery and Imaging Applications

Balaji Maddiboyina[1], Ramya Krishna Nakkala[1], and Gandhi Sivaraman[2]

[1] *NRK & KSR Gupta College of Pharmacy, Department of Pharmacy, Guntur, Andhra Pradesh, India*
[2] *Gandhigram Rural Institute Deemed University, Department of Chemistry, Dindigul, Tamil Nadu, India*

6.1 Introduction

Carbon nanomaterials are still a developing focus in the field of material science. Because of their extensive enactment in various contexts, they are required to receive the expected pronounced consideration. Since they have an unusual composition, possessing distinct physiochemical properties and biological behaviors, such as high chemical inertness/mechanical constancy, exceptional electrical conductivity, and adequate biocompatibility [1]. Graphene, fullerene, carbon nanotubes, and carbon dots (CDs) are among the sp2 carbon-based constituents that have attracted widespread interest for their potential implementation in the biomedical arena, including drug delivery systems (DDSs). These therapies have unique properties in terms of chemistry and physics, such as chemophotothermal synergistic therapy, gene transfection, and in vivo real-time imaging [2–4]. A few examples include a reworked surface, increased photothermal conversion capacity, supramolecular p–p stacking, and biological behaviors like excellent compatibility, high adsorption efficiency, and theranostic utility [5]. The good news is that even at high doses, these sp2-based carbon nanomaterials are completely safe and biocompatible. Preclinical biodistribution, biocompatibility, and hemocompatibility experiments were used to confirm these findings. For a long time, biocompatible inorganic mesoporous materials have intrigued biomedical researchers [6]. Mesoporous structures with large surface areas, high pore volumes, and variable pore sizes provide large reservoirs for guest molecules and show drug release profiles that are explicit in design [7].

Mesoporous carbon nanoparticles (MCNs) assimilate the advantages comprising:

- Enormous surface area and pore volume that are promising for drug loading;
- Adaptable pore structure permits enhanced control of drug release;
- Certainly revised surface potency advance controlled and targeted DDS to enhance the therapeutic efficacy and lessen side effects;
- Exceptional heat translation capacity in the near-infrared (NIR) region might deliver potentials for photothermal therapy (PTT);

Biomass-Derived Carbon Materials: Production and Applications, First Edition.
Edited by Alagarsamy Pandikumar, Perumal Rameshkumar, and Pitchaimani Veerakumar.

- Admirable biocompatibility and physicochemical stability;
- Supramolecular p–p stacking facilitates the high drug loading capacity and sustained drug release to load aromatic drugs;
- In vivo real-time monitoring and theranostic functions are made possible by extraordinary optical qualities and a peaceful interaction with luminous composites;
- Additional requests for PTT, synergistic therapy, cell labeling, exclusion of hazardous components, collection of peptides from biosamples, and biosensing. Consequently, MCNs are debated like the following cohort of platforms for biological applications [8].

This chapter highlights the present biomedical applications of MCNs, which include medication delivery, PTT, synergistic therapy, and delivery of biotherapeutic agents. The extraction of peptides from biosamples, as well as biosensing and bioimaging, is facilitated by mesoporous carbon.

6.2 Drug Delivery Systems Based on MCNs

Drug release rates can be reduced by MCN nanosized mesopores, resulting in persistent release behavior. Because of this, MCNs with adjustable particle size and spherical morphology will be able to travel through the veins. Since the carbonaceous outline of MCNs can practise explicit supramolecular connections through aromatic drug molecules, MCNs are plentiful and seemly for the encapsulation and distribution of hydrophobic medicinal proxies. As a result, MCNs are an excellent source of encouragement for DDS professionals.

6.2.1 Immediate-release DDS

As a result of their increased porosity and powerful adsorption facility, MCNs have a greater capacity for drug loading than other cell types. In order for chemical medications requiring large dosages to fulfill therapeutic needs, this is critical. A hollow mesoporous carbon (HMC) fixated with $Eu3+/Gd3+$-EDTA increased oral bioavailability of the insoluble prototype medicine carvedilol (CAR), which was improved by 2.2 and 6.5 times. As a result, oral bioavailability was greatly altered [9].

6.2.2 Sustained-release DDS

Sustained release by deliberately liberating the encapsulated drugs over a sustained period of time, DD methods can exert a prolonged therapeutic effect, potentially lowering the incidence of supervision, stabilizing drug concentration intensities in the blood, reducing side effects, and increasing patient amenability. To account for persistent release, the carbonaceous environment of MCNs can create supramolecular p–p stacking surfaces with aromatic drug fragments [10]. MCNs demonstrated a good capacity for loading doxorubicin (DOX) due to hydrophobic contacts and supramolecular p–p stacking between DOX and MCNs, as well as a prolonged release [11].

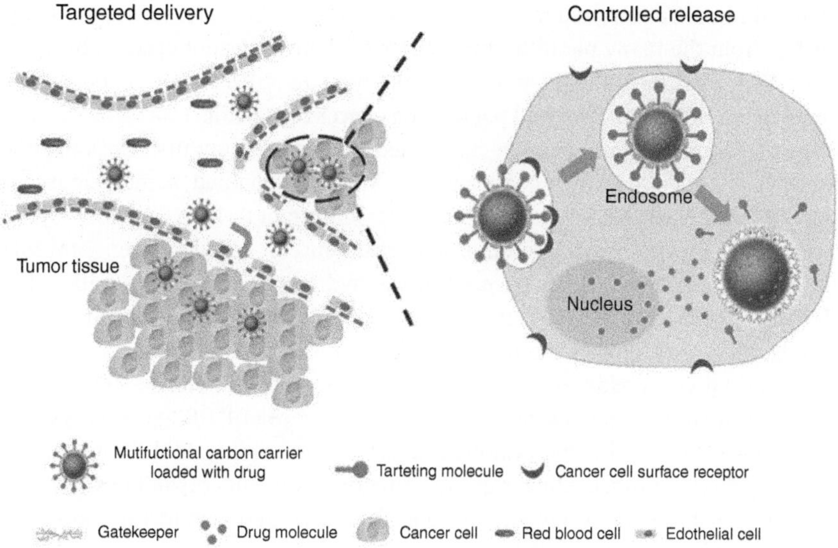

Figure 6.1 Schematic illustration for the in vivo process of controlled and targeted DDS based on MCNs [12]. Source: Ref. [12] / with permission of Elsevier.

6.2.3 Controlled/Targeted DDS

There is growing concern about the multifunctional DDS derived from carbon. This can deliver the drug to the desired location and then release it in a controlled manner to maximize their cellular absorption via the tiniest impulsive release prior to reaching the desired location (Figure 6.1). The carbon-based multifunctional DDS strategy, which retains the ability to transport the drug to the desired place and distribute it in a regulated manner, increases cellular absorption by minimizing premature release prior to achieving the specified site. The surface modification of MCN with hyaluronic acid (HA) over disulfide bonds prepared the scheme is subtle to both intracellular hyaluronidase-1 and GSH, hence controlling drug release. HA could be used as a targeting agent for the CD44 receptor on articulated cancer cells, enhancing the colloidal stability and biocompatibility of MCNs, and anticipating drug release caused by hyaluronidase-1 [12].

6.3 Photothermal Therapy

PTT has proven extremely effective in treating cancer by utilizing NIR-resonant nanoparticles that absorb NIR light and convert it to cytotoxic heat to kill malignant cells while minimizing the intrusive damage to normal cells. It is widely established that direct thermal ablation of cancer cells requires a high temperature (e.g. greater than 50 °C), which may cause damage to nearby normal tissues and cells [13]. PTT significantly enhances the photothermal effects of the tumor by raising the temperature in the tumor range of 43–45 °C and maintaining the hyperthermia for a few minutes. Tumor ablation could be persuaded effectively. MCNs are distinguished from the many photothermal agents by their persistent optical absorption in the NIR. Demonstrating its potential efficacy as an NIR-resonant nano-agent in converting NIR light to heat and persuading cancer cells to die. [14].

Tumor targeting with peptide-conjugated core–shell graphitic carbon@silica nanospheres via dual-ordered mesopores (MMPS) remained a marvelous concoction. Advanced MMPS is a synergistic antitumoral DOX delivery system (MMPSD) for synergistic photothermal treatment of breast cancer. MMPSD's upright water dispersibility is ensured by the hydrophilic mesoporous silica shell. The hydrophobic graphitic mesoporous carbon core enables high hydrophobic drug loading, rapid contact between the drug and photothermal hotspots, and high NIR photothermal conversion efficiency. The SP13 peptide facilitates MMPSD for the targeted and enhanced delivery of DOX in the presence of HER2-positive SK-BR-3 breast cancer cells, while PEGylation ensures biocompatibility. MMPSD technology showed superior drug loading capacity, superior targeting capabilities, sensitive NIR/pH-responsive DOX release, prolonged release, and remarkable anticancer effectiveness in combination [15].

PTT enhances therapeutic efficiency by expending hollow carbon nanospheres (HCSs) as a carrier. The photothermal effects proceed by:

- Slight blisters persuaded by the laser through great power density might rupture the cell membrane and indications to cell death accordingly;
- Lysosomal membrane permeation (LMP) might be improved that would consent for the discharge of DOX@HCSs from lysosomes and enable the penetration of DOX into the nuclei.
- The laser-converted heat could condense DOX certainly detach from the carbon matrix, consequently preclude inadequate drug release at the diseased site (Figure 6.2a).

While laser irradiation increased DOX accumulation beneath the nucleus, NIR irradiation increased DOX accumulation beneath the nucleus due to increased expression of an HSF-1 gene, which may encode a large number of HSF-1 protein homotrimers to suppress resistance-related pathways (Figure 6.2b and c); high levels of ROS may be persuaded by MCNs during NIR irradiation to prepare the MCF-7/ADR cells to DOX. As a result, HCSs continue to be employed to combat chemo-resistance by combining chemotherapy with laser irradiation stimuli to generate heat and free radicals (Figure 6.2d). These HCSs exhibited high load efficiency and were capable of delivering a large number of medicines into cells. Additionally, laser irradiation of HCSs can not only induce photothermal effects but also disrupt the intracellular redox state, which results in the production of stubborn-free radicals. The free radicals resolve to promote HSF-1 gene expression and protein synthesis in order to suppress resistance-related pathways. The usefulness of resistant cells was harmed as a result of the discovery that laser activation of DOX@HCSs may not only release DOX but also operate as a ROS generator by generating a large number of free radicals [16].

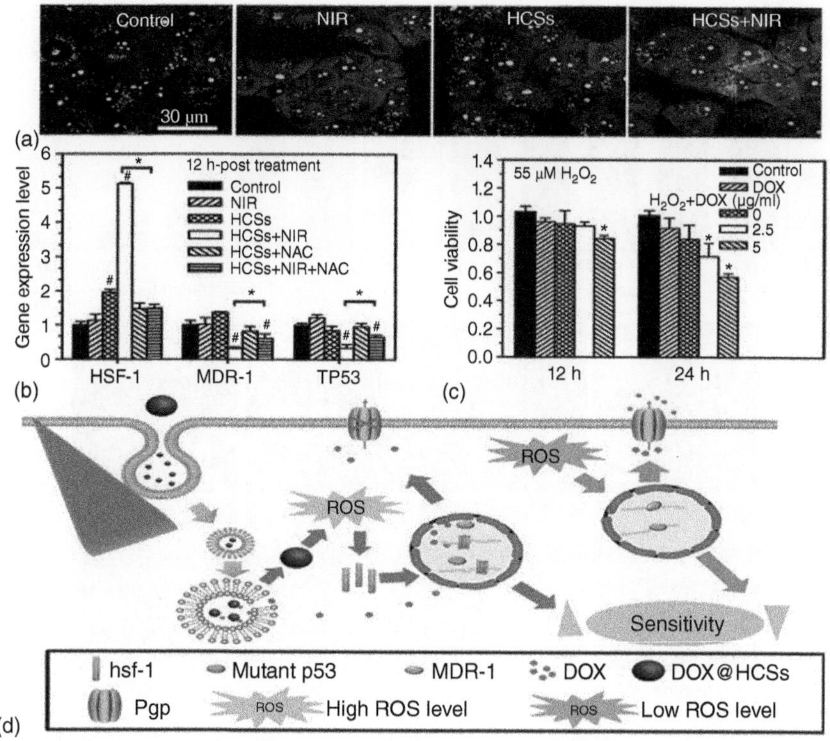

Figure 6.2 (a) Changes in the lysosomal membrane permeation induced by the photothermal effects determined by AO staining. (b) Impact of intracellular ROS levels on the expression of HSF-1, MDR-1, and TP53 genes. The asterisk (*) shows significant differences between test samples and the sample under laser irradiation (HCSsþNIR) ($p < 0.05$); the pound symbol (#) indicates significant differences between test samples and the control. (c) Change in the DOX sensitivity of MCF-7/ADR cells treated with 55 μM H_2O_2. Asterisk (*) indicates significant differences between control and test samples. (d) Combatting the chemotherapeutic resistance of cancer using HCSs under NIR laser irradiation. Source: Ref. [16] / with permission of American Chemical Society.

Permutation of carbon and silica elements may result in the formation of mesoporous amalgamated constituents with enhanced photothermal adaptability and regulated drug release enactment. Additionally, it has a high capacity for medication loading and a well-regulated drug release pattern. The semi-graphitized carbon revealed sundry encouraging properties:

- Great drug loading ability owing to the sp2-hybridized context;
- Enhanced photothermal conversion ability owing to the hotspots of the graphitic pore walls.

Additionally, the mesoporous silica shell could be modified to validate hydrophilicity and DDS targeting. In situ familiarization of semi-graphitized carbon with the mesopores of MSNs for photothermal treatment and DDS was performed [17].

In vitro PTT of MCNs was chosen because it ensured that no cytotoxicity occurred when the cells were cultured at a high concentration of 200 μg mL^{-1} for 24 hours.

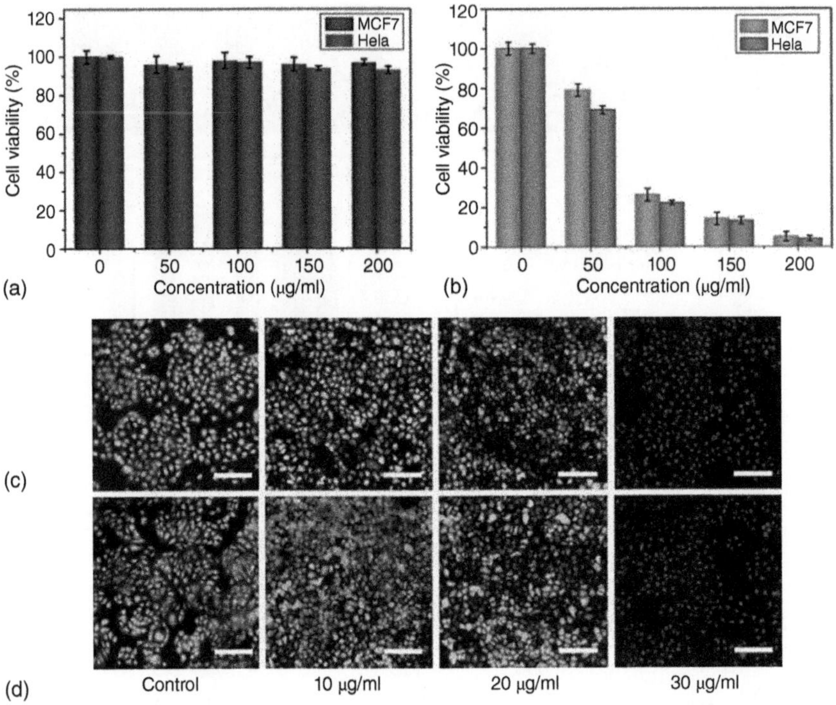

Figure 6.3 In vitro cytotoxicity and photothermal therapy effect. (a) Relative viabilities of MCF7 and HeLa cells after incubation through several concentrations of Meso-CNs for 24 hours. (b) Relative viabilities of MCF7 and HeLa cells after incubation with diverse concentrations of Meso-CNs and ensuing exposure to 650-nm light at a power density of 0.3 W cm^{-2} for 30 minutes. Confocal images of Calcein AM/PI-stained MCF7 cancer cells later incubation with diverse concentrations of Meso-CNs and irradiation with (c) an 808-nm laser at 1.0 W cm^{-2} for 10 minutes and (d) a 1120-nm laser at 0.5 W cm^{-2} for 20 minutes. Scale bar: 150 μm. Source: Ref. [18] / with permission of Ivyspring International Publisher.

The practicalities of MCF7 and HeLa cells were both greater than 90%, implying a remarkable biocompatibility via MCNs (Figure 6.3a). The results demonstrate that following NIR light-induced PTT with MCNs, MCF7, and HeLa cells were killed in a concentration-dependent manner. Both cell types demonstrated strangely diminished capacities when MCN concentrations were as low as 30 μg mL^{-1} (Figure 6.3b). Calcine-AM (living cells, green fluorescence) and propidium iodide (PI; dead cells, red fluorescence) co-staining was used to distinguish live/dead cells using fluorescence confocal microscopy [18]. After irradiation with light at 808 and 1120 nm, the staining results indicated a dose-dependent PTT effect. Almost all cells were killed following incubation with an MCN concentration of 30 μg mL^{-1}, as determined by PTT management (Figure 6.3c and d). These findings indicate that MCNs exhibited exceptional photothermal effects that may be readily observed upon exposure to both NIR-I and NIR-II light.

In vivo, photothermal treatment was considered using H22 tumor-bearing ICR mice. For 20 minutes, numerous groups of mice were treated with the 808- or

1120-nm laser at a power density of $0.5\,W\,cm^{-2}$. As illustrated in Figure 6.4a, the local temperature of tumors injected via Meso-CNs and Meso-CNs loaded with DOX rapidly increased to 65 °C. In comparison, tumors injected with PBS displayed only a minor increase in size when exposed to the same irradiation conditions. Within 10 days, there were no statistically significant changes in the weight deviations between these groups (Figure 6.4b). Surprisingly, tumor development was significantly suppressed in mice receiving DOX-loaded Meso-CNs following NIR laser irradiation as a result of the combination chemo-photothermal therapy (Figure 6.4c).

Numerous tumors in the management groups were completely eradicated (Figure 6.4d). The animals were surrendered for histological examinations using hematoxylin and eosin (H&E) staining. The staining results of the tumor sections demonstrate unequivocally that the cells in the tumors injected with Meso-CNs (or Meso-CNs + DOX) and then exposed to laser irradiation at both 808 and 1120 nm were significantly damaged. In comparison, control group tumor cells mostly engaged their individual morphologies with usual membrane and nuclear components. Additionally, histological examination revealed no injury or inflammation in the respective group's primary organs (Figure 6.4e) [18].

Another work used copper sulfide (CuS) nanoparticles to cap MCNs with a diameter of 150–200 nm (NPs). CuS was a well-known p-type semiconductor material that can be used to demonstrate NIR absorption capacity and photothermal conversion capability [19].

6.3.1 Synergistic Therapy

The development of MCNs for operational dual-triggered synergistic cancer therapy is progressing. MCNs used in the modification act not only as NIR absorbers but also as highly efficient medication transporters. The alteration of MCN's surface with biomacromolecules HA over the disulfide unit renders the scheme susceptible to both intracellular hyaluronidase-1 and GSH, allowing the medication to be released. HA on MCN serves as both a biocompatibility indicator and a cancer cell targeting agent. The MCN-HA compromises the ensuing smart sorts:

- Adsorption of antitumor drug DOX with a great loading efficiency for chemotherapy;
- Exceptional dual stimuli-responsive (Hyal-1/redox) release features;
- Proficient heat conversion for PTT;
- Optical imaging capacity that might aid to track cellular uptake of MCN.

Numerous activities have been ingeniously focused on a single nanomaterial, which has the potential to significantly improve therapeutic efficacy while avoiding undesirable side effects. As a result, MCN-HA is expected to be widely employed in biomedical activities, particularly for cancer therapy [20].

6.3.2 Cell Labeling

Using a conventional electrostatic attraction technique, identical mesoporous carbon spheres (UMCS) were functionalized with HA. This HA modification validated

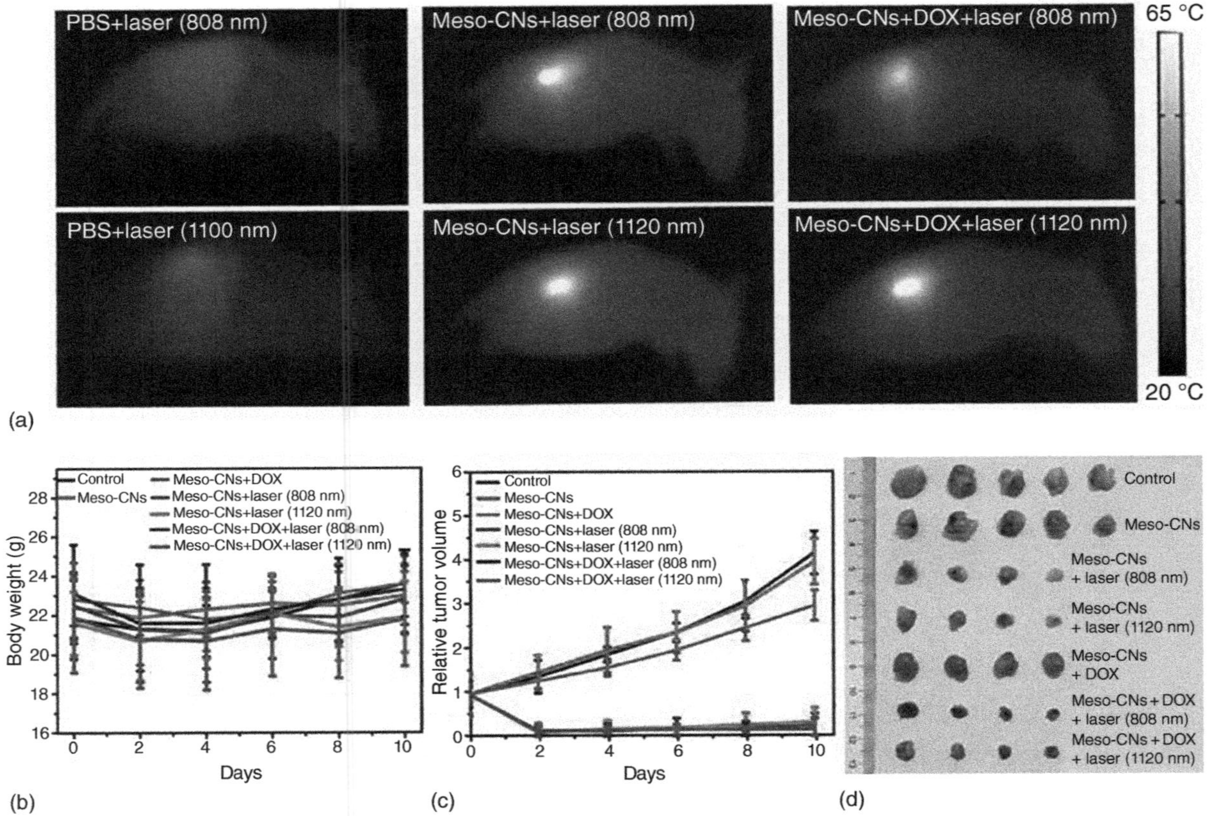

Figure 6.4 In vivo photothermal therapy. (a) IR thermal images of tumor-bearing mice ensuing intratumoral injection of PBS, Meso-CN, or DOX-loaded Meso-CN solutions under 808 or 1120-nm laser irradiation at a power density of $0.5\,W\,cm^{-2}$ for 15 minutes. (b) Average body weight of mice after many treatments. Five mice were used in each group. (c) H22 tumor growth curves of mice after various treatments. The relative tumor volumes were normalized to their initial sizes. (d) Photograph of the tumors collected from different groups of mice on the 10th day posttreatment. Note that in the laser treatment groups, some tumors were completely eliminated after treatment. (e) H&E-stained tumor and major organ tissue slices collected from mice post various treatments. Scale bar: 200 μm. Source: Ref. [18] / with permission of Ivyspring International Publisher.

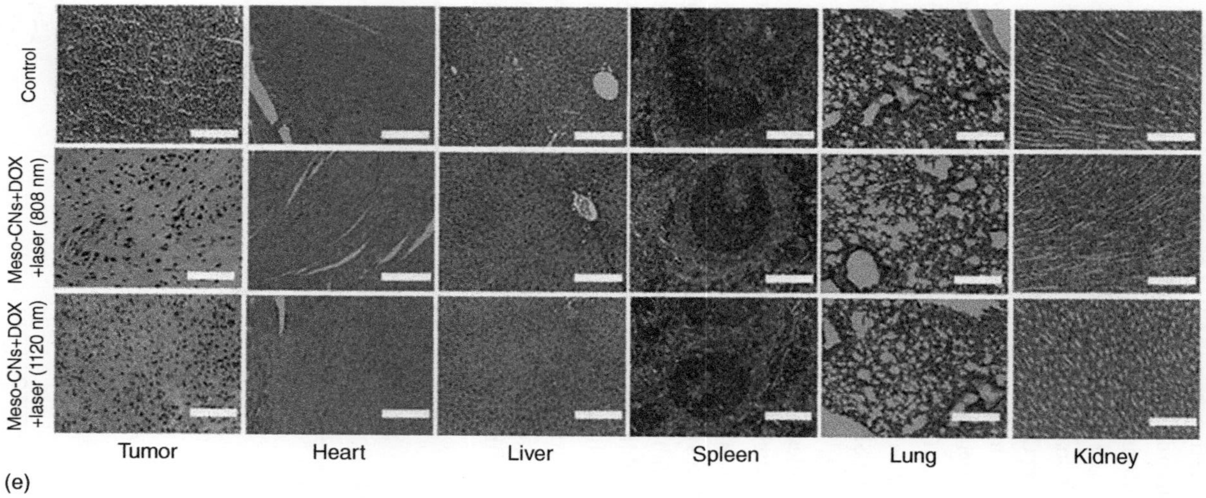

Figure 6.4 (*Continued*)

the unflinching encapsulation of drugs in MCNs in an extracellular environment by increasing colloidal stability, biocompatibility, cell targeting capacity, and controlled cargo release. Cellular adhesion studies using fluorescently labeled MCNs with or without HA functionalization demonstrated that HA-UMCS are capable of explicitly targeting cancer cells overexpressing CD44 receptors. Additionally, DOX and verapamil (VER) loaded cargo displayed dual pH and hyaluronidase-1 responsive release in the tumor microenvironment. Additionally, VER/DOX/HA-UMCS demonstrated higher therapeutic efficacy in vivo against an HCT-116 tumor in BALB/c nude mice. HA-UMCS is likely to jeopardize an innovative method for co-delivery of medicines to tumors overexpressing CD44 receptors [21].

The fluorescence used to label cells was photostable and displayed more enactment restoration than organic fluorescent dyes. Fabricated fluorescent porous carbon nanocapsules (FPC-NCs) are embedded with fluorescent CDs. The assimilated CDs may be considered as an option for optical imaging contrast in confocal laser scanning microscopy (CLSM) and two-photon fluorescence cell imaging [22].

By covalently linking DNA to CDs, it is possible to detect fluorescently DNA-specific biomolecules.

The assembly of a sandwich nanostructure for C-dots-based fluorescent thrombin exposure was driven by a target (Figure 6.5). Two thrombin aptamers were enhanced exclusively on the surfaces of C-dots and silica nanoparticles, resulting in the formation of two distinct assembling units. Due to the precise aptamer–thrombin contact in the presence of thrombin, the smaller unit C-dots involved roughly the apparent size of the superior unit silica nanoparticles. Following centrifugation and redispersion, the sandwich structure's product can be quantified using fluorescence emission spectra or fluorescence microscopy pictures, from which the amount of thrombin can be determined. By substituting additional modified aptamers and antibodies

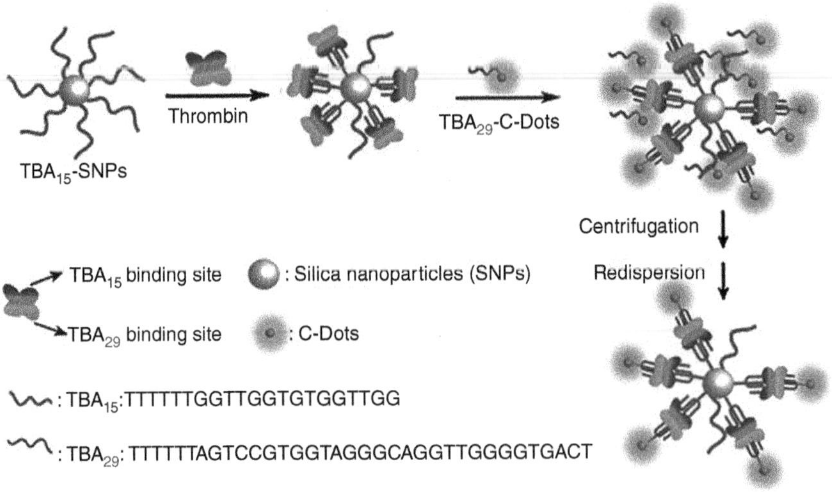

Figure 6.5 Nanostructure for C-dots-based fluorescent exposure of thrombin. Source: Ref. [23] / with permission of Royal Society of Chemistry.

for the thrombin-binding aptamers, this technology may be applicable to the detection of a variety of proteins in biological, pharmacological, and nanomechanical applications [23].

6.3.3 Removal of Toxic Substances

A copolymer-templated nitrogen-doped mesoporous carbon (CTNC) has been primed via the carbonization of well-defined polyacrylonitrile-block-poly (*n*-butyl acrylate) (PAN-b-PBA) block copolymer structures. The adsorbent CTNC demonstrated a high exclusion efficiency for bilirubin and bile acid from human plasma due to its pronounced exact extent, hierarchical open-porous edifice, and framework of nitrogen atoms in the graphitic sp2 network. The bilirubin exclusion rate was greater than 50.7% with a minor loss of albumin. Meanwhile, more than 95% of bile acid has been eliminated. It was noted that albumin had an effect on the adsorption kinetics of bilirubin. The result indicated that the adsorption rate of bilirubin bound to BSA was slower than that of free bilirubin. However, the adsorption capacity was quite high, reaching 39.8 $mg\,g^{-1}$ within two hours. Additionally, the effect of porosity and nitrogen fillings on bilirubin adsorption capacity and blood compatibility was extensively investigated. The large decrease in nitrogen concentration indicated only a negligible hemolysis activity. As a result, the nitrogen-doped mesoporous carbon developed in this study is likely to find application in blood purification for the effective elimination of bilirubin [24].

6.3.4 Transmembrane Delivery

A fundamentally well-organized MCN significant of the CMK-1 type was wonderfully synthesized using a mesoporous silica nanoparticle as a template. MCN's structure was continued to be examined using a variety of techniques, including scanning and transmission electron microscopy, powder X-ray diffraction, and N2 sorption research. The feasibility of using these fundamentally well-organized MCNs to deliver membrane-impermeable chemical proxies to cells other than eukaryotic cells, as well as the cellular uptake ability and biocompatibility of MCN with human cervical cancer cells (HeLa), was investigated. The results demonstrate that MCN's inhibitory concentration (IC50) is extremely high (>50 $\mu g\,mL^{-1}$ per million cells), indicating that it is relatively biocompatible in vitro. Similarly, a membrane-impermeable fluorescent dye, Fura-2, was delayed until it reached the MCN mesoporous matrix. It was established that MCN's high potency unquestionably serves as a transmembrane carrier for carrying Fura-2 across the cell membrane in order to release these molecules in the absence of live HeLa cells. Subsequent breakthroughs in this MCN area are expected to result in the development of a distinct class of nanodevices for transmembrane delivery and intracellular release applications [25].

6.3.5 Photoacoustic Imaging

Photoacoustic (PA) imaging is a widely used diagnostic imaging technique that utilizes a pulsed laser as the energy source and ultrasonic waves as the signal.

When compared to conventional optical imaging techniques, PA imaging has a higher spatial resolution and sensitivity, as well as a deeper tissue penetration. These expansions are sparked in part by the discovery of phonons, a collection of vibrational vibrations, as an auxiliary of photons, energy packets. Phonons, unlike photons, cannot be dispersed in tissues [26].

Under 808-nm laser stimulation, the PA characteristics of Meso-CNs and advanced contrast agents (graphene, SWCNTs, and GNR808) were determined. The animal injected with Meso-CNs exhibited significantly more advanced PA signals than the mice injected with the other ingredients, as demonstrated by the phantom experiment results. As a result, the capable PA production by Meso-CNs represents a significant advancement in high-sensitivity PAI, with potential clinical diagnostic applications [18].

Carbon-based nanomaterials may also be obligated, as they require exceptional NIR absorption, biocompatibility, and great photostability. By combining the two elements, the Si/C NPs retain their capability as a biodegradable PTT carrier that exhibits a strong PA signal. Using PA imaging, it was possible to discern the accumulation of PEG-Si/C-DOX in tumor tissue. Through a combination of chemo-thermal therapy and PEG-Si/C-DOX NPs, tumors were eventually eradicated in vivo [27].

6.3.6 Therapeutic Biomolecule Delivery

Gene therapy is expected to remain a viable and safe strategy for overpowering oncogenes and containing the spread of resistant tumors via the evolution of exogenous nucleic acids as therapeutic agents. PEI-treated oxidized mesoporous carbon nanospheres (OP) were used for combined photothermal and gene therapy (Figure 6.6a). The synthesized OP possessed a three-dimensional spherical assembly via an undeviating diameter, well-ordered mesopores via graphitic domains, exceptional water dispersibility, and upright biocompatibility. Plasmid DNA encoding ING4 or pING4 was recommended for loading into OP in order to overwhelm cancer cell growth and angiogenesis. Through electrostatic attraction, the primed OP was subdued as both the photothermal carrier with robust NIR light absorption capability and the gene vector. The median survival data (Figure 6.6c) indicated that mice treated with a combination of photothermal and gene treatment had a median survival period of 76 days, which was much longer than mice treated with PTT (48 days) or gene therapy alone (62 days). Additionally, the tumors of certain mice completely disappeared following the combinational medication and remained living till the experiment was completed. As a result, combining PTT with gene therapy may induce widespread and complete apoptosis of tumor cells, significantly suppress tumor growth, and prolong the life of tumor-bearing animals [28].

6.3.7 Biosensing

The self-assembly of CDs into aggregates is commonly complicated by fluorescence quenching, which provides the impetus for the development of

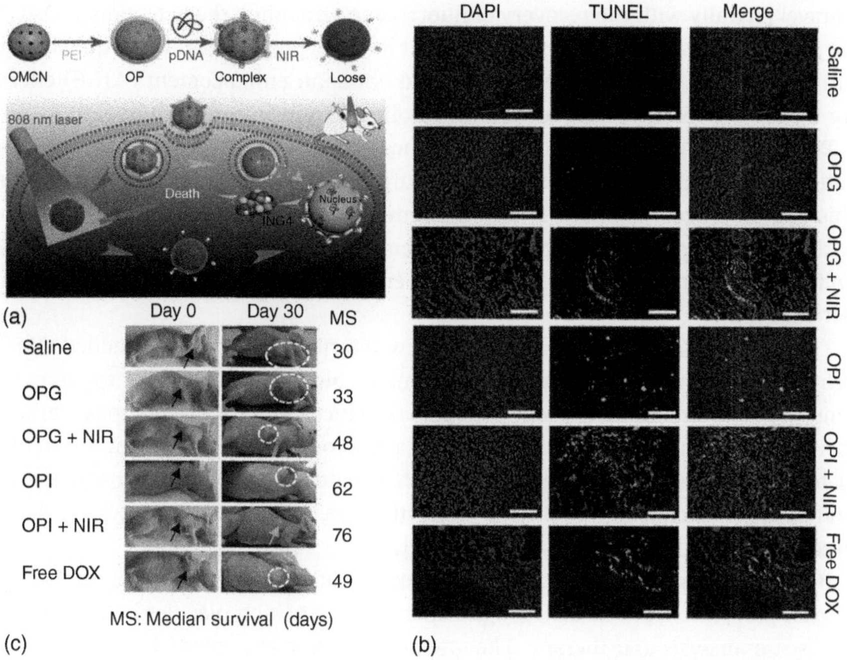

Figure 6.6 (a) Illustration of the synthesis and combined photothermal and gene therapy achieved by PEI-grafted OMCN. (b) Apoptosis results on 14th day post-injection with different treatments based on TUNEL assay. Blue: DAPI-stained nucleus; green: FITC-labeled apoptosis cells. (c) Images of mice on 30th day post-injection and median survival data with different treatments. Black arrows and white dotted circles indicated the tumor sites before and after treatments, respectively. Source: Ref. [28] / with permission of Elsevier.

biosensing technologies. Using phenylboronic acid as the sole precursor, boronic acid-functionalized C-dots for glucose sensing in blood samples, fluorescence retorts could be used to accurately quantify glucose in the range of 9–900 M, which is 10–250 times more intense than previous boronic acid-based fluorescent nanosensing systems. Remaining on a "inert" surface, the C-dots may effectively fight invasions from several biomolecules, revealing an exceptional level of fussiness. The predicted sensing system has been successfully applied to the determination of glucose levels in human serum. Due to the specific interaction between boronic acid species and glucose, C-dots accumulated strictly in the presence of glucose and the fluorescence was switched off. Due to its simplicity and efficacy, it demonstrates enormous potential as a platform for blood glucose sensing [29].

Using electrostatic self-assembly, we integrated anionic C-dots and oppositely charged biopolymers into hybrid components with the goal of developing label-free biosensors. Depending on the biopolymer, ionic self-assembly generated a succession of hierarchical assemblies that displayed C-dot fluorescence quenching caused by aggregation. After viable glycosaminoglycans (GAGs) are bound to biopolymers or the protein backbone is hydrolyzed enzymatically, the hybrid structures unravel laterally with the recovery of fluorescence, enabling the detection of GAGs (heparin) and protease (trypsin). In contrast to the aggregation-induced quenching of fluorescence, a new aggregation-induced emission enhancement (AIEE) effect for C-dots was recently discovered and exploited for sensing applications [30].

Pt-containing mesoporous carbon enhances glucose oxidase's electron translocation and redox capacity. A modest exposure frontier was chosen for establishing this source glucose biosensor [31]. 3D-ordered graphitized mesoporous carbon with 6-nm pores enhanced pore-size-dependent enzymatic stability, bioactivity, and direct electron transfer of entrapped enzymes designed to detect hydrogen peroxide [32].

Through high selectivity, stability, and low interference, NiFex-embedded OMC provided a high-performance electrochemical biosensing platform for amperometric exposure to hydrogen peroxide or glucose [33]. Mesoporous carbon exhibited similar diversity when used as a microelectrode with an ionic liquid (1-butyl-3-methylimidazolium hexafluorophosphate) and protein (glucose oxidase) and demonstrated enhanced electrocatalytic activity and sensitivity, as well as exceptional stability for glucose biosensing [34].

Biomolecules play critical roles in all aspects of lifetime augmentation, including disease prevention, and so a thorough understanding of biomolecules is detrimental to disease analysis and therapy. The graphene-based components have to tolerate being utilized to create numerous biosensors that utilize a variety of sensing modalities, including optical and electrochemical signaling [35]. Graphene-enhanced electrode for determining dopamine's fortitude (DA). Dopamine would adsorb on the electrode surface due to its phenyl structure and the p–p stacking interaction with graphene. Subsequently, DA will mostly certainly be selectively increased and controlled.

6.3.8 Magnetic Resonance (MR) Imaging

It is a widely used biomedical instrument capable of obtaining anatomic evidence noninvasively via high spatial and temporal resolution [36]. MCN imaging has frequently been accomplished by embedding precise inorganic NPs, such as adolinium (Gd) chelates, Fe_3O_4, and manganese oxide, within the carbonaceous architecture. Emergence of an intelligent stimuli-responsive nanosystem is based on MCNs to improve the resolution and specificity of investigative imaging and the efficacy of cancer therapy [37]. Superparamagnetic Fe_3O_4 nanoparticles are continually used as a T2 contrast agent in MRI because they can generate local magnetic field inhomogeneity and, as a result, a decrease in regional signal intensity due to the processing spins dephasing more rapidly (shorter T2 relaxation time). Amino dextran-coated Fe_3O_4 nanoparticles were fixed onto GO to improve biocompatibility and cellular MRI. The complexes demonstrated exceptional physiological stability and low cytotoxicity and were capable of being internalized by HeLa cells. In comparison with isolated Fe_3O_4 nanoparticles, the Fe_3O_4–GO composites displayed a statistically significant increase in cellular MRI signal [38].

6.4 Conclusion and Future Perspectives

We detailed the significant advancements of MCNs in terms of DDS and biomedical solicitations in this chapter. Due to their huge surface area and pore volume, amendable pore structure, and easily modified surface, MCNs have enormous potential as a drug carrier to control drug release and comprehend spatial–temporal DDS. MCNs have high potential in PTT, synergistic therapy, fluorescence labeling, bioadsorption of hazardous pathogenic components, peptide separation, and biosensing due to their peculiar mesoporous structure, carbonaceous shape, and high biocompatibility. Additionally, because MCNs can adapt NIR light to heat, a combination of PTT and chemotherapy could be performed to enhance the clinical therapeutic impact. MCNs can be employed for biodetection and real-time imaging when combined with fluorescent dyes or CDs. As a result, MCNs are being considered as the next generation of DDS and biological stages.

The residues of surface modification that inspired modern oxidation of MCNs can endow the carrier with some specific organic groups for further modification. Primarily, surface modification is required to provide drug carriers with targeted DDS and sustained/controlled drug release capabilities. Apart from DDS, conventional mesoporous silica-based biomaterials have proven to be extremely advantageous for gene transfer, photodynamic therapy, antibacterial management, and even tissue engineering. MCNs are expected to garner considerable attention in advanced biological fields due to their unique structure, composition, and physicochemical features. MCNs with assimilated functional modules are likely to provide diagnostic imaging capabilities such as magnetic resonance (MR) imaging, ultrasonography, computed tomography, and positron emission computed tomography, among others.

Systematic biosafety evaluations of MCNs are required to ensure their clinical translation in the near future, and these evaluations will be highly dependent on the growth of ways to get expected MCNs. Biosafety assessments should place a premium on biodistribution, biodegradation, excretion, and other specific toxicities such as neurotoxicity, reproductive toxicity, and embryonic toxicity. Quantitative measurement of MCNs in vivo may be puzzling due to the influence of carbon in living systems on their carbonaceous configuration. Radiolabeling the potency of MCNs alleviates this problem for an upcoming biosafety assessment.

References

1 Allen, M.J., Tung, V.C., and Kaner, R.B. (2010). Honeycomb carbon: a review of graphene. *Chem. Rev.* 110: 132–145.

2 Doane, T.L. and Burda, C. (2012). The unique role of nanoparticles in nanomedicine: imaging, drug delivery and therapy. *Chem. Soc. Rev.* 41: 2885–2891.

3 Wei, A., Mehtala, J.G., and Patri, A.K. (2012). Challenges and opportunities in the advancement of nanomedicines. *J. Control. Release* 164: 236–246.

4 Master, A., Livingston, M., and Gupta, A.S. (2013). Photodynamic nanomedicine in the treatment of solid tumors: perspectives and challenges. *J. Control. Release* 168: 88–102.

5 Wang, K., Yao, H., and Meng, Y. (2015). Specific aptamer-conjugated mesoporous silica–carbon nanoparticles for HER2-targeted chemo-photothermal combined therapy. *Acta Biomater.* 16: 196–205.

6 Zhao, Q., Wang, T., and Wang, J. (2012). Fabrication of mesoporous hydroxycarbonate apatite for oral delivery of poorly water-soluble drug carvedilol. *J. Non-Cryst. Solids* 358: 229–235.

7 Chen, Y., Chen, H.R., and Shi, J.L. (2014). Construction of homogenous/heterogeneous hollow mesoporous silica nanostructures by silica-etching chemistry: principles, synthesis, and applications. *Acc. Chem. Res.* 47: 125–137.

8 Zhaoa, Q., Lina, Y., Hanb, N. et al. (2017). Mesoporous carbon nanomaterials in drug delivery and biomedical application. *Drug Delivery* 24 (2): 94–107.

9 Liu, J., Zhao, Y., and Cui, Y. (2016). A Eu^{3+}/Gd^{3+}-EDTA-doped structurally controllable hollow mesoporous carbon for improving the oral bioavailability of insoluble drugs and *in vivo* tracing. *Nanotechnology* 27: 315101.

10 Chen, Y., Xu, P., and Wu, M. (2014). Colloidal RBC-shaped, hydrophilic, and hollow mesoporous carbon nanocapsules for highly efficient biomedical engineering. *Adv. Mater.* 26: 4294–4301.

11 Zhu, J., Liao, L., and Bian, X. (2012). pH-Controlled delivery of doxorubicin to cancer cells, based on small mesoporous carbon nanospheres. *Small* 8: 2715–2720.

12 Zhou, L., Dong, K., and Chen, Z. (2015). Near-infrared absorbing mesoporous carbon nanoparticle as an intelligent drug carrier for dual-triggered synergistic cancer therapy. *Carbon* 82: 479–488.

13 Yang, K., Feng, L., and Liu, Z. (2016). Stimuli responsive drug delivery systems based on nano-graphene for cancer therapy. *Adv. Drug Deliv. Rev.* 19: 316–317.

14 Xu, G., Liu, S., Niu, H. et al. (2014). Functionalized mesoporous carbon nanoparticles for targeted chemo-photothermal therapy of cancer cells under near-infrared irradiation. *RSC Adv.* 4: 33986–33997.

15 Wang, Y., Wang, K., Zhang, R. et al. (2014). Synthesis of core-shell graphitic carbon@silica nanospheres with dual-ordered mesopores for cancer-targeted photothermo chemotherapy. *ACS Nano* 8: 7870–7879.

16 Wang, L., Sun, Q., Wang, X. et al. (2015). Using hollow carbon nanospheres as a light-induced free radical generator to overcome chemotherapy resistance. *J. Am. Chem. Soc.* 137: 1947–1955.

17 Wang, Y., Wang, K., Yan, X., and Huang, R. (2014). A general strategy for dual-triggered combined tumor therapy based on template semi-graphitized mesoporous silica nanoparticles. *Adv. Healthc. Mater.* 3: 485–489.

18 Zhou, L., Jing, Y., Liu, Y. et al. (2018). Mesoporous carbon nanospheres as a multifunctional carrier for cancer theranostics. *Theranostics* 8 (3): 663–675.

19 Zhang, L., Li, Y., and Jin, Z. (2015). Mesoporous carbon/CuS nanocomposites for pH-dependent drug delivery and near-infrared chemo-photothermal therapy. *RSC Adv.* 5: 93226–93233.

20 Zhou, L., Dong, K., Chen, Z. et al. (2015). Near-infrared absorbing mesoporous carbon nanoparticle as an intelligent drug carrier for dual-triggered synergistic cancer therapy. *Carbon* 8 (2): 479–488.

21 Wan, L., Jiao, J., Cui, Y. et al. (2016). Hyaluronic acid modified mesoporous carbon nanoparticles for targeted drug delivery to CD44-overexpressing cancer cells. *Nanotechnology* 27 (13): 135102.

22 Wang, C., Xu, L., and Liang, C. (2014). Immunological responses triggered by photothermal therapy with carbon nanotubes in combination with anti-CTLA-4 therapy to inhibit cancer metastasis. *Adv. Mater.* 26: 8154–8162.

23 Xu, B., Zhao, C., Wei, W. et al. (2012). Aptamer carbon nanodot sandwich used for fluorescent detection of protein. *Analyst* 137: 5483–5486.

24 Rong, Y., Yang, S., Chengling, W. et al. (2020). Preparation of nitrogen-doped mesoporous carbon for the efficient removal of bilirubin in hemoperfusion. *ACS Appl. Bio Mater.* 3 (2): 1036–1043.

25 Tae-Wan, K., Po-Wen, C., Igor, I.S. et al. (2008). Structurally ordered mesoporous carbon nanoparticles as transmembrane delivery vehicle in human cancer cells. *Nano Lett.* 8 (11): 3724–3727.

26 Pu, K., Shuhendler, A.J., and Jokerst, J.V. (2014). Semiconducting polymer nanoparticles as photoacoustic molecular imaging probes in living mice. *Nat. Nanotechnol.* 9: 233–239.

27 Wang, H., Sun, Y., and Yi, J. (2015). Fluorescent porous carbon nanocapsules for two-photon imaging, NIR/pH dual-responsive drug carrier, and photothermal therapy. *Biomaterials* 53: 117–126.

28 Meng, Y., Wang, S., and Li, C. (2016). Photothermal combined gene therapy achieved by polyethyleneimine-grafted oxidized mesoporous carbon nanospheres. *Biomaterials* 100: 134–142.

29 Shen, P. and Xia, Y. (2014). Synthesis-modification integration: one-step fabrication of boronic acid functionalized carbon dots for fluorescent blood sugar sensing. *Anal. Chem.* 86 (11): 5323–5329.

30 Lin, Y., Chapman, R., and Stevens, M.M. (2015). Integrative self-assembly of graphene quantum dots and biopolymers into a versatile biosensing toolkit. *Adv. Funct. Mater.* 25 (21): 3183–3192.

31 You, C., Li, X., Zhang, S. et al. (2009). Electrochemistry and biosensing of glucose oxidase immobilized on Pt-dispersed mesoporous carbon. *Micro Chim. Acta* 167: 109–116.

32 Lu, X., Xiao, Y., Lei, Z. et al. (2009). A promising electrochemical biosensing platform based on graphitized ordered mesoporous carbon. *J. Mater. Chem.* 19: 4707–4714.

33 Xiang, D., Yin, L., Ma, J. et al. (2015). Amperometric hydrogen peroxide and glucose biosensor based on NiFe2/ordered mesoporous carbon nano composites. *Analyst* 140: 644–653.

34 Sun, W., Guo, C.X., Zhu, Z., and Li, C.M. (2009). Ionic liquid/mesoporous carbon/protein composite microelectrode and its biosensing application. *Electrochem. Commun.* 11: 2105–2108.

35 Liu, Y., Dong, X., and Chen, P. (2012). Biological and chemical sensors based on graphene materials. *Chem. Soc. Rev.* 41 (6): 2283–2307.

36 Hricak, H. (2005). MR imaging and MR spectroscopic imaging in the pretreatment evaluation of prostate cancer. *BJR* 78: S103-11.

37 Zhang, S., Qian, X., and Zhang, L. (2015). Composition–property relationships in multifunctional hollow mesoporous carbon nano-systems for PH-responsive magnetic resonance imaging and on-demand drug release. *Nanoscale* 7: 7632–7643.

38 Chen, W., Yi, P., Zhang, Y. et al. (2011). Composites of aminodextran-coated Fe_3O_4 nanoparticles and graphene oxide for cellular magnetic resonance imaging. *ACS Appl. Mater. Interfaces* 3 (10): 4085–4091.

7

Mesoporous Carbon Synthesized from Biomass as Adsorbent for Toxic Chemical Removal

Babu Cadiam Mohan[1], Srinivasan Vinju Vasudevan[2], Ramkumar Vanaraj[3], Sundaravel Balachandran[4], and Selvamani Arumugam[3]

[1] *National University of Singapore, Department of Chemical and Biomolecular Engineering, Singapore 117585, Singapore*
[2] *Jiangsu University, School of Agricultural Equipment Engineering, Zhenjiang 212013, Jiangsu Province, PR China*
[3] *CSIR-Central Leather Research Institute, Chennai 600020, Tamil Nadu, India*
[4] *Kalasalingam Academy of Research and Education, School of Advanced Sciences, Department of Chemistry, Krishnankoil 626 126, Tamil Nadu, India*

7.1 Introduction

The developed industrialization in the modern world leads to increasing industrial origin pollution, which generates a potential threat to ecosystems from the evacuation of toxic chemicals such as heavy metals, organic dyes, and phenolic compounds [1–4]. These poisonous chemicals come mainly from chemical fertilizers, mining, steel, machine industries, textile industries, and refineries [5]. Even if all the industries have their in-house waste management systems, the release of waste chemicals is unavoidable. Hence, the researchers' attention is more on developing proper methods and materials to remove toxic chemicals from the contaminated ones. Physical separation and chemical and biological treatment are the important methods for toxic chemical removal, including chemical precipitation, adsorption, ion exchange, oxidation, reduction, evaporation, neutralization, filtration, membrane process, electrodeposition, and high-temperature incineration [6–9]. Each technology or method has some merits and demerits so that a constant effort must be developed to product the environment. Among the various methods, adsorption has been recognized as attractive because of its low-cost process, high removal efficiency, and easy operation.

Already thousands of different types of adsorbents have been reported for toxic chemical removals such as metal oxides [10, 11], carbon nanotubes [12], graphene oxides [13, 14], metal–organic framework [15], chitosan [16], clay minerals [17], zero-valent iron ions [18], biochar, and activated carbons. All the materials have different properties and can be used to remove different types of pollutants. Among, carbon-based materials derived from biomass (agricultural, biodegradable materials, and household residues) have been widely used to repossess hazardous

Biomass-Derived Carbon Materials: Production and Applications, First Edition.
Edited by Alagarsamy Pandikumar, Perumal Rameshkumar, and Pitchaimani Veerakumar.

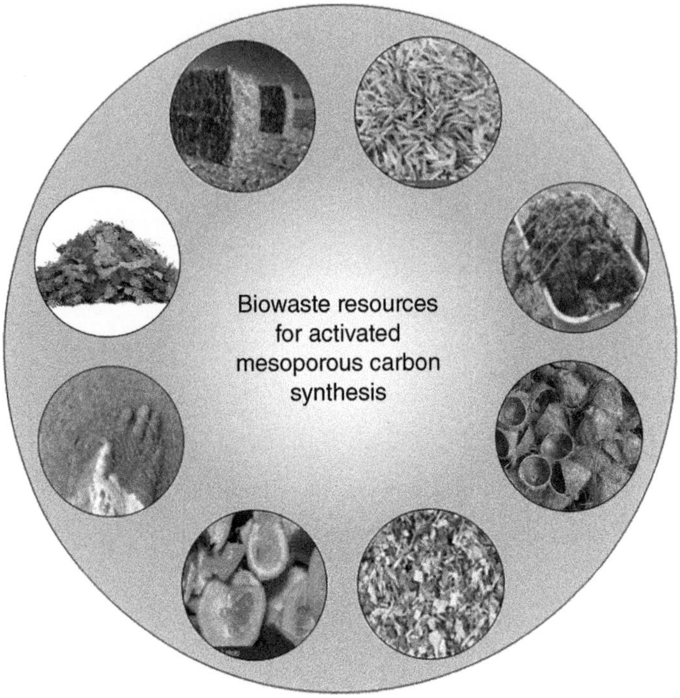

Figure 7.1 General resources to synthesize mesoporous carbon from biomass. Source: OlegD / Adobe Stock; Lilya / Adobe Stock.

substances from the environment due to their distinctive qualities of mechanical integrity and regeneration. In this sequence, more attention has been paid on mesostructured carbon due to their versatility role in catalysis as support and standing out highly efficient adsorbents [19, 20]. Moreover, the use of biomass as a precursor for the development of mesostructured carbon materials is cost-effective and attracted considerable attention because of its unique properties such as renewable precursors, high surface area, low density, high chemical stability, and good electrical conductivity [21, 22]. The mesoporous materials have significant role in the adsorption process and also act as the main transport channels for large molecules. Thus, developing activated carbon with high mesoporosity is supposed to be more useful for the adsorption of large molecules. The general resources of biowaste for the synthesis of mesoporous carbon have been illustrated in Figure 7.1. In this book chapter, we highlighted the synthesized methods of mesoporous carbons from biowaste resources and their applications in the removal of toxic chemicals such as dyes, metal ions, and phenolic compounds. Problems and future trends of these promising materials are also presented based on the literature report.

7.2 Synthesized Methods of Mesoporous Carbons from Biowaste or Biomass

The synthesis of porous carbon materials from cheap, naturally occurring biodegradable resources is a hot topic in modern materials science research due to low manufacturing costs and environmental impact. The mesoporous carbon materials can be divided into two types as ordered pore and irregular pore according to their structural order. The ordered mesoporous carbon from raw biomass is tough to synthesize. In principle, the preparation of porous carbon materials from biomasses can easily be synthesized by the activation process at elevated temperatures by physical, chemical, physicochemical, and microwave-assisted activation [23]. The micro-, meso-, and macroporous materials are formed according to the activation method and activating agent.

The physical activation method is considered inexpensive and processes due to the chemical-free preparation. It has two steps that start with the carbonization of the raw material of biomass precursor in an inert atmosphere followed by activation at the high-temperature range of 800–1100 °C [23]. Even though this process is green, it has disadvantages like prolonged activation time, the low adsorption capacity of prepared activated carbon, and high energy consumption [24]. Chemical activation is a one-step process that involves impregnating acidic/basic solutions or inorganic metal salts with the carbonaceous material and heat treatment at a temperature of 450–600 °C under an inert atmosphere, and it has more advantages than physical activation, such as a one-step process, high yield in carbon, low energy cost [25]. The activated carbon can also be prepared through the physicochemical method in which the biomass precursors or char is soaked with an appropriate activating agent and then heated in the oxidant atmosphere (CO_2, steam gas or O_2) flow [25]. The advantage of the physicochemical method is the formation of high surface area, pore volume, and porosity by the simultaneous activation process [26]. The high-temperature requirements, a low percentage of carbon yield, and extended process time of the physicochemical process make it expensive. The microwave-assisted process is the same as chemical activation. It is a fast and alternative method to prepare activated mesoporous carbon. Further, the uniform temperature distribution, raising the carbon yield, and energy saving are attractive over conventional thermal methods. In this process, the dried raw biowaste precursor is soaked with activating agents and then heated by microwave radiation in the range of 300–1400 W to the synthesis of mesoporous activated carbon materials [25, 27]. The radiation power and activating agent influence the porosity, surface area, and pore volumes. However, a significant problem is the lack of large-scale industrial demonstration units for microwave technology.

7.3 Application of Mesoporous Activated Carbons

7.3.1 Removal of Dyes

Plastic, cosmetic, pharmaceutical, textile, food, paper, and printing industries utilize natural and synthetic coloring materials called dyes to impart color to the finished goods. These dyes are generally water-soluble and discharge into aquatic ecosystems or wastewater streams, leading to pollution [2, 3, 28]. Hence, they are considered as potential pollutants for the environment. These dyes are harmful and produce adverse effects on the biotic and abiotic components of the aquatic ecosystem. The researchers have adopted different methods to remove these compounds from the aquatic phase. A standard parameter observed in these approaches is the adsorption of dyes by a suitable adsorbent. The researchers evaluated various adsorbent materials and suggested that an adsorbent material with good adsorption capacity and low production cost could be used to remove dyes from the aqueous phase. Especially, mesoporous carbon surpasses other adsorbents in terms of adsorption capacity, stability, and ease of use. The major hurdle in synthesizing mesoporous carbon materials is their production cost. Even though, many researchers worldwide made successful attempts to synthesize mesoporous carbon from biomass resources. This part will discuss the various biomass utilized by the researchers to synthesize mesoporous carbon and their adsorption behavior over cationic and anionic dyes. In general, rice husk (RH), wastes from fruit/food processing industries, hydrochars, cattail biomass, sawdust, agricultural biomass like fruit and vegetable peels, fungi residue, plant wastes, shells of aquatic animals or nuts, and coconut coir dust are the primary choice for the researchers to synthesize activated mesoporous carbons. Here, some of the activated carbon prepared from biomass resources have been discussed below.

7.3.1.1 GWAC as an Adsorbent for Methylene Blue and Metanil Yellow

Vitis vinifera (grapevine) is a crop cultivated for its fruit in various parts of the world. A large amount of solid waste is formed during the manufacture of wine from grapes. Sayğılı et al. reported $ZnCl_2$ activation method to synthesize activated mesoporous carbon from these solid wastes [29]. The authors achieved this by activating the solid waste with $ZnCl_2$ in 6 : 1 ratio for 60 minutes at 600 °C. The authors have chosen grape waste since it has lower ash content and a remarkable amount of carbon. The $ZnCl_2$ activation reduces the char formation and increases the porosity of the resulting activated carbon. Thus, synthesized activated carbon possesses 1455 $m^2\ g^{-1}$ specific surface area and 2.318 $cm^3\ g^{-1}$ pore volume. The activated carbon synthesized from GWAC possessed micropores as well as mesopores which were evidenced from the nitrogen sorption isotherm. The nitrogen sorption isotherm pattern varies with activation time and impregnation ratio. The researchers thoroughly investigated the isotherm pattern and established a method to fine-tune the porosity of the GWAC. The authors also claimed that the activated carbon production using GW by this method is economically feasible and it is two to four times lesser than other methods. GWAC was tested for the removal of methylene blue (MB) and metanil yellow (MY) from aqueous solutions at natural pH conditions. The equilibrium adsorption behavior was fitted with Freundlich and Langmuir linear isotherm equations, and it is reported that the Freundlich constant (K) for MB and MY is close to 160 which can be attributed to the ease of these dyes adsorption on the GWAC substrate in the given experimental conditions. The GWAC has a greater affinity for cationic dyes compared to anionic dyes which is inferred from the adsorption intensity ($1/n$) value of MB (0.126) and MY (0.139). However, the adsorption of MB and MY adsorption over GWAC obeys Langmuir adsorption isotherm which is noted in the correlation coefficient (R^2) value. The R^2 value is close to 1 for Langmuir model and deviates to a low value in the case of the Freundlich model. The separation factor (R_L) for MB and MY over GWAC lies between 0 and 1. Hence, the adsorption process is favorable and can be reversed by choosing appropriate conditions. GWAC showed a maximum adsorption capacity of 417 $mg\,g^{-1}$ for MB and 386 ($mg\,g^{-1}$) for MY which is comparably large with other known activated carbon synthesized from other biowaste. Similarly, the authors also reported the synthesis of mesoporous activated carbon from tomato processing wastes and applied it for MB and MY removal from aqueous solutions. In this sequence, Sayğılı and Güzel researchers were also utilized GWAC-derived mesoporous activated carbon to adsorb malachite green (MG) and congo red (CR) from the aqueous solutions [30]. The GWAC has more number of mesopores (close to 95%) which is responsible for the high adsorption capacity of the material. The researchers reported that the sorption of MG and CR over GWAC occurred readily at the expense of thermal energy. The maximum sorption capacity calculated for MG and CR over GWAC is found to be 667 and 455 $mg\,g^{-1}$, respectively. The sorption kinetics followed pseudo-second-order kinetics, and the researchers established that the intra-particle diffusion is not only the factor deciding the rate of sorption but also the initial concentration of dyes decides the rate of sorption process. The researchers were found in their previous study as GWAC has a greater affinity for cationic dyes [29]. This statement is well justified in the effect of GWAC dose on MG and CR sorption study. The study showed that the optimum dose for removal of MG from aqueous solution is 0.01 g/50 mL, whereas for CR is 0.02 g/50 mL. Since the sorption of MG or CR dye over GWAC is an endothermic process, the sorption capacities of the GWAC increase with an increase in temperature. It is also noted that in the given temperature conditions, the cationic dye MG shows a higher affinity than CR. Thus, mesoporous activated carbon synthesized from grapefruit waste showed excellent sorption properties over a range of dyes and can be a promising sorbent for dyes removal.

7.3.1.2 Rice Husk (RH)-Derived Mesoporous Activated Carbon (AC) for Methylene Blue (MB) Dye Removal

Rice husk or rice hull (RH) is one of the major agricultural wastes found in abundance in rice-producing countries mainly around the Asian region. RH has been reported to consist mainly of organic lingo-cellulosic biomass and ash which can be effectively utilized for the synthesis of porous carbon through precarbonization and chemical activation. Lin et al. reported the synthesis of mesoporous activated carbons from RH in concentrated alkali medium (NaOH) [31]. The authors reported that the mesoporous activated carbon (AC) pertaining surface area above 1000 $m^2\,g^{-1}$ and total pore volume equal to 0.75 $cm^3\,g^{-1}$ can be synthesized from RH using 2 M NaOH as an activating agent. The researchers extracted one part of clean, dry RH powder with seven parts of aqueous 2.0 M NaOH at 100 °C for 4 hours to yield a solid residue rich in carbonaceous materials. Then, the solid RH residue obtained was carbonized at 700 °C for 1.5 hours in the presence of a nitrogen atmosphere. The researchers mixed the char obtained from the carbonization furnace with 2.0 M aqueous NaOH solution maintaining 1 : 4 weight/volume ratios and heated at 800 °C for 90 minutes. Thus, prepared activated carbon showed type IV nitrogen sorption isotherm. Lin et al. reported spontaneous adsorption of MB dye over the mesoporous AC derived from RH is an endothermic process and follows a pseudo-second-order kinetics. The adsorption capacity of AC increases with an increase in pH of the solution. About 1 g of material is capable of adsorbing 390 mg of MB dye from the aqueous solution of pH close to 12 at 293 K within 30 minutes of contact time.

7.3.1.3 Activated Carbon from Rattan Waste for Methylene Blue (MB) Removal

During the manufacture of furniture from rattan (Lacosperma secundiflorum), a large quantity of wastes is generated which imposes a threat to the environment. Islam et al. found a way to utilize the rattan wastes to synthesize activated mesoporous carbon [32]. The researchers carbonized the biomass in an autoclave at 200 °C for 5 hours to obtain hydrochar. Thus, produced hydrochar was dried in an oven at 105 °C for 24 hours and mixed with aqueous NaOH solution in 3 : 1 (w/w) ratio and allowed to stand for overnight. The hydrochar impregnated with NaOH is then dried in an oven at 105 °C. Further activation of modified hydrochar was achieved by heating it in an electric furnace at 600 °C in a nitrogen atmosphere. The sample obtained was washed with hot double-distilled water until the pH of filtrate drops down to a range of 6–7. Thus, obtained mesoporous activated carbon had surface area of 1135 $m^2\ g^{-1}$ and mesopore volume equal to 0.44 $cm^3\ g^{-1}$. The adsorption of MB over the mesoporous activated carbon best fitted with the Langmuir isotherm with a correlation coefficient equal to 0.92, and the kinetic study revealed that the sorption process follows the pseudo-second-order model. Hence, rattan wastes can be a promising source for the synthesis of mesoporous activated carbon and utilized for adsorption of cationic dye MB from aqueous solution.

7.3.1.4 Activated Carbon from Cattail Biomass (CAC) for Malachite Green (MG) Removal

Yu et al. reported that the synthesis of carbon using the cattail biomass by a double-stage thermal treatment [33]. The carbon monoliths were activated using carbon dioxide, resulting in mesoporous activated carbon having an average pore diameter of less than 2.09 nm. The major advantage of this method is the activation of carbon by CO_2 gas. Commercial-grade activated carbon (CAC) material showed type IV nitrogen sorption isotherm with H4 hysteresis loop and specific surface area equal to 441.2 $m^2\ g^{-1}$. Similar to other adsorbent materials, CAC also followed pseudo-second-order kinetics and perfectly fit in the Langmuir adsorption model. Since MG is a cationic dye, hence the sorption of this dye over CAC must occur at higher pH rather in the lower pH values. The CAC material showed a maximum adsorption capacity of 210.18 $mg\,g^{-1}$ for the MG dye.

7.3.1.5 Wood Sawdust Waste Activated Carbon (WACF-P) for Xylenol Orange (XO) Removal

Minghui et al. attempted to eliminate the usage of alkali solutions as activators in the mesoporous carbon synthesis and reported the synthesis of wood-derived activated carbon fiber using phosphoric acid as an activating agent [34]. The researcher dispersed the carbon fibers obtained from wood sawdust in 10% phosphoric acid solution and activated at 900 °C to obtain WACF-P materials and produced mesoporous volume equal to 0.607 $cm^3 g^{-1}$ and specific surface area equal to 1037 $m^2 g^{-1}$. The researchers reported that the sorption rate of xylenol orange (XO) over WACF-P material at various activation temperatures depends on the pore volume diffusion and also reported that XO adsorption increased with an increase in mesopore volume of WACF-P materials.

7.3.1.6 Mesoporous Activated Carbon from Agricultural Waste for Methylene Blue Removal

An interesting study on MB removal was reported by Mahmoodi et al. [35]. The researcher utilized agricultural wastes like cucumber peel (CP), kiwi peel (KP), and potato peel (PP) to synthesize mesoporous activated carbon (AC) by adopting the procedure reported by Sayğılı and Güzel [30]. Mahmoodi et al. performed an interesting study on MB removal from aqueous solutions in the presence of other dyes like MG and rhodamine B (RhB). The major highlight of the work is that the researchers demonstrated the sorption of MB dye both theoretically and experimentally. The researchers reported the artificial neural network (ANN) model for dye removal. With the aid of the model, various parameters that influence the dye sorption, including dye concentration, adsorbent dosage, solution pH, temperature, and process time, were extensively studied. The results produced by ANN model were in good agreement with the practically determined values. The adsorbent preferentially adsorbs MB from an aqueous solution containing MB and MG dyes. Also, the preferential adsorption of MB dye is not affected when the aqueous solution contains RhB along with MG dye. Hence, Mahmoodi et al. proved theoretically and practically that CP, PP, and KP can be utilized to manufacture mesoporous AC with high surface area and good sorption properties.

7.3.1.7 Mesoporous Activated Carbon from Edible Fungi Residue (EFR-AC) for Reactive Black 5 Removal

Xiao et al. adopted microwave-assisted activated carbon synthesis from edible fungi residue (EFR) [36]. Potassium carbonate was mixed with EFR in 1 : 0.8 ratio and irradiated with 520 W microwave radiations for 16 minutes to obtain EFAR-AC with BET surface area equal to 683.76 $m^2 g^{-1}$ and mesopore volume of 0.334 $cm^3 g^{-1}$. In EFR-AC, micropores contribute to 43.5% of the total pore volume. Like other activated carbons, EFR-AC also follows pseudo-second-order kinetics and the sorption process is well fitted with Langmuir isotherm. The maximum adsorption capacity of the EFR-AC was determined using iodine as a probe molecule and the maximum iodine number value reaches 654.83 $mg g^{-1}$. EFR-AC effectively adsorbs reactive black 5 at low pH conditions.

7.3.1.8 Mesoporous Activated Carbon from Plant Wastes for Methylene Blue (MB) Removal

Açıkyıldız et al. reported the synthesis of mesoporous activated carbon from plant wastes like pine sawdust (PS), rose seed (RS), and cornel seed (CS) [37]. All the biomass sources yielded activated carbon with the surface area close 1000 $m^2 g^{-1}$

and MB sorption index close to 300 mg g^{-1}. The researchers also designed a mathematical model to determine the surface area and sorption behavior of synthesized activated carbon. The researchers performed a nonlinear regression analysis to study the effect of carbonization temperature, impregnation ratio, and impregnation time on sorption properties like surface area and MB index using SPSS 11.5 package and reported that the empirical equations derived were well in agreement with the values reported. In these series, a variety of biomass like Jerusalem artichoke stalk, coconut shells, pinewood chips, banana peels, and alginate beads were converted to mesoporous activated carbon using $ZnCl_2$ activation method and tested for MB dye adsorption [38–40].

7.3.1.9 Mesoporous Activated Carbon from *Corozo oleifera* Shell for Methylene Blue (MB) Removal

Kang et al. utilized Camellia oleifera shell as a starting material for the synthesis of mesoporous activated carbon [41]. An interesting report about this study is that the researchers used water vapor gasification followed by a phosphoric acid modification to achieve the high mesoporous volume of 1.17 cm^3 g^{-1} and specific surface area equal to 1608 m^2 g^{-1}. The researchers carbonized the shell at 500 °C followed by activation at 820 °C for 120 minutes. The materials synthesized by adopting these conditions showed a specific surface area equal to 1076 m^2 g^{-1}, but the material's micropore volume was almost double the mesopore volume. The number of mesopores in the material is increased by heating the sample in a temperature range between 600 and 800 °C in the presence of phosphoric acid. During the phosphoric acid modification, the sample to phosphoric acid ratio was maintained at 3 : 1. When the modification temperature was 800 °C, the micropore to the mesopore ratio becomes 1 : 3. Thus, modified activated carbon had a maximum adsorption capacity of 330 mg g^{-1} for MB dye sorption and 1326 mg g^{-1} for iodine.

7.3.1.10 Mesoporous Activated Carbon from Coconut Coir Dust for Methylene Blue (MB) and Remazol Yellow (RY) Removal

Fooand Hameed also adopted the chemical activation method to prepare mesoporous activated carbon from coconut coir dust [42]. The coconut coir dust and zinc chloride were taken in 1 : 3 (w/w) ratio heated to 800 °C in a nitrogen atmosphere and kept at the same temperature for 2 hours in a carbon dioxide atmosphere. Then the sample was cooled to room temperature in a nitrogen atmosphere. It was washed with aqueous HCl and then with distilled water to remove the chloride ions and dried at 100 °C for 2 hours. The synthesized materials showed type IV nitrogen sorption isotherm and pore size range between 2 and 4 nm. The amount of Remazol Yellow (RY) dye adsorbed at equilibrium was close to 160 mg g^{-1}, whereas the amount of MB dye adsorbed at equilibrium was close to 15 mg g^{-1}. It was also observed that these values did not change appreciably in the pH range 4.0–8.0. The adsorption kinetics study also revealed that the adsorption mechanism follows pseudo-second-order kinetics.

7.3.1.11 Mesoporous Activated Carbon from Macadamia Nut Shell (MNS) Waste for Methylene Blue (MB) Removal

Wongcharee et al. compared the adsorption behavior of commercial-grade activated carbon with activated carbon (MAC) prepared from macadamia nut shell (MNS) waste [43]. Dry MNS waste was subjected to anaerobic carbonization at 400 °C for 2 hours to yield char. The char obtained was heated to 900 °C at a heating rate of 30 °C min^{-1} in a tubular furnace under nitrogen gas flow (0.3 L min^{-1}). Then the material was activated by passing carbon dioxide gas at a flow rate of 0.3 L min^{-1} for 1 hour. Thus, synthesized MAC had a specific surface area of 830 m^2 g^{-1} and mesopore volume of 0.614 cm^3 g^{-1}. Even though the CAC material had a higher surface area than MAC, the CAC contains 42% of mesopores which was 22% lesser than the MAC. The MB dye adsorption reached a maximum value within 4 hours of contact time. A variety of kinetic models like pseudo-first-order, pseudo-second-order, fractal-like pseudo-first-order, fractal-like pseudo-second-order, and elovich model were examined for the sorption mechanism and concluded that the sorption data best fitted the FL-PFO model. The maximum dye adsorption capacity for MB dye over MAC was 135 mg g^{-1}.

7.3.1.12 Mesoporous Activated Carbon from Neobalanocarpus Heimii Wood Sawdust (WSAC) for Methylene Blue (MB) Removal

Foo and Hameed converted chengal (Neobalanocarpus heimii) wood sawdust to mesoporous activated carbon by a microwave-assisted potassium carbonate activation method [42]. Chengal wood sawdust of particle size 1–2 mm was carbonized at 700 °C in a tubular furnace under nitrogen gas flow to obtain the char. The char was mixed with the calculated quantity of potassium carbonate in 1 : 1.25 ratios and irradiated with 600-W microwaves for 6 minutes. Nitrogen gas was allowed to flow into the activation chamber at the rate of 300 mL min^{-1} throughout the activation process. Thus, the prepared material was washed with 0.1 M HCl and rinsed with distilled water until the filtrate showed neutral pH. The synthesized materials (WSAC) showed excellent textural properties. The surface area of the materials was almost 1500 m^2 g^{-1}. It had an average pore size of 2.31 nm and the total pore volume of 0.864 mL g^{-1}. The adsorption of MB dye over WSAC material showed a monolayer adsorption capacity equal to 462.10 mg g^{-1}. The researchers also reported that the sorption of MB over WSAC follows a film diffusion-controlled mechanism. The overall results of the discussed work have been presented in Table 7.1 with their textural properties.

7.3.2 Removal of Metal Ions

For decades, heavy metal contamination of water has been a challenge for many countries around the world. It has attracted the attention of researchers due to its severe effects on human health and the environment [44]. The primary hazardous heavy metals are chromium (Cr), iron (Fe), selenium (Se), vanadium (V), copper (Cu), cobalt (Co), nickel (Ni), cadmium (Cd), mercury (Hg), arsenic (As), lead (Pb), and zinc (Zn). Conventional methods such as chemical precipitation, ion

Table 7.1 The textural properties of activated carbon discussed in this section.

S.no	Adsorbent source (Biowaste)	Surface area ($m^2\ g^{-1}$)	Total pore volume ($cm^3\ g^{-1}$)	Mesopore volume ($cm^3\ g^{-1}$)	Sudied dyes	Maximum adsorption capacity ($mg\ g^{-1}$)	References
1	Grape waste	1455	2.318	–	MB MY MG CR	417 386 667 455	[29, 30]
2	Rice husk or rice hull	1015	0.75	0.75	MB	390	[31]
3	Rattan waste	1135	0.61	0.44	MB	359	[32]
4	Cattail biomass	441.2	0.33	—	MG	210.18	[33]
5	Wood sawdust	1037	0.825	0.607	XO	140	[34]
6	Agricultural biowastes	1113	0.3004	0.0938	MB	350	[35]
7	Edible fungi residue	683.76	0.591	0.334	RB 5	172.43	[36]
8	Plant wastes		—	—	MB		[37]
	Pine sawdust (PS)	—				—	
	Rose seed (RS)	986				292	
	Cornel seed (CS)	1238				298	
9	Camellia oleifera shell	1608	1.17	0.71	MB	330	[41]
10	Coconut coir	—	—	—	RY MB	160 15	[42]
11	Macadamia nut shell	830	0.830	0.614	MB	135	[43]
12	Jerusalem artichoke stalk	582	0.32	0.10	MB MO	630 540	[40]

exchangers, chemical oxidation/reduction, reverse osmosis, electrodialysis, and ultrafiltration have been used to remove heavy metals [45]. However, these techniques have some limitations, such as low efficiency, sensitive operating conditions, and the production of secondary sludge, which increases the cost. These metals should be removed through economically viable roots. Notably, the use of biomass to remove heavy metals from wastewater has attracted much attention due to its economic advantages and high removal efficiency, and the biomass-derived carbon materials are attributed to different functional groups. The sorption mechanism of biomass can consist of several steps, including chemisorption, complexation, adsorption on the surface, diffusion through pores, and ion exchange. A simple, high-efficiency method that utilizes locally available agricultural materials or other biomass is needed to prepare low-cost and environmental-friendly materials and also economically viable method [46]. Preparing biomass-derived mesoporous activated carbons to remove heavy metals through the adsorption process could be a better way. Here are discussed some of the biomass-derived carbon materials for heavy metal removals.

7.3.2.1 Use of Chicken Feather and Eggshell to Synthesize a Novel Magnetized Activated Carbon for Sorption of Heavy Metal Ions

Keeping environment and sustainability concept in view with the preparation of new sorbents, two waste by-products from the poultry industry, i.e. feather and eggshell, were used for the synthesis of a new magnetic-activated carbon for sorption of heavy metal ions. The prepared activated carbon was magnetized for easy separation from water media, and iron oxide-magnetized ESCFC (IOM-ESCFC) was comprehensively examined for removing some heavy metallic ions (Pb^{2+}, Cd^{2+}, Cu^{2+}, Zn^{2+}, and Ni^{2+}) from water. The typical carbon material was prepared as follows: first, the chicken eggshells were washed with a detergent solution and deionized water, respectively. The material was oven-dried at 100 °C for 4 hours and, then, it was converted to a fine powder by grinding in mortar for 20 minutes. Then, different amounts (1–5 g) of eggshells were dissolved in 2.5-mL portions of concentrated HNO_3. After filtration of the undissolved constituents of eggshell, utilizing a 0.45-μm cellulose acetate membrane, the supernatant solutions were diluted to 50 mL and stored to be used for the activation process (Figure 7.2). Batch-wise sorption experiments were accomplished to assess the sorption behavior of the interested metallic ions onto the newly prepared sorbent (IOM-ESCFC). For the whole metal ion sorption experiments, 50-mL aliquots of Cd^{2+}, Cu^{2+}, Pb^{2+}, Ni^{2+}, or Zn^{2+} aqueous solutions with the prespecified initial concentrations were shaken at 150 rpm and constant temperature with a certain amount of sorbent in 150-mL conical flasks. All the studies showed that the new sorbent had a high sorption capacity for the interested metal ions, and the sorption process was fast and spontaneous. In summary, the sorbent showed useful features in the treatment plants of wastewaters contaminated by heavy metal ions [47].

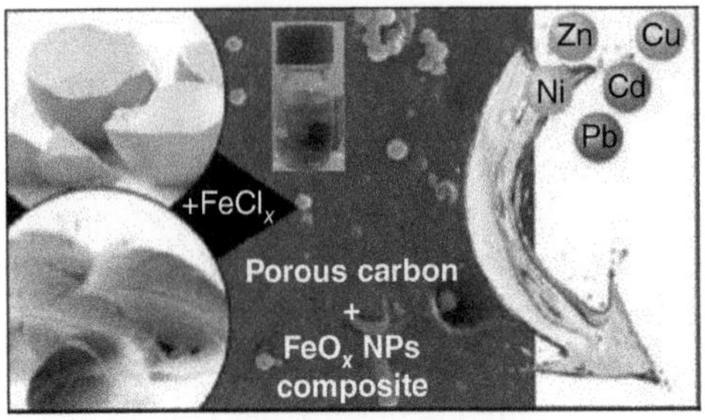

Figure 7.2 Removal of heavy metal ions using eggshell and chicken feather.

7.3.2.2 Meso/micropore-Controlled Hierarchical Porous Carbon Derived from Activated Biochar as a High-Performance Adsorbent for Copper Removal

High-quality meso/micropore-controlled hierarchical porous carbon (HPC) was synthesized by a hard template method utilizing RH biochar and then used to adsorb copper ions from an aqueous solution. The preparation procedure was included two main steps: base leaching and physicochemical activation. The improvement is ascribed to the high specific surface area and favorable hierarchical structure. In a typical synthesis, the cleaned RHs were dried at 105 °C for 24 hours to reduce their moisture content before further use. Then, the RHs were pyrolyzed at 500 °C for 1 hour under nitrogen to prepare the RH biochar (denoted RHP). In this work, the authors successfully synthesized a series of HPC materials derived from RHP through alkaline leaching and physicochemical activation processes. These steps are crucial to achieving the high-performance adsorption of Cu(II) ions onto HPCs. This work demonstrates that the specific surface area, total pore volume, and mesoporosity can be well controlled by using different KOH impregnation ratios, activation temperatures, and activation times under a CO_2 atmosphere. As demonstrated, the HPC 3-0.5-800 had a high specific surface area of 2330 m^2 g^{-1} with an 81% mesopore to total specific surface area ratio. The most prominent finding to emerge from this study was that HPCs could obtain a high adsorption capacity and fast removal of cooper ions synthesized from activated biochar [48]. Figure 7.3 represents the overall activity of HPC from RH biowaste for the removal of metal ions.

All the biowaste resources can be utilized to prepare mesoporous activated carbon with suitable activation and activating agents. Heavy metal adsorption using inexpensive and efficient biosorbents from agricultural waste materials has been reported as a promising replacement for existing conventional systems [49–55]. Several reports on biomass-derived porous carbon materials and their maximum adsorption capacity have been listed in Table 7.2 [56].

7.3.3 Removal of Phenolic Compounds

Phenolic compounds have attracted great interest and are used to prepare adhesives, insecticides, dyes, explosives, resin manufacturing, plastic, wood products, etc. Discharging phenolic effluents after industrial use without treatment may lead to serious health risks for humans, animals, and aquatic systems [57]. Phenolic

Table 7.2 Different types of heavy metals removal from the biowaste materials [56].

S. no.	Toxic metal ions	Biowaste	Biomass source	Adsorption capacity (mg g^{-1})
1	Cd(II)	Rice husk	Natural rice husk	73.96
		Wheat waste	Wheat bran	15.82–22.78
		Coconut waste	Puresorbe	285.70
		Peel	Orange peel	47.60
		Seeds	Raw date pit	35.90
		Coffee waste	Raw coffee powder	15.65
2	Cr(VI)	Wheat	Wheat bran	310.58
		Fruit	Bael fruit	17.27
		Husk	Groundnut husk	7.00
		Shell	Almond shell	3.40
		Bark	Pinus roxburghii bark	4.15
3	Cu(II)	Peel	Mango peel	46.09
		Hull	Peanut hull	9.00–21.25
		Seed	Cicerarientinum	18.00
		Shell	Chestnut shell	12.56
		Bark	*Casuarina equisetifolia* bark	16.58
		Tea	Tea waste	8.64–48.00
4	Co(II)	Coconut	Coir pith	12.82
		Peel	Lemon peel	22.00
5	Pb(II)	Wheat	Wheat bran	87.00
		Coconut	Coir pith waste	263.00
		Tea	Spent black tea	129.90
		Peel	Mango peel	99.05
		Bark	*Moringa oleifera* bark	34.60
		Shell	Hazelnut shell	28.18
6	Ni(II)	Bark	Acacia leucocephala bark	294.10
		Peel	Pomegranate peel	52.00
		Seed	Guava seed	18.05
		Coconut	Coir pith	15.95
		Tea	Tea waste	73.00
7	Zn(II)	Seed	Cicerarientinum	20.00
		Wheat	Wheat bran	16.40
		Tea	Tea waste	8.90
		Peel	Mango peel	28.21

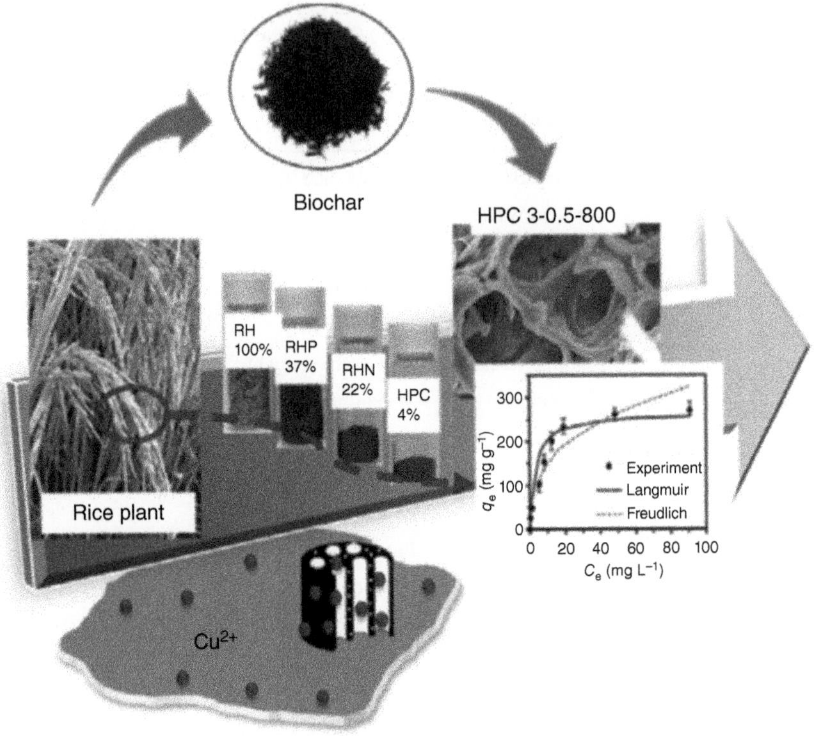

Figure 7.3 Procedure of the hard template method for HPC synthesis from rice husk biochar.

compounds have not been generated only by human activity, but they are formed naturally also due to the degradation of humic substances, tannins, and lignins. Therefore, these compounds produce contamination in groundwater. Thus, there is tremendous environmental interest in the removal of phenolic compounds. The World Health Organization (WHO) has urged an acceptable phenolic concentration of 0.1 μg L^{-1} in potable water. In contrast, the European Union (EU) has set a maximum level of 0.5 μg L^{-1} of total phenols in drinking water [58]. Further, the US Environmental Protection Agency (EPA) said that phenolic compounds are priority pollutants with high toxicity even at low concentrations [59]. EPA proposed 11 phenolic compounds as priority pollutants which are shown in Figure 7.4. Therefore, removing or diminishing these organics compounds to reach the permitted levels before discharging becomes challenging.

Different types of technologies have been developed to remove phenols from wastewater, including distillation, coagulation, chemical oxidation, electrochemical oxidation, pervaporation, solvent extraction, membrane process, and adsorption. Mainly, adsorption technology has been used to remove phenolic compounds from water which is effective from low concentrations to high concentrations. For phenolic adsorption processes, carbon-based materials are economical and one of the most investigated liquid-phase applications as adsorbents [44]. On that

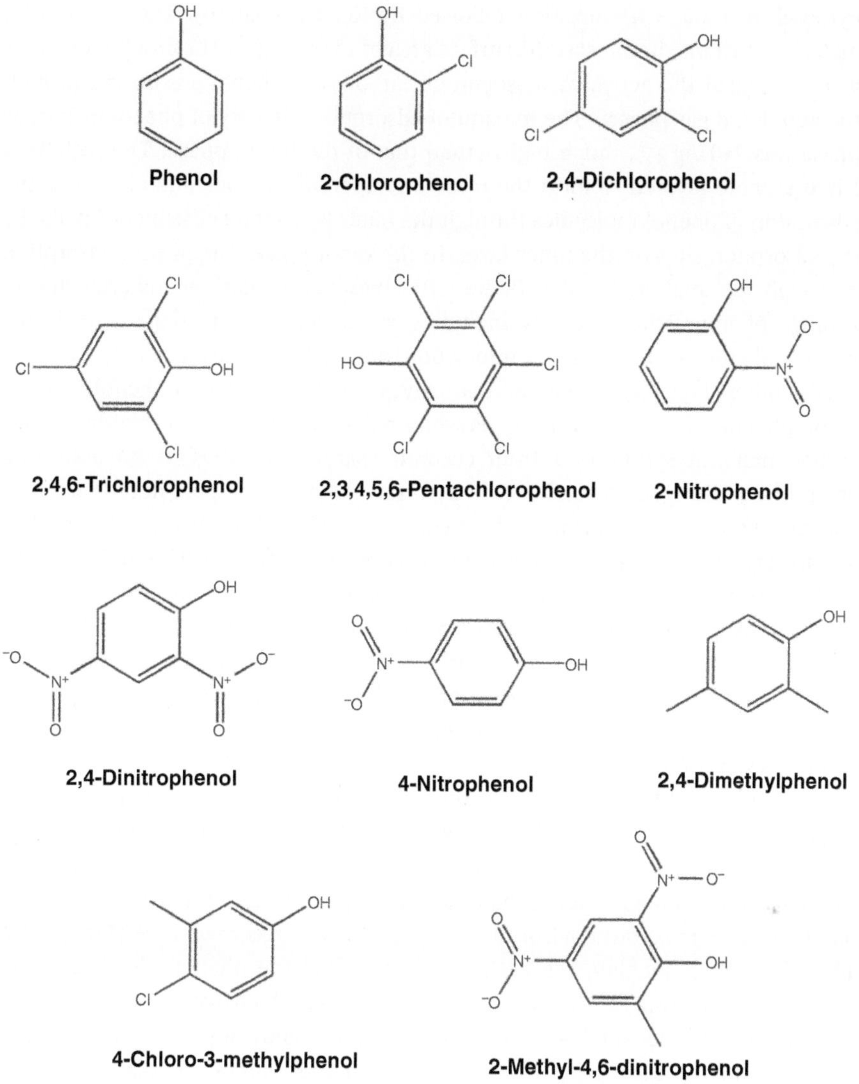

Figure 7.4 Structures of 11 phenolic compounds considered priority pollutants by US EPA [60].

basis, we have discussed some recent and important biomass-derived mesoporous carbons for effective removals of phenolic compounds. For example, Sekaran et al. were prepared porous carbon from RH through carbonization followed by chemical activation using phosphoric acid at 900 °C. The prepared carbon material had a maximum total surface area of 438.9 m^2 g^{-1} and mesopore surface area of about 224 m^2 g^{-1} which was employed for phenol adsorption study. At the optimum condition, the maximum phenol adsorption for RH-derived mesoporous carbon catalyst was 22 mg g^{-1} at a pH 2.7 [61]. Further, Yafei Shen et al. were synthesized activated mesoporous carbon from unaltered and pelletized RH by a two-step pyrolysis method (carbonization followed by KOH activation). The synthesized material had a maximum specific surface area of 1818 $m^2 g^{-1}$. The group memebers were evaluated the activated mesoporous carbons for phenol adsorption in both the liquid and gas phases. The maximum adsorption capacity of phenol in the gas phase was 740 $mg g^{-1}$, much higher than that of the liquid phase. The pelletized RH was more favored to form the meso-microporous carbons and enhanced the adsorption of phenol molecules through the outer layer and subsequent uptake by the adsorption sites on the inner layer. In the vapor phase, the phenol adsorption was high for unaltered and pelletized RH mesoporous carbon material due to a range of interactions, mainly including electrostatic interaction, pore filling, functional groups effect (e.g. π–π interaction) and hydrophobic effect [62].

The other types of phenolic derivatives, such as 2-chlorophenol and 4-chlorophenol, were analyzed by Prashanthakumar et al. with highly porous carbon material synthesized from coconut spathe and KOH as an activating agent under nitrogen atmosphere via pyrolysis method. The pore development in coconut spathe was significant at temperature 800 °C and reached maximum BET surface area (1705 $m^2 g^{-1}$) and maximum pore radius (2 nm). The authors were performed adsorption of phenolic derivatives using the coconut spathe-derived mesoporous carbon materials in an aqueous medium with the effect of various adsorption parameters such as adsorbent dosage, initial concentration of the adsorbate, contact time, pH, and reusability. The uptake of phenol, 2-chlorophenol, and 4-chlorophenol was found to be 120, 225, and 275 $mg g^{-1}$, respectively, within 15 minutes of contact time [63].

Another kind of biomass wastes is the fallen leaves which have been considered as potential sources to synthesize porous carbon. In this category, Zhang et al. were used Toona sinensis leaves to prepare the porous carbon by NaOH activation and evaluated the obtained mesoporous carbon for phenol adsorption in an aqueous solution. The maximum adsorption capability was recorded as 325 $mg g^{-1}$ of phenol at 25 °C. Further, the authors reported that the phenol adsorbed dominantly as monolayer on the porous carbon surface, which was explained by the pseudo-second-order and Langmuir models. The thermodynamic analysis of carbon materials for the adsorption process confirmed the phenol monolayer physical adsorption mechanism in mesoporous carbon materials based on π–π dispersion interaction [64].

As a natural plant, cattail fiber is abundant, cheaper, biodegradable, and environment-friendly which is a good precursor for developing carbon-based adsorbent materials. Zhang et al. researched cattail fiber and successfully produced a large specific surface area of 890 $m^2 g^{-1}$ using H_3PO_4 as an activating agent with well-developed porous structures. This material had a high adsorption capacity for 2,4-dichlorophenol (2,4-DCP) and 2,4,6-trichlorophenol from the aqueous solution over a wide range of concentrations. The adsorption ability of both phenolic compounds was found to increase when increased the agitation time. The authors mentioned the good adsorption property of cattail fiber due to the high surface area and average pore diameter (3.88 nm) of the material [65].

The adsorption of *p*-chlorophenol (PCP) and 2,4-DCP were studied by Zaharaddeen and Rahima using high surface area and efficient mesoporus adsorbent synthesized from *Prosopis africana* seed hulls (PASH-AC). They reported that sodium acetate (CH_3COONa) as an activating agent was very promising in the synthesis of mesoporous carbon. The maximum adsorption capacity of 347 and 380 mg g^{-1} were achieved for PCP and 2,4-DCP respectively. Langmuir model best described the PASH-AC adsorption capacity of both the PCP and 2,4-DCP, while the kinetics obeyed the pseudo-second-order model. Results of the study confirmed that PASH can be used as a low-cost and effective precursor for activated carbon preparation [66].

Pavan et al. were prepared activated mesoporous containing carbon from *Eragrostis plana* Nees (EPN) biomass by chemical activation method using microwave-induced pyrolysis. For mesopores formation, ZnCl2 and the biomass were made as a paste with a 1 : 1 ratio. Then the material was dried and carbonized in a microwave oven. Further, a leaching-out procedure with HCl 6.0 mol L^{-1} was carried out to eliminate the utterly present inorganics on the carbonized sample. The activated carbon was designated as MWEPN and the surface area was 1251 m^2 g^{-1} with a high total pore volume of 0.7296 g cm^{-3}. The BJH pore size distribution curve of the activated carbon material showed 80.49% mesopores regions. TGA analysis reveals that MWEPN exhibited higher thermal stability even after 750 °C. The experiment, Boehm titration, well explained the predominance of acid groups on the surface of MWEPN adsorbent. The adsorption efficiency of MWEPN was examined for a 2-nitrophenol with a contact time of 30 minutes. The maxima adsorption capacity (Qmax) was 255.8 mg g^{-1} for 2-nitrophenol at optimized adsorption conditions. The investigation on EPNs showed that it could efficiently be used as a carbon precursor to produce activated carbon by chemical activation [67]. Further, Umpierres et al. used tucumã seeds wastes (Astrocaryum aculeatum) as raw material for activated mesoporous carbon synthesis through a microwave heating process using different ratios of $ZnCl_2$ as an activating agent. The produced activated carbon exhibited a high specific surface area, up to 1318 m^2 g^{-1} for the ratio of 1.0 : 2.0 (tucumã seed: $ZnCl_2$). Surface characterization revealed that activated carbon has hydrophilic surfaces and has predominantly acidic groups on their surfaces. The obtained porous activated carbon from tucumã seeds was used to remove 2-nitrophenol from water and exhibited high values for maximum adsorption amounts (Qmax). The Qmax values of 2-nitrophenol at 50 °C was 1191, 1355, and 1382 mg g^{-1} for TMAC-1.0, TMAC-1.5, and TMAC-2.0, respectively. Here, the designated TMCA is tucumã seed-derived carbon material with the different ratios of $ZnCl_2$. The general-order kinetic and Liu isotherm models have explained the adsorption kinetics of 2-nitrophenol onto the mesoporous activated carbon [68].

Many biomass-derived mesoporous carbons are efficiently involving in phenolic compounds removal. The discussed work for phenol compound adsorption and their maximum adsorption capacity has been listed in Table 7.3.

Table 7.3 Adsorption capacities of biowaste-derived mesoporous carbons for the removal of phenolic compounds.

S. no.	Adsorbent source	Activating agents	Total surface area ($m^2\,g^{-1}$)	Type of phenolic compounds	Maximum adsorption ($mg\,g^{-1}$)	Reference
1	Rice husk	H_3PO_4	439	Phenol	22	[61]
2	Pelletized rice husk	KOH	1320	Phenol	179 (liq. phase) 740 (gas phase)	[62]
3	Toona sinensis leaves	NaOH	2719	Phenol	325 $mg\,g^{-1}$	[64]
5	Coconut spathe	KOH	1705	Phenol 2-Chlorophenol 4-Chlorophenol	120 225 275	[63]
6	Cattail fiber	H_3PO_4	890	2,4-Dichlorophenol 2,4,6-Trichlorophenol	124 172	[65]
7	*Prosopis africana* seed hulls	CH_3COONa	1085	*p*-Chlorophenol 2,4-Dichlorophenol	347 380	[66]
8	*Eragrostis plana* Nees	$ZnCl_2$, Microwave 1300 W	1250	2-Nitrophenol	255.8	[67]
9	Tucumã seeds	$ZnCl_2$, Microwave 1200 W	1318	2-Nitrophenol	1382	[68]

7.4 Conclusion and Future Outlooks

In this chapter, an extensive range of activated mesoporous carbon derived from biomass materials has been discussed. The adsorption potential of the biomass-derived carbon material has been observed according to the nature of the surface area, porosity, and functional group. Biomass-derived carbon materials reduce the environmental and management-related problem of the wastes and provide a sustainable alternative way to produce cost-effective value-added materials for the benefit of society. The main interaction of dyes and phenolic compounds on the activated carbon surface is occurred through π–π interactions due to their aromatic characters. The kinetics and adsorption capacity of all the dyes, metals, and phenolic compounds are directly influenced by the basic and structural characteristics of the carbon materials. It is mandatory to note that the system of adsorbent–adsorbate is unique, and therefore, to make a safe comparison, the adsorption conditions must be the same. Otherwise, the comparison is faulty.

Future Outlooks

Mostly, the prepared carbon materials from biomass resources have a more microporous structure and considerable mesoporous structure. The synthesis of pure mesoporous carbon and ordered mesoporous carbon from biomass resources still is a challenge. Among the various techniques used to activate carbon, chemical activation plays a crucial role in creating mesoporous structures and increasing the surface area. This chemical activation again leads to the discharge of undesirable materials into the aquatic ecosystem. Hence, a more reliable, cost-effective, and efficient physical activation method should be developed. We hope that the biomass-derived mesoporous carbon could have great development and practical applications in the near future in the toxic chemical removal and energy conversion and storage application.

References

1 Ali, H. and Khan, E. (2017). Environmental chemistry in the twenty-first century. *Environmental Chemistry Letters* 15: 329–346.

2 Bashir, I., Lone, F.A., Bhat, R.A. et al. (2020). Concerns and threats of contamination on aquatic ecosystems. *Bioremediation and Biotechnology* 1–26.

3 Lellis, B., Fávaro-Polonio, C.Z., Pamphile, J.A., and Polonio, J.C. (2019). Effects of textile dyes on health and the environment and bioremediation potential of living organisms. *Biotechnology Research and Innovation* 3: 275–290.

4 Wan, X., Yang, X., Wen, Q. et al. (2020). Sustainable development of industry-environmental system based on resilience perspective. *International Journal of Environmental Research and Public Health* 17: 645.

5 Desore, A. and Narula, S.A. (2018). An overview on corporate response towards sustainability issues in textile industry. *Environment, Development and Sustainability* 20: 1439–1459.

6 Camargo, F.P., Sérgio Tonello, P., dos Santos, A.C.A., and Duarte, I.C.S. (2016). Removal of toxic metals from sewage sludge through chemical, physical, and biological treatments – a review. *Water, Air, & Soil Pollution* 227: 433.

7 Dermont, G., Bergeron, M., Mercier, G., and Richer-Laflèche, M. (2008). Soil washing for metal removal: a review of physical/chemical technologies and field applications. *Journal of Hazardous Materials* 152: 1–31.

8 Ma, M., Ying, H., Cao, F. et al. (2020). Adsorption of congo red on mesoporous activated carbon prepared by CO_2 physical activation. *Chinese Journal of Chemical Engineering* 28: 1069–1076.

9 Schnell, A., Steel, P., Melcer, H. et al. (2000). Enhanced biological treatment of bleached kraft mill effluents – I. Removal of chlorinated organic compounds and toxicity. *Water Research* 34: 493–500.

10 Babu, C.M., Palanisamy, B., Sundaravel, B. et al. (2013). A novel magnetic Fe_3O_4/SiO_2 core-shell nanorods for the removal of arsenic. *Journal of Nanoscience and Nanotechnology* 13: 2517–2527.

11 Nagpal, M. and Kakkar, R. (2019). Use of metal oxides for the adsorptive removal of toxic organic pollutants. *Separation and Purification Technology* 211: 522–539.

12 Mashkoor, F., Nasar, A., and Inamuddin (2020). Carbon nanotube-based adsorbents for the removal of dyes from waters: a review. *Environmental Chemistry Letters* 18: 605–629.

13 Babu, C.M., Vinodh, R., Sundaravel, B. et al. (2016). Characterization of reduced graphene oxide supported mesoporous Fe_2O_3/TiO_2 nanoparticles and adsorption of As(III) and As(V) from potable water. *Journal of the Taiwan Institute of Chemical Engineers* 62: 199–208.

14 Baig, N., Ihsanullah, M., and Sajid, T.A. (2019). Saleh, graphene-based adsorbents for the removal of toxic organic pollutants: a review. *Journal of Environmental Management* 244: 370–382.

15 Ramanayaka, S., Vithanage, M., Sarmah, A. et al. (2019). Performance of metal–organic frameworks for the adsorptive removal of potentially toxic elements in a water system: a critical review. *RSC Advances* 9: 34359–34376.

16 Vakili, M., Deng, S., Cagnetta, G. et al. (2019). Regeneration of chitosan-based adsorbents used in heavy metal adsorption: a review. *Separation and Purification Technology* 224: 373–387.

17 Yadav, V.B., Gadi, R., and Kalra, S. (2019). Clay based nanocomposites for removal of heavy metals from water: a review. *Journal of Environmental Management* 232: 803–817.

18 Zou, Y., Wang, X., Khan, A. et al. (2016). Environmental remediation and application of nanoscale zero-valent iron and its composites for the removal of heavy metal ions: a review. *Environmental Science & Technology* 50: 7290–7304.

19 Deng, J., Li, M., and Wang, Y. (2016). Biomass-derived carbon: synthesis and applications in energy storage and conversion. *Green Chemistry* 18: 4824–4854.

20 Hu, Z., Srinivasan, M.P., and Ni, Y. (2000). Preparation of mesoporous high-surface-area activated carbon. *Advanced Materials* 12: 62–65.

21 Singh, G., Lakhi, K.S., Kim, I.Y. et al. (2017). Highly efficient method for the synthesis of activated mesoporous biocarbons with extremely high surface area for high-pressure CO_2 adsorption. *ACS Applied Materials & Interfaces* 9: 29782–29793.

22 Üner, O., Geçgel, Ü., and Bayrak, Y. (2019). Preparation and characterization of mesoporous activated carbons from waste watermelon rind by using the chemical activation method with zinc chloride. *Arabian Journal of Chemistry* 12: 3621–3627.

23 Bedia, J., Peñas-Garzón, M., Gómez-Avilés, A. et al. (2018). A review on the synthesis and characterization of biomass-derived carbons for adsorption of emerging contaminants from water. *C—Journal of Carbon Research* 4: 63.

24 Yahya, M.A., Al-Qodah, Z., and Ngah, C.W.Z. (2015). Agricultural bio-waste materials as potential sustainable precursors used for activated carbon production: a review. *Renewable and Sustainable Energy Reviews* 46: 218–235.

25 Ukanwa, K.S., Patchigolla, K., Sakrabani, R. et al. (2019). A review of chemicals to produce activated carbon from agricultural waste biomass. *Sustainability* 11: 6204.

26 Hu, Z. and Srinivasan, M.P. (2001). Mesoporous high-surface-area activated carbon. *Microporous and Mesoporous Materials* 43: 267–275.

27 Cheng, C., Liu, H., Dai, P. et al. (2016). Microwave-assisted preparation and characterization of mesoporous activated carbon from mushroom roots by phytic acid (C6H18O24P6) activation. *Journal of the Taiwan Institute of Chemical Engineers* 67: 532–537.

28 Shahid, M., Shahidul, I., and Mohammad, F. (2013). Recent advancements in natural dye applications: a review. *Journal of Cleaner Production* 53: 310–331.

29 Sayğılı, H., Güzel, F., and Önal, Y. (2015). Conversion of grape industrial processing waste to activated carbon sorbent and its performance in cationic and anionic dyes adsorption. *Journal of Cleaner Production* 93: 84–93.

30 Sayğılı, H. and Güzel, F. (2015). Performance of new mesoporous carbon sorbent prepared from grape industrial processing wastes for malachite green and congo red removal. *Chemical Engineering Research and Design* 100: 27–38.

31 Lin, L., Zhai, S.-R., Xiao, Z.-Y. et al. (2013). Dye adsorption of mesoporous activated carbons produced from NaOH-pretreated rice husks. *Bioresource Technology* 136: 437–443.

32 Islam, M.A., Ahmed, M.J., Khanday, W.A. et al. (2017). Mesoporous activated carbon prepared from NaOH activation of rattan (Lacosperma secundiflorum) hydrochar for methylene blue removal. *Ecotoxicology and Environmental Safety* 138: 279–285.

33 Yu, M., Han, Y., Li, J., and Wang, L. (2017). CO_2-activated porous carbon derived from cattail biomass for removal of malachite green dye and application as supercapacitors. *Chemical Engineering Journal* 317: 493–502.

34 Wenjing, L., Ximing, W., and Minghui, Z. (2017). Preparation of highly mesoporous wood-derived activated carbon fiber and the mechanism of its porosity development. *Holzforschung* 71: 363–371.

35 Mahmoodi, N.M., Taghizadeh, M., and Taghizadeh, A. (2018). Mesoporous activated carbons of low-cost agricultural bio-wastes with high adsorption capacity: preparation and artificial neural network modeling of dye removal from single and multicomponent (binary and ternary) systems. *Journal of Molecular Liquids* 269: 217–228.

36 Xiao, H., Peng, H., Deng, S. et al. (2012). Preparation of activated carbon from edible fungi residue by microwave assisted K_2CO_3 activation – application in reactive black 5 adsorption from aqueous solution. *Bioresource Technology* 111: 127–133.

37 Açıkyıldız, M., Gürses, A., and Karaca, S. (2014). Preparation and characterization of activated carbon from plant wastes with chemical activation. *Microporous and Mesoporous Materials* 198: 45–49.

38 Jain, A., Jayaraman, S., Balasubramanian, R., and Srinivasan, M.P. (2014). Hydrothermal pre-treatment for mesoporous carbon synthesis: enhancement of chemical activation. *Journal of Materials Chemistry A* 2: 520–528.

39 Kong, W., Zhao, F., Guan, H. et al. (2016). Highly adsorptive mesoporous carbon from biomass using molten-salt route. *Journal of Materials Science* 51: 6793–6800.

40 Yu, L. and Luo, Y.-m. (2014). The adsorption mechanism of anionic and cationic dyes by Jerusalem artichoke stalk-based mesoporous activated carbon. *Journal of Environmental Chemical Engineering* 2: 220–229.

41 Kang, S., Jian-chun, J., and Dan-dan, C. (2011). Preparation of activated carbon with highly developed mesoporous structure from Camellia oleifera shell through water vapor gasification and phosphoric acid modification. *Biomass and Bioenergy* 35: 3643–3647.

42 Foo, K.Y. and Hameed, B.H. (2012). Mesoporous activated carbon from wood sawdust by K_2CO_3 activation using microwave heating. *Bioresource Technology* 111: 425–432.

43 Wongcharee, S., Aravinthan, V., Erdei, L., and Sanongraj, W. (2018). Mesoporous activated carbon prepared from macadamia nut shell waste by carbon dioxide activation: comparative characterisation and study of methylene blue removal from aqueous solution. *Asia-Pacific Journal of Chemical Engineering* 13: e2179.

44 Singh, R., Gautam, N., Mishra, A., and Gupta, R. (2011). Heavy metals and living systems: an overview. *Indian Journal of Pharmacology* 43: 246–253.

45 Tchounwou, P.B., Yedjou, C.G., Patlolla, A.K., and Sutton, D.J. (2012). Heavy metal toxicity and the environment. *Experientia Supplementum* 101: 133–164.

46 Hegazi, H.A. (2013). Removal of heavy metals from wastewater using agricultural and industrial wastes as adsorbents. *HBRC Journal* 9: 276–282.

47 Rahmani-Sani, A., Singh, P., Raizada, P. et al. (2020). Use of chicken feather and eggshell to synthesize a novel magnetized activated carbon for sorption of heavy metal ions. *Bioresource Technology* 297: 122452.

48 Cuong, D.V., Liu, N.-L., Nguyen, V.A., and Hou, C.-H. (2019). Meso/micropore-controlled hierarchical porous carbon derived from activated biochar as a high-performance adsorbent for copper removal. *Science of the Total Environment* 692: 844–853.

49 El-Azazy, M., El-Shafie, A.S., Issa, A.A. et al. (2019). Potato peels as an adsorbent for heavy metals from aqueous solutions: eco-structuring of a *green* adsorbent operating Plackett–Burman design. *Journal of Chemistry* 2019: 4926240.

50 Farooq, U., Kozinski, J.A., Khan, M.A., and Athar, M. (2010). Biosorption of heavy metal ions using wheat based biosorbents – a review of the recent literature. *Bioresource Technology* 101: 5043–5053.

51 Liu, Z. and Zhang, F.-S. (2011). Removal of copper (II) and phenol from aqueous solution using porous carbons derived from hydrothermal chars. *Desalination* 267: 101–106.

52 Noor, N.M., Othman, R., Mubarak, N.M., and Abdullah, E.C. (2017). Agricultural biomass-derived magnetic adsorbents: preparation and application for heavy metals removal. *Journal of the Taiwan Institute of Chemical Engineers* 78: 168–177.

53 Xia, D., Li, H., Chen, Z. et al. (2019). Mesoporous activated biochar for As(III) adsorption: a new utilization approach for biogas residue. *Industrial & Engineering Chemistry Research* 58: 17859–17870.

54 Yang, J., Hou, B., Wang, J. et al. (2019). Nanomaterials for the removal of heavy metals from wastewater. *Nanomaterials (Basel)* 9: 424.

55 Yun, Y.-S., Park, D., Park, J.M., and Volesky, B. (2001). Biosorption of trivalent chromium on the brown seaweed biomass. *Environmental Science & Technology* 35: 4353–4358.

56 Alalwan, H.A., Kadhom, M.A., and Alminshid, A.H. (2020). Removal of heavy metals from wastewater using agricultural by products. *Journal of Water Supply: Research and Technology-AQUA* 69: 99–112.

57 Sun, X., Wang, C., Li, Y. et al. (2015). Treatment of phenolic wastewater by combined UF and NF/RO processes. *Desalination* 355: 68–74.

58 World Health Organization (ed.) (2008). Guidelines for drinking-water quality: second addendum. In: *Recommendations*, 3rde, vol. 1. Geneva: World Health Organization.

59 Villegas, L.G.C., Mashhadi, N., Chen, M. et al. (2016). A short review of techniques for phenol removal from wastewater. *Current Pollution Reports* 2: 157–167.

60 Mahugo Santana, C., Sosa Ferrera, Z., Padrón, M.E.T., and Rodríguez, J.J.S. (2009). Methodologies for the extraction of phenolic compounds from environmental samples: new approaches. *Molecules* 14: 298–320.

61 Kennedy, L.J., Vijaya, J.J., Kayalvizhi, K., and Sekaran, G. (2007). Adsorption of phenol from aqueous solutions using mesoporous carbon prepared by two-stage process. *Chemical Engineering Journal* 132: 279–287.

62 Shen, Y., Zhou, Y., Fu, Y., and Zhang, N. (2020). Activated carbons synthesized from unaltered and pelletized biomass wastes for bio-tar adsorption in different phases. *Renewable Energy* 146: 1700–1709.

63 Prashanthakumar, T.K.M., Kumar, S.K.A., and Sahoo, S.K. (2018). A quick removal of toxic phenolic compounds using porous carbon prepared from renewable biomass coconut spathe and exploration of new source for porous carbon materials. *Journal of Environmental Chemical Engineering* 6: 1434–1442.

64 Kong, X., Gao, H., Song, X. et al. (2020). Adsorption of phenol on porous carbon from Toona sinensis leaves and its mechanism. *Chemical Physics Letters* 739: 137046.

65 Ren, L., Zhang, J., Li, Y., and Zhang, C. (2011). Preparation and evaluation of cattail fiber-based activated carbon for 2,4-dichlorophenol and 2,4,6-trichlorophenol removal. *Chemical Engineering Journal* 168: 553–561.

66 Garba, Z.N. and Rahim, A.A. (2016). Evaluation of optimal activated carbon from an agricultural waste for the removal of para-chlorophenol and 2,4-dichlorophenol. *Process Safety and Environmental Protection* 102: 54–63.

67 Cunha, M.R., Lima, E.C., Cimirro, N.F.G.M. et al. (2018). Conversion of Eragrostis plana Nees leaves to activated carbon by microwave-assisted pyrolysis for the removal of organic emerging contaminants from aqueous solutions. *Environmental Science and Pollution Research* 25: 23315–23327.

68 Umpierres, C.S., Thue, P.S., Lima, E.C. et al. (2018). Microwave-activated carbons from tucumã (Astrocaryum aculeatum) seed for efficient removal of 2-nitrophenol from aqueous solutions. *Environmental Technology* 39: 1173–1187.

8

Biomass-derived Carbon as Electrode Materials for Batteries

P. Vengatesh[1,3], C. Karthik Kumar[2], T.S. Shyju[1,3], and M. Paulraj[4]

[1] *Centre for Nanoscience and Nanotechnology, Sathyabama Institute of Science and Technology, Chennai 600119, Tamil Nadu, India*
[2] *Department of Chemistry, Vinayaka Mission's Kirupananda Variyar Arts and Science College, Vinayaka Mission's Research Foundation (Deemed to Be University), Salem 636308, Tamil Nadu, India*
[3] *Centre of Excellence for Energy Research, Sathyabama Institute of Science and Technology, Chennai 600119, Tamil Nadu, India*
[4] *University of Concepcion, Department of Physics, Faculty of Physical and Mathematical Sciences, Post Box 160-C, Concepcion, Chile*

8.1 Introduction

Energy is the basic need for the existence as well as for the development of human society. The rapid development in the advanced energy technology and the emergence of the high-quality energy techniques has revolutionized the world in the energy research aspect. Hence, people are more concern toward the utilization of conventional fossil fuels (termed as nonrenewable energy resources) for fulfilling the global energy demand. However, the usage of conventional fossil fuels, such as coal, petroleum, and natural gas in the past several decades, leads to the rapid exhaustion of these resources along with the severe damage to the environment. Therefore, the renewable resources are given more importance in recent years in order to satisfy the depletion of the nonrenewable fossil fuels.

The development of renewable and green energy technologies is believed to be the great potential for future energy generation, storage, and their usage to the society. In this context, the electrical energy storage devices are becoming increasingly important to develop clean energy technologies, particularly for grid-scale storage of the energy derived from the renewable energy resources, in which it can address the problem of continuous supply of the energy to the world. Among them, electrochemical energy storage devices (commonly denoted as EES) and electrochemical capacitors (supercapacitors) have received tremendous interest due to their high energy conversion efficiency and of precise interest to the energy research community in order to confront the energy shortage in the upcoming years [1]. Supercapacitors possess high cycle and high power density, but they have limited energy density,

Biomass-Derived Carbon Materials: Production and Applications, First Edition.
Edited by Alagarsamy Pandikumar, Perumal Rameshkumar, and Pitchaimani Veerakumar.

whereas the rechargeable batteries (RBs) have high energy density depending on the charge–discharge rate and cycle life which will be discussed in detail in Section 8.4. Hence, the electrochemically active materials with high energy density have been emerged in order to fulfill the next-generation electronic devices [2]. Therefore, in this chapter, we particularly focus on the electrode materials for the RBs and their performance.

8.1.1 Batteries

A battery is an electrical energy storage device that consists of large number of electrochemical cells connected in series or parallel to achieve the required voltage and current. It acts as a primary power source for any electronic gadgets. The conversion of chemical energy directly into the electrical energy by means of electrochemical (redox) reactions is taking place in a typical electrochemical cell with the usage of electrochemically active materials. There are three main components present in the battery cells which are anode, cathode, and electrolyte. The redox reactions occur at specified sites in the battery, i.e. at the anode and cathode with negative free energy change that drives the battery to function. This is a spontaneous reaction[1].

8.1.2 Classification of Batteries

Batteries are generally classified into two major categories: (i) primary (non-rechargeable) batteries and (ii) secondary (rechargeable) batteries. Primary batteries are the one which cannot be recharged once it gets depleted. They are made up of the electrochemical cells where the electrochemical reactions are irreversible. One such example for primary batteries is alkaline batteries which possess high specific energy and are environmental-friendly and cost-effective.

Secondary batteries or RBs are the batteries which consist of electrochemical cells whose chemical reactions are reversible, and hence the electroactive materials spent during the discharging reactions are restored by passing the electric current. Once the energy has been drained, it can be recharged for further usage of energy. Such kinds of batteries are further classified into several categories based on the characteristics of batteries. However, some of the secondary batteries that pay increased attention in recent years are shown in Figure 8.1.

8.1.3 Characteristics of Batteries

The terminologies to represent the performance of a typical battery are (i) nominal cell voltage, (ii) internal resistance, (iii) charging/discharging efficiency, (iv) gravimetric energy density, (v) power density, (vi) specific capacity of electrodes, (vii) open-circuit voltage, (viii) closed-circuit voltage, (ix) cycle life, and (x) shelf life. Although each characteristic plays a significant role, energy density, specific capacity of electrodes, charging/discharging phenomenon, and cycle life are given

1 https://batteryuniversity.com/learn/article/secondary_batteries.

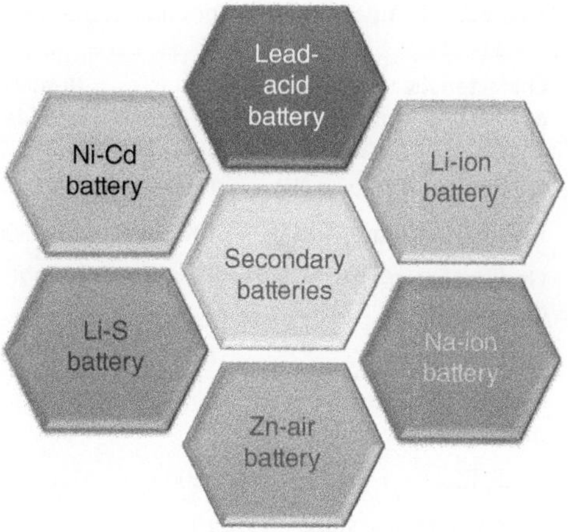

Figure 8.1 Classification of rechargeable batteries.

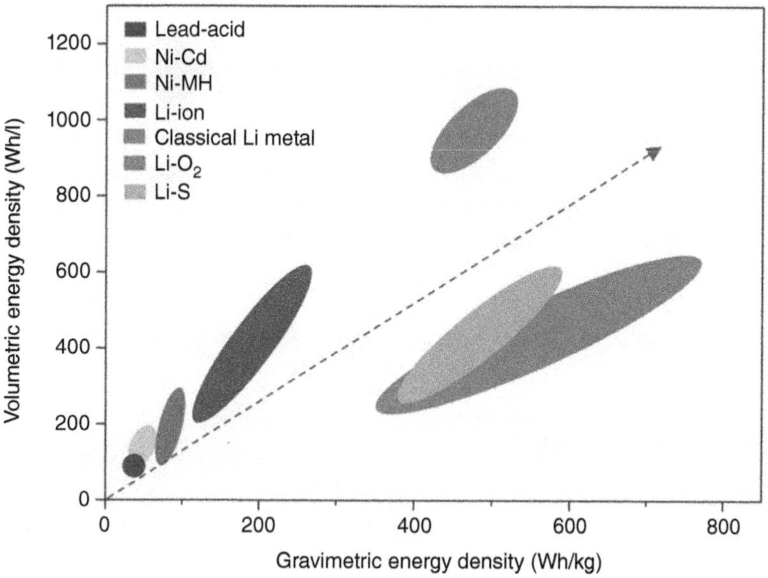

Figure 8.2 Comparison of energy density of some selected RBs. Source: Ref. [2] / with permission of Springer Nature.

more preferences in order to have excellent RBs. Figure 8.2 shows the comparison of energy densities of various RBs where it is inferred that the smaller size and lighter weight of lithium-based batteries have high energy density compared to lead-acid and Ni-Cd batteries. Table 8.1 outlines the characteristics of some of the RBs. It is observed from this table that the lithium-based batteries have high nominal cell voltage and high energy density. However, the RBs such as lithium-ion (Li-ion) batteries (LIBs), sodium-ion (Na-ion) batteries (SIBs), lithium-sulfur batteries (Li-S), and zinc-air batteries (ZABs) possess several advantages and disadvantages depending on the electrode characteristics that are discussed in detail with mechanism in

Table 8.1 Comparison of major characteristics of various RBs.

Characteristic of a battery	Ni-Cd	Ni-metal hydride	Lead-acid	Li-ion	Li-ion polymer
Nominal cell voltage (V)	1.25	1.25	2	3.6	3.6
Internal resistance (mΩ)	100–200	200–300	<100	150–250	200–300
Gravimetric energy density (Wh kg^{-1})	45–80	60–120	30–50	110–160	100–130
Fast charge time	1 h	2–4 h	8–16 h	2–4 h	2–4 h
Self-discharge rate/month	20%	30%	5%	10%	~10%
Cycle life	1500	300–500	200–300	500–3000	300–500
Commercial use since	1950	1990	1970 (sealed lead acid)	1991	1999

Section 8.2. Moreover, the specific capacity of electrodes is considered as a key factor in the assembly of batteries[2].

One has to be keen on the aforementioned battery parameters for the proper utilization of the devices. Therefore, the selection of battery for the energy devices should be based on the considerations of the characteristics such as energy density, power density, charging/discharging efficiency, cycle life, stability, and cost.

8.2 Role of Carbon with Mechanism of Rechargeable Batteries (RBs)

8.2.1 Li-Ion Batteries (LIBs)

Among the metals, lithium is a promising material for electrodes in batteries due to its lightweight with high voltage and high energy density. As lithium has the lowest electronegativity in existing metals with the standard cell potential of −3.045 V, the anode material most easily gives electrons to form positive ions [3]. The lithium ion intercalates must move easily back and forth between the cathode and anode intercalator in the RBs based on the concept of "rocking chair" and the concept of "shuttle cock."

LIBs consist mainly of a carbonaceous anode and a lithium transitional metal oxide, such as $LiCoO_2$, $LiNiO_2$, and $LiMn_2O_4$, as cathode. The anode on Cu foil and the cathode on Al foil are made into spiral- or plate-folded shapes, each giving the US 18650 cylindrical cell shape and rectangular cells. A porous polymer separator of approximately 25-mm-thick polyolefin, made of polyethylene (PE) and polypropylene (PP), is placed between these two electrodes [4]. An electrolyte is an organic liquid, such as Propylene carbonate (PC), Ethylene carbonate (EC1)

2 http://web.mit.edu/evt/summary_battery_specifications.pdf.

Diethyl carbonate (DEC), and a newly created, high-voltage gel-like polymer which is stable under high voltages. In the electrolyte, lithium salts such as $LiClO_4$, $LiBF_4$, and $LiPF_6$ are dissolved.

The theoretical lithium storage capacity of a Li-ion secondary battery graphite anode, corresponding to the first LiC-GIC stage, was estimated to be 372 mA h g^{-1}. The total reactions to the charging/dismantling and the Li^+ intercalation/deintercalation anode reaction are illustrated as follows:

$$LiCoO_2 + yC \underset{\text{Discharge}}{\overset{\text{Charge}}{\rightleftarrows}} Li_{1-x}CoO_2 + Li_xC_y \tag{8.1}$$

$$yC + xLi + xe^- \underset{\text{Discharge}}{\overset{\text{Charge}}{\rightleftarrows}} Li_xC_y \tag{8.2}$$

In contrast, the theoretical capacity was overreported for disordered carbons with Li storage capacities. This phenomenon is still hard to explain and new systems should be needed to be put in place. Experimentally and theoretically, many types of carbonaceous materials have been studied, ranging from highly ordered graphite to disordered carbon. Investigations are focused on improving energy storage devices' specific capacities, cyclic efficiency, and cyclic lifetime.

8.2.2 Li-S Batteries (Li-S)

The Li-S battery proof-of-concept analysis is dated back to the 1960s with the direct application of primary sulfur and metallic lithium as cathode and anode, respectively. The overall reaction within a Li-S battery follows the simple Li and S combination for Li_2S:

$$16Li + S_8 \leftrightarrow 8Li_2S \tag{8.3}$$

Li-S batteries are composed of a composition of positive elemental sulfur compound with a conductive additive, a polymer-binding agent (cathode), a negative lithium metal anode, and a porous separator, impregnated with organic fluid electrolyte (Figure 8.3a), close to the configuration of other secondary batteries (Figure 8.3a) [6]. In this case, the battery consisting of a composite sulfur cathode with a lithium metal anode is already in "charged state" when it is placed together and assembled. For certain other situations, when a Li_2S composite cathode is used, lithium and sulfur must be charged to the relatively electrode in the first position. The intermediate state between charge and discharge may be called when the "catholyte" of lithium polysulfide (usually Li_2S_6 or Li_2S_8) is chosen. The difference in the electric chemistry of these systems lies only in the first charging or discharging process, despite the different starting electrode materials; in later cycles, they are all subject to the battery reaction as depicted in Eq. (8.3). The electrochemical method is also complex and multistep, with soluble and insoluble polysulfide intermediates (Figure 8.3b), given the apparently simple reaction in

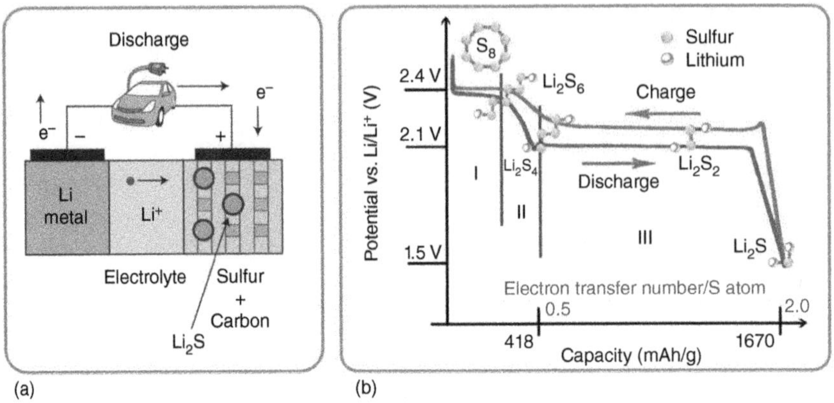

Figure 8.3 (a) A typical Li-S battery and (b) charge–discharge voltage profiles of the cathode and the intermediate products in a Li-S battery. Source: Ref. [5] / with permission of Elsevier.

Li-S batteries. Throughout Li-S discharge processes, the reduction of elemental sulfur to lithium sulfide (Li_2S) can typically be split into three stages with the two separate output voltage plateau at approximately 2.4 and 2.1 V versus Li/Li^+ on an intermediate discharge profile that lead to intermediate solubility.

A fundamental necessity is necessary to use sulfur in a Li-S battery that leads to the poor conductivity of the electron. The combination of sulfur with various carbon forms is also used to accomplish this. In order to fulfill its usefulness in a Li-S battery, carbon must be in close contact with sulfur. Sulfur is, therefore, the principal factor in the incorporation of carbon which in turn a major factor affecting the battery performance [5]. The conductive agent in a sulfur cathode can be used to promote the electron's transport due to its poor electronic conductivity (5×10^{-30} S cm^{-1}). After the initial production of Li-S batteries, carbon materials have also been utilized to improve its conductivity. The electronic conductivity of sulfur/carbon composites depends largely on the carbon added and on the type of contact between its building blocks such as carbon nanotubes (CNTs), carbon fibers, or other microscopic compounds (Figure 8.4).

8.2.3 Na-Ion Batteries (SIBs)

Recent advances in the area of post-lithium batteries are SIBs. They seek to address the resource availability, cost, or efficiency limitations of existing LIBs. Following that, a drop-in technology, based on similar electrical processes, materials, and manufacturing processes, they are considered to be the same working principle as the LIBs (Figure 8.5).

Over the past few years, significant efforts have been made to investigate the effective electrode materials to use the low-cost rechargeable sodium battery chemistry [7, 8]. Graphite is a possible anode source for SIBs. Earlier, it was proved that a tiny amount of Na ions could be intercalated into the graphite structure (i.e. 30 mA h g^{-1}), which is attributed to the thermodynamic stability of the compound

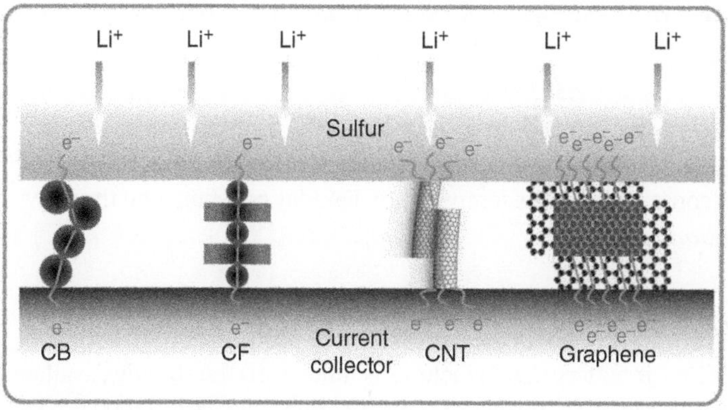

Figure 8.4 Schematic representation of various carbon (conductive) additives along with the contact modes between sulfur and the current collector. Source: Ref. [5] / with permission of Elsevier.

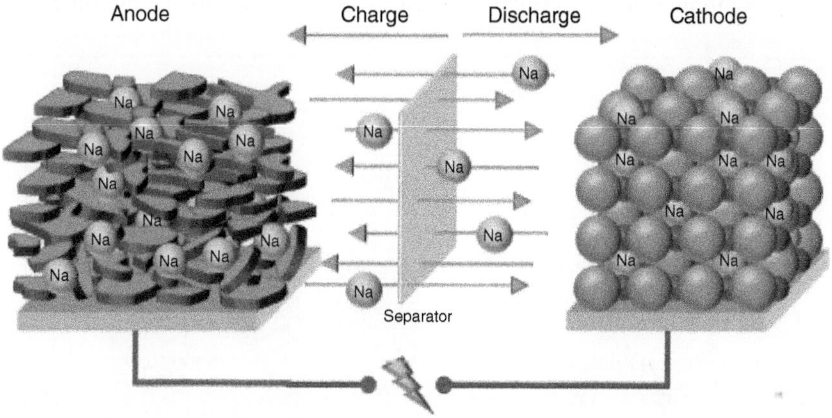

Figure 8.5 Working principle of sodium-ion battery (SIB). Source: Ref. [7] / MDPI / CC BY 4.0.

based on experimental tests, which was intercalated with Na-intercalated binary graphite compound (b-GIC) [9]. In the meantime, Jache and Adelhelm [10] and Kim et al. [11] showed that sodium can be contained in graphite reversibly by co-intercalation, and the sodium ions form a ternary graphite intercalation compound (t-GIC) into the graphite galleries. Co-intercalation reactions provided with remarkable reversibility of graphite electrodes and rapid sodium intercalation kinetics, which pave the way to exploit graphite as a promising anode for advanced SIBs. In recent years, there have been extensive investigations into the electrochemical performance of a graphite anode and its mechanism of intrinsic reaction. The electrode reactions for a half-cell of graphite || 1 M $NaPF_6$ diethylene glycol dimethyl ether (G_2) || sodium can be described as follows:

$$\text{Anode: } \mathrm{Na} + (\mathrm{G_2}) \leftrightarrow [\mathrm{Na} - \mathrm{G_2}]^+ + \mathrm{e^-} \tag{8.4}$$

$$\text{Cathode: } x\mathrm{C} + [\mathrm{Na} - \mathrm{G_2}]^+ + \mathrm{e^-} \leftrightarrow [\mathrm{Na} - \mathrm{G_2}]\mathrm{C}_x \tag{8.5}$$

Based on the Eqs. (8.4) and (8.5), the co-intercalation reaction voltage can be expressed as:

$$V = \left(E^0_{\text{t-GIC}} - E^0_{\text{Na}}\right) + \frac{2.303RT}{nF}\log a_{\text{G2}} \tag{8.6}$$

where $E^0_{t\text{-GIC}}$, E^0_{Na}, R, T, F, and a_{G2} denote the standard electrode potentials of t-GIC and Na metal, gas constant, absolute temperature, Faraday constant, and the activity of the solvent, respectively.

8.2.4 Zn-Air Batteries (ZABs)

The advantages of Zn-air battery (ZAB) include as follows: (i) the stability, availability, and low price of Zn metal; (ii) the energy density of ZAB device is >1000 Wh kg^{-1} (five times higher than LIBs); and (iii) the inherent protection feature of the Zn-air safety property. ZAB performs two slothful, fundamental electrochemical reactions during discharge and charge procedures: oxygen reduction response (ORR) and oxygen evolution reaction (OER), respectively (Figure 8.6a shows the operating

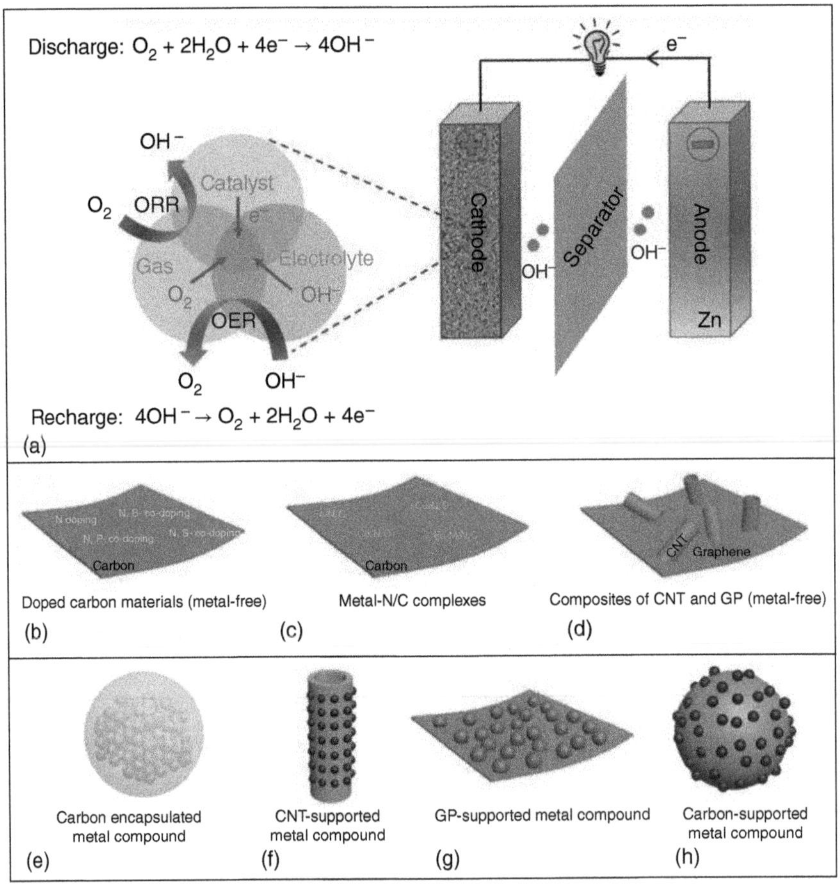

Figure 8.6 (a) Schematic representation of a typical ZAB along with the oxygen bifunctional electrocatalysts. (b–h) Various carbon-based electrocatalysts for rechargeable ZABs. Source: Ref. [12] / with permission of Elsevier.

theory of ZAB in accordance with ORR and the OER scheme). Specifically, Zn is oxidized following during the discharge: on the anode [12]:

$$Zn + 4OH^- \rightarrow Zn(OH)_4^{2-} + 2e^- \tag{8.7}$$

$$Zn(OH)_4^{2-} \rightarrow ZnO + H_2O + OH^- \tag{8.8}$$

On the cathode, O_2 is reduced to the OH^- as follows:

$$O_2 + 2H_2O + 4e^- \rightarrow 4OH^- \tag{8.9}$$

Theoretically, the reactions are reversible and allow a ZAB to be recharged electrically. During charging on the anode, ZnO is reduced back to:

$$ZnO + H_2O + OH^- \rightarrow Zn\,(OH)_4^{2-} \tag{8.10}$$

$$Zn\,(OH)_4^{2-} + 2e^- \rightarrow Zn + 4OH^- \tag{8.11}$$

On the cathode, the OH^- is oxidized to O_2 as follows:

$$4OH^- \rightarrow O_2 + 2H_2O + 4e^- \tag{8.12}$$

The cathode (ORR and OER) reactions are the speed-restricting step for ZAB reactions. Therefore, for the development of high-performance ZABs, efficient ORR and OER catalysts are very important. In the recent research progresses, carbon-based ORR/OER electrocatalysts (Figure 8.6b–h) should be satisfied in order to obtain better functionally and performance on the future development of carbon-based catalysts for advanced ZABs.

8.3 Biomass-derived Carbonaceous Materials

Environmental deterioration and increasing energy demand have urged researchers to find facile, low cost, and green routes for the synthesis of novel-advanced materials from natural resources. Although several materials have been explored, carbon-based nanomaterials show significant interest to the scientific community because of their intriguing properties such as low cost, high abundance, high conductivity, good electrochemical activity, good biocompatibility, low toxicity, high stability, and environmental friendliness. Apart from the focus of the carbonaceous materials to the many scientific fields, they are also attentive to the energy storage and conversion devices such as batteries, supercapacitors, and solar cells [13, 14].

The large-scale commercialized batteries should have long cycle life, low cost, and high safety. Hence, the electrode materials could be selected based on the stability, durability, nontoxicity, and abundance of the materials. Among the reported anode materials for EES, the carbonaceous materials have gained more interest due to their excellent stability, cost-effectiveness, and high abundance. Some of the carbonaceous materials that have been utilized are graphite, graphene, carbon spheres, carbon fibers, porous carbons, carbon sheets, carbon tubes, and amorphous carbons [15].

Biomass, an efficient renewable energy resource alternative to the commercially available materials, has been widely utilized to derive various forms of carbonaceous materials [16]. People are more focused toward the discovery and innovation of

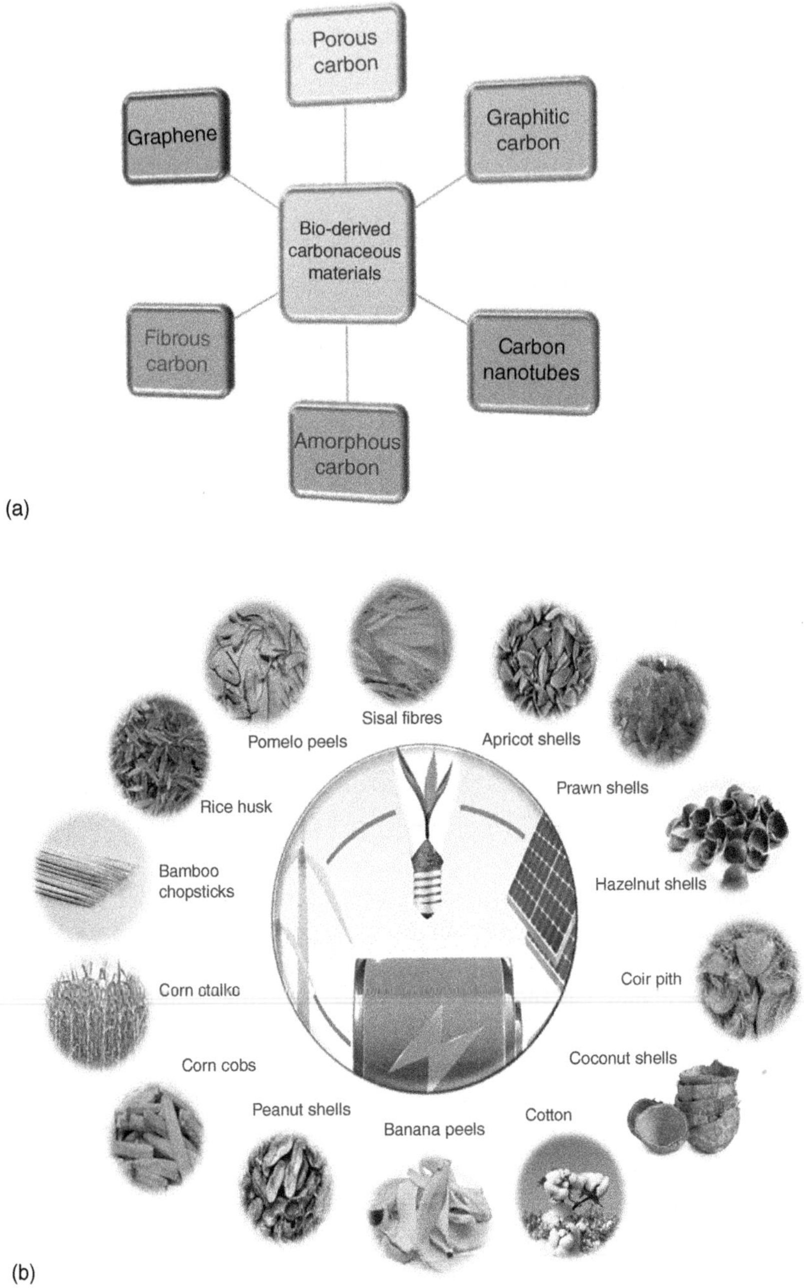

Figure 8.7 (a) Biomass-derived carbonaceous materials for RBs and (b) various biomasses used for the preparation of electrode materials for RBs. Source: Abimagestudio / Adobe Stock; Gargonia / Adobe Stock; Thapakorn / Adobe Stock.

new electrodes using the carbonaceous materials derived from biomass to enhance the electrochemical performances of RBs (as shown in Figure 8.7a). Since biomass-derived carbonaceous materials offer several advantages such as structural versatility, tunable physical/chemical properties, environmental friendliness, and low cost, they have been employed as efficient electrodes in RBs to improve their storage efficiency [17]. Increasing demand of batteries urged researchers to work toward the improvement of the battery performance and their stability which could be fulfilled by the utilization of aforementioned biomass-derived carbonaceous materials (Figure 8.7b) either as such or as nanocomposites.

The aforementioned carbon materials could in turn easily be derived from the biomass which enhances the performances of the batteries when they are utilized as anode materials. The biomass-derived carbon could be achieved by several ways of activation such as physical and chemical activation (the detailed synthetic procedures for biomass-derived carbon could be witnessed in Chapter 1). Some of them are listed as follows:

- Activation
- Template method
- Hydrothermal carbonization
- Molten salt carbonization

8.4 Electrochemical Performances of RBs using Biomass-derived Carbon Electrodes

8.4.1 Li-Ion Batteries (LIBs)

LIBs are crucial category of RBs. As mentioned earlier, for commercial-type LIBs, graphite is the most preferred anode. Since their theoretical capacity is limited to ~372 mA h g^{-1}, there are more preambles to explore the variety of low-cost and well-sourced carbon anode materials for LIBs that can boost up the specific capacity and charging–discharging efficiency. In particular, deep attention has been paid toward the biomass-derived carbon materials which are economically viable route to obtain green and sustainable electrodes. Biomass-derived carbonaceous materials can usually trigger the specific capacity while being employed for the preparation of electrode materials for LIBs [17]. Materials derived from biomass also have anticipated molecular architectures in which they favor the efficient charge storage and transport.

8.4.1.1 Biomass-derived Undoped Carbon Electrodes

Sun et al. explored a simple, efficient, and economical route for the preparation of carbonaceous material by the direct pyrolysis of spongy pomelo peels, which are rich in pectin and cellulose constituents [18]. The resulted carbon powder was used to fabricate anode for LIBs that exhibited high capacity of 452 mA h g^{-1} at the current density of 90 mA g^{-1} after 200 cycles. They suggested that the electrode prepared from spongy pomelo peels for LIBs maintained a steady discharge capacity at higher current density. The Coulombic efficiency was found to be 99.5% which indicated the excellent cycling stability. They also added that the high electrochemical performance was due to the unique structures and intrinsic properties of the biomass-derived electrode.

Unur et al. developed nanoporous carbons from hazelnut shells through hydrothermal carbonization method followed by the three different approaches, namely sole heat treatment, KOH activation, and environmentally benign MgO templating in order to introduce porosity. They also prepared seven types of porous materials with different surface areas and pore volumes. When they are tested in Li cells as Li-intercalation hosts, the material obtained by MgO templating (with surface area of $150\,m^2\,g^{-1}$) showed the best cycling performance ($307\,mA\,h\,g^{-1}$ at cycle 100 at 1 C) and the material obtained by successive hydrothermal and thermal treatments at 600 °C (with surface area of $250\,m^2\,g^{-1}$) showed the best overall electrochemical performance. They attributed this enhancement due to the properties such as minimum surface functionality, maximum aromaticity and structural order, optimal specific surface area, and well-organized micro- and mesoporous network [19].

Rice husk (RH), which is a well-known biomass material, is also being utilized for the preparation of electrodes for LIBs. SiO_2, cellulose (38%), hemicelluloses (18%), and lignin (22%) are the major ingredients present in RH. When these RHs are pyrolyzed under inert atmosphere, they could yield different types of carbonaceous materials based on the preparative techniques. The pyrolytic carbons were prepared using RH as the precursor in the presence of strong acids or bases by Prof. Fey's group. Thus, obtained carbons were utilized as a negative electrode in LIBs. The carbon obtained from RH treated with 0.3 M NaOH showed highest insertion and deinsertion capacities such as 819 and $463\,mA\,h\,g^{-1}$, respectively [20]. Followed by this work, Wang et al. obtained a porous fibrous carbon from the same RH but with different approach [21]. The RHs are subjected to acid pretreatment and hydrothermal carbonization followed by calcination and SiO_2 removal to get the porous carbon fibers which were then used as anodes in LIBs. Thus, prepared electrode delivered superior electrochemical performance with the initial discharge capacity of $789\,mA\,h\,g^{-1}$ and irreversible capacity during lithium desertion of $396\,mA\,h\,g^{-1}$. They reported that the increased electronic conductivity as well as the hierarchical (mesoporous and macroporous) fibrous network of the porous carbon derived from the RH were attributed to the enhanced electrochemical performance. Li et al. also employed RH as the biomass precursor and obtained disordered carbon by pyrolysis technique followed by the base treatment with 2 mol L NaOH solution at 150 °C for 3 hours [22]. The initial discharge capacity was found to be $1647\,mA\,h\,g^{-1}$ at 0.2 C with high reversible capacity of $502\,mA\,h\,g^{-1}$ after 100 cycles with the same current density.

Graphitic carbon fibers were also developed by Jiang et al. from the disposable bamboo chopsticks by hydrothermal process at alkaline conditions followed by calcination at 800 °C [23]. Figure 8.8 displays the preparation of carbon microfibers from bamboo chopsticks along with the Scanning electron microscope (SEM) images, X-ray diffraction (XRD), Raman spectrum, and pore size distribution curve.

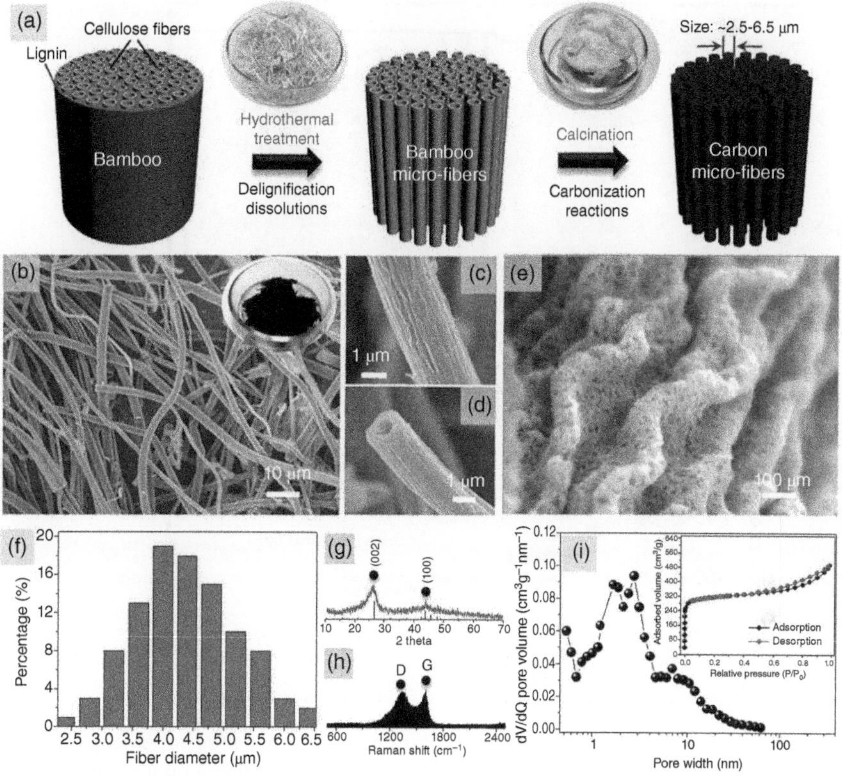

Figure 8.8 (a) Overall view of the evolution of carbon fibers from bamboo chopsticks. (b–e) SEM and optical images, (f) distribution curve, (g) and (h) XRD and Raman analysis, and (i) pore size distribution curve of the carbon fibers. Source: Ref. [23] / with permission of Royal Society of Chemistry.

Although the evolved carbon fibers were showed superior electrochemical behavior when tested as anode, the hybrid electrode C/MnO_2 @carbon fibers (which possess synergetic core–shell structure) showed a good cyclic performance (reversible capacity of ~710 mA h g^{-1} for 300 cycles) and rate performance.

Amorphous carbon was derived from the sisal fibers by Yu et al. through pyrolysis and hydrothermal activation with high specific surface area of 616.4 m^2 g^{-1}. The Li-ion insertion capacity of 646 mA h g^{-1} was obtained for the amorphous carbon electrodes at the first cycle [24]. The same pyrolysis technique was also employed by Adams et al. to obtain amorphous carbon from loblolly pine woodchips [25]. The KOH-activated and -pyrolyzed woodchips gave amorphous carbon with Brunauer–emmett–teller (BET) surface area of 1580 m^2 g^{-1}. The resulted anode for LIBs enabled better lithiation capacity in defect and interfacial sites due to the micro- and mesoporosity as well as high surface area. They also obtained a stable capacity of 700 mA h g^{-1} at C/10 rate and rate capacity of 350 mA h g^{-1} at 1 C rate (as shown in Figure 8.9). The fabricated electrodes showed stable cycling with 99.6% Coulombic efficiency after 100 cycles for a C/5 rate.

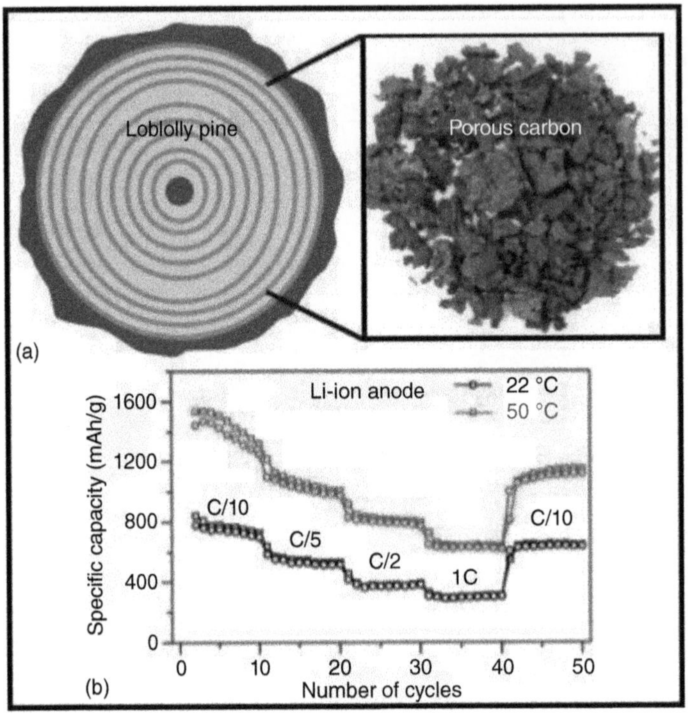

Figure 8.9 (a) Image of loblolly pine woodchips and successive preparation of porous carbon from it and (b) electrochemical performance (rate studies) of porous carbon derived from woodchips. Source: Ref. [25] / with permission of American Chemical Society.

Mullaivananathan et al. developed microporous carbon which contains randomly oriented graphene sheet from KOH activated carbon derived from yet another biomass precursor, i.e. coir pith [26]. The coir pith-derived carbon (CPC) was demonstrated as an excellent LIB anode because of its high surface area of 2500 m^2 g^{-1}. CPC obtained from carbonization process was subjected to the various KOH activation temperatures such as 800, 850, and 900 °C in order to obtain microporous carbon. Among all the CPC-based anode materials activated at different temperatures, CPC-850 exhibited a steady state capacity of 837 mA h g^{-1} at 100 mA g^{-1}. Figure 8.10 depicts the cyclic voltammogram (CV) curves of CPC electrodes activated at different temperatures and also charge–discharge behavior at the current density of 100 mA g^{-1}.

Fromm et al. reported the preparation of carbon materials from five different biomass precursors (different types of bamboo wood) and the comparison of electrochemical performance of the respective electrodes made out of these carbon materials along with the petroleum coke-derived carbon as a reference material [27]. They also studied the correlation between the structural changes of carbon with temperature dependence and their corresponding anodic behavior in LIBs.

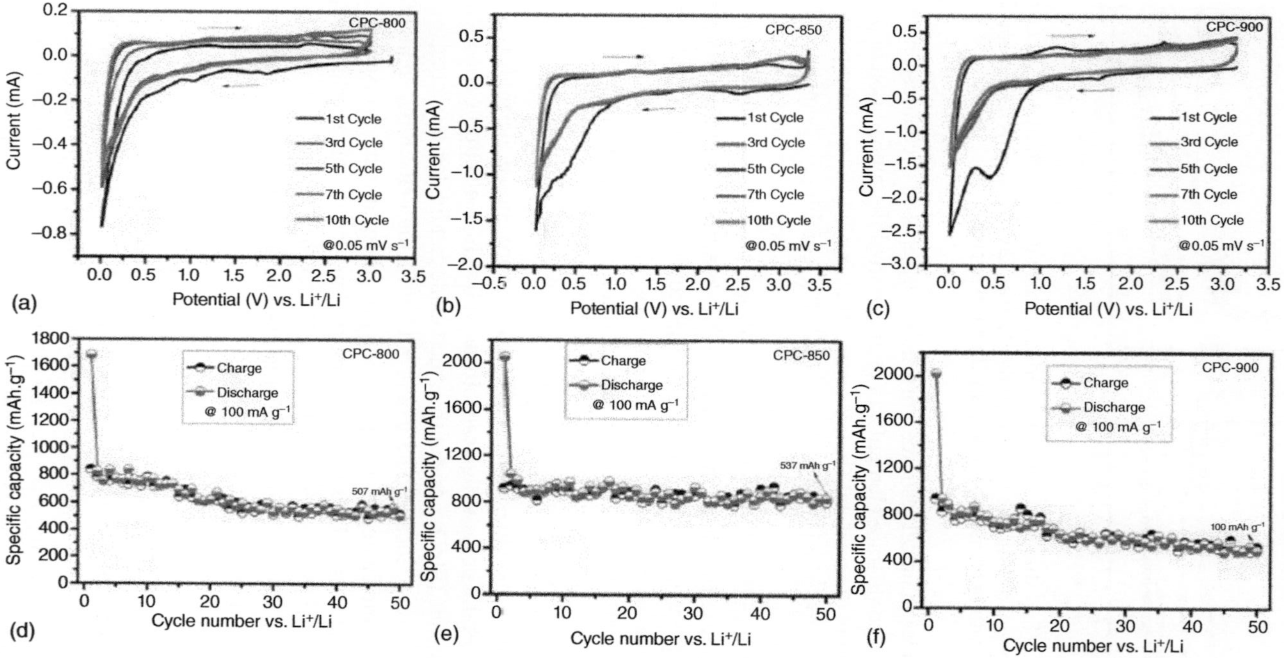

Figure 8.10 (a–c) Cyclic voltammogram of carbon electrodes derived from coir pith at different activation temperatures such as 800, 850, and 900 °C recorded at scan rate of 0.05 mV s^{-1}. (b–f) Charge–discharge behavior of CPC anodes at the current density of 100 mA g^{-1}. Source: Ref. [26] / with permission of Elsevier.

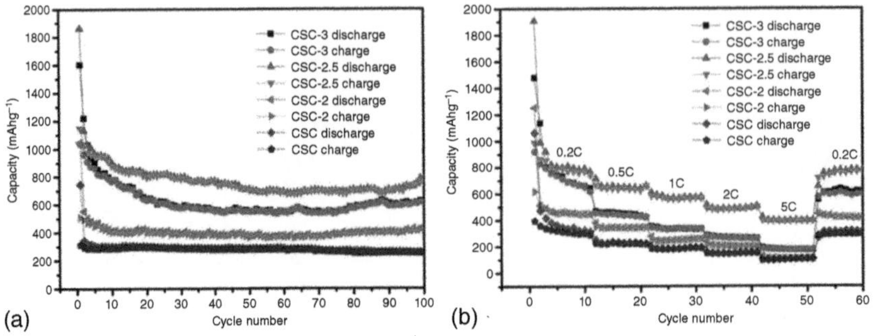

Figure 8.11 (a) Cycling studies of cornstalk-derived carbon (CSC) electrodes at 0.2 C and (b) rate performance of the various CSC electrodes such as CSC-3, CSC-2.5, CSC-2, and CSC. Source: Ref. [28] / with permission of American Chemical Society.

As an important biological resource, corn stalk is composed of cellulose, hemicellulose lignin, and some trace elements. Li et al. reported the preparation of porous carbon derived from the corn stalk biomass via carbonization and activation of $CaCl_2$ which is a cheap and recyclable activator [28]. The combined advantages of less harmful effects of corn stalk and adjustment of pore size and distribution by controlling the amount of $CaCl_2$ additive are responsible for the enhanced electrochemical performance of the cornstalk-derived carbon (CSC) electrodes for LIBs. The surface area of 370.6 $m^2\,g^{-1}$ with average pore size of 9.65 nm was obtained for the electrodes which showed the specific capacity of 783 $mA\,h\,g^{-1}$ after 100 cycles. Figure 8.11 shows the cyclic performance profiles for CSC electrodes at different weight ratios of corn stalk powder and $CaCl_2$.

8.4.1.2 Metal Oxides @ Biomass-derived Carbon Nanocomposite Electrodes

Metal oxides generally possess high capacity as well as high electronic conductivity. Molybdenum dioxide (MoO_2), one of the metal oxide candidates, is of precise interest due to its high theoretical capacity (838 $mA\,h\,g^{-1}$), high electronic conductivity (>1×10^4 $S\,cm^{-1}$), and facile ion transport. However, there are certain disadvantages during the cycle process of LIBs such as lethargic kinetics and large volume expansion [29]. Hence, tuning the nanostructures of MoO_2 and design and synthesis of nanocomposites of MoO_2@C are being employed as suitable processes to prepare anode materials for LIBs in order to encounter the aforementioned problems.

One such work was reported by Che et al. (2016), where they developed MoO_2@ C aerogel nanocomposite electrode by a simple one step in situ method using sodium alginate (seeweed biomass) as the carbon precursor (SAC) [29]. Briefly, the aqueous solution containing ammonium molybdate tetrahydrate (AMM) and sodium alginate was poured into HCl solution to obtain conductive hydrogel networks via ion exchange process (replacing Na^+ ions by H^+ ions). The obtained hydrogels were subjected to freeze-drying process for dehydration in order to generate AMM aerogels. Upon further stabilization in air and subsequent carbonization under nitrogen atmosphere, they achieved three different aerogels, namely MoO_2 @ SAC –X (where X denoted the different stabilization temperatures such as 350, 375, and 400 °C).

They also found BET surface area for these aerogels which are 210.1, 196.7, and 25.22 $m^2 g^{-1}$, respectively. The precipitation of more carbon in the redox reaction and destroy of carbon network were the reasons attributed to the decrease in surface area upon increase of stabilization temperature. Among the three different electrodes, the MoO_2 @ SAC-375 electrode displayed better electrochemical performance with a discharge capacity of 438 mA h g^{-1}. The same electrode exhibited best cycle performance with a reversible capacity of 490 mA h g^{-1} at a current density of 200 mA g^{-1} after 120 cycles. The brilliant Li storage property of these electrodes was attributed to (i) porous structure of carbon aerogel matrix that enabled fast charge flow and could accommodate the volume expansion during cycling process and (ii) synergistic effect between MoO_2 and carbon aerogel matrix.

Cheng et al. demonstrated an easy way to synthesize interconnected carbon nanoflakes-incorporated MnO nanoparticles (NPs) by biomass soaking method using the auricularia as a biomass precursor [30]. Since auricularia possess excellent swelling effects and the complexation with metal ions and chitin, guest-active molecules could easily be incorporated. Hence, they obtained MnO NPs immobilized in the 3D carbon nanoflake networks (MnO@C hybrid) using ammonia as precipitation agent in metal ions and subsequent carbonization. Thus, prepared MnO@C hybrid electrode, when tested as anode for LIBs, retained capacity of 868 mA h g^{-1} at 0.2 A g^{-1} over 300 cycles and 668 mA h g^{-1} at 1 A g^{-1} over 500 cycles, indicating a high Li storage capacity and superior cycling stability. They reported that this type of cross-linked nanoflake network structure not only benefited the fast ion/electron transportation but also improved the conductive connectivity and lowered the barrier for lithium intercalation. The combination of highly conductive nanoflake network and well-dispersed MnO NPs with porous hybrid structure were attributed to the enhanced electrochemical performance of the hybrid electrode.

Iron-based oxides such as α-Fe_2O_3 and Fe_3O_4 have also been extensively studied as alternate anode materials for LIBs due to their high theoretical capacity of ~1007 mA h g^{-1} (α-Fe_2O_3) and 926 mA h g^{-1} (Fe_3O_4). However, the same drastic volume expansion during cycling process causes severe structural pulverization and also voltage hysteresis along with low conductivity decline in the electrochemical behavior. Therefore, in order to address these issues, nanocomposites with highly conducting carbon were introduced to enhance the electrochemical aspects [31]. The carbon materials derived from biomass could help in preparing green and environmental-friendly nanocomposites with iron oxides. In this scenario, Wu et al. combined carbon with α-Fe_2O_3 to construct a graphitic carbon-encapsulated α-Fe_2O_3 (α-Fe_2O_3@GC) nanocomposite via an "absorption-catalytic graphitization-oxidation" process [31]. Utilizing degreasing cotton as a cost-effective biomass material to derive carbon, they formed onion-like carbon shell-structured graphitic carbon-encapsulated α-Fe_2O_3 which was distributed in a fibrous carbon matrix. The anodic performances of α-Fe_2O_3 @GC was compared with α-Fe_2O_3 electrodes where the former electrode revealed a Coulombic efficiency of 99% along with a reversible capacity of 1070 mA h g^{-1} after 430 cycles. The authors correlated the performance with the well-protected Fe_2O_3 NPs using the graphitic carbon matrix that resulted in the minimal structural damage, agglomeration, and pulverization. Similarly, Huang et al designed and fabricated an amorphous Fe_2O_3-coated carbon (A- Fe_2O_3@C) electrode for LIBs [32]. The amorphous Fe_2O_3 was tightly coated onto the 3D bacterial cellulose-derived carbon nanofibers as a shell in a highly interconnected nanofibrous structure. At a current density of 400 mA g^{-1}, the A-Fe_2O_3@C anode retained a high capacity of ~1135 mA h g^{-1} with a Coulombic efficiency of nearly 100% over 200 cycles.

Li et al. achieved Fe_3O_4 NPs/carbon nanocomposite (Fe_3O_4 NP/C) from $Fe(NO_3)_3{\cdot}9H_2O$ by infiltrating it with the biochar which in turn derived from inner pomelo pericarp, one of the biomass carbon sources [33]. The resulted nanocomposite revealed outstanding cycling performance (1003.3 mA h g^{-1} at 100 mA g^{-1} and 634.6 mA h g^{-1} at 500 mA g^{-1}). They also prepared acid-treated carbon nanocomposite in addition to the pyrolyzed carbon nanocomposite and compared the anodic performances. Although the specific surface area of acid-treated carbon was high (254.8 m^2 g^{-1}), when compared to Fe_3O_4 NP/C (207.6 m^2 g^{-1}), it showed poor cyclic performance. The enhanced performance of pyrolyzed carbon composite was demonstrated by the higher electronic conductivity (due to the presence of mineral substances such as KCl and $CaCO_3$) than the acid-treated carbon.

SnO_2 were also utilized as anode materials in LIBs due to their theoretical capacity around 1490 mA h g^{-1}. In this regard, wheat straw-derived 1D porous carbon-anchored-nanosized amorphous SnO_2 (SnO_2/C) electrodes were obtained by Kong et al., where the porous carbon derived from wheat straw provided a reversed-free expansion of SnO_2/C during the cycling process [34]. At the current density of 80 mA g^{-1}, the electrode exhibited an initial capacity of 517.6 mA h g^{-1} after 100 cycles. The capacity retention of this electrode was compared with pure SnO_2 and graphene-coated SnO_2, where they found the enhanced capacity retention from 15 to 84.1% after 300 cycles. The other metal oxides such as Co_3O_4 and CuO were also been engaged for the preparation of biomass-derived carbon nanocomposite electrodes [17].

8.4.1.3 Metal Sulfides @ Biomass-derived Carbon Nanocomposite Electrodes

Transition metal sulfides also showed profound interest in the energy storage devices because of their high theoretical capacity and superior electrochemical responses, similar to that of transition metal oxides. Hence, the transition metal sulfides such as MoS_2, CoS_2, FeS, FeS_2, and bimetallic sulfides were also been utilized effectively as anode materials in LIBs [17]. However, there are certain discrepancies such as volume expansion, agglomeration, and poor electronic conductivity which could easily be overcome by employing the nanocomposite strategy with carbon materials.

MoS_2 (with the theoretical capacity of 670 mA h g^{-1}) possess two-dimensional-layered structure similar to that of graphene sheets which enable facile intercalation of Li ions [35]. Moreover, it is an attractive candidate upon considering electrochemical performances. But still, there are some practical challenges such as the low electronic conductivity and poor cyclic stability that tend the material to be more tuned. One robust research direction is to combine MoS_2 with high conductive carbon-nanostructured materials that can facilitate the rapid electron transfer. In this direction, Wang et al. used carbonization process to incorporate MoS_2

nanosheets with a few layers into biomass-derived carbon (MoS_2/C) [35]. They also prepared different electrodes from 54 wt% MoS_2/C composite and 79 wt% MoS_2/C composite. However, the 54 wt% MoS_2/C composite was used in the electrochemical measurements using an assembled LIB. It was found at 200 mA g^{-1} that the 54 wt% MoS_2/C anode exhibited an improved specific capacity (707.4 mA h g^{-1}) compared with the commercial MoS_2 (580.2 mA h g^{-1}) and the corresponding hierarchical porous carbon (215.5 mA h g^{-1}).

Ma et al. demonstrated the vertical growth of MoS_2 nanosheets on the inner and outer surfaces of carbonized corn stalks (MoS_2/CCS nanocomposite) by simple hydrothermal reaction [36]. The MoS_2/CCS electrode delivered initial specific capacity of 1409.5 mA h g^{-1} with Coulombic efficiency of 72.06% which are endorsed to the formation of solid–electrolyte interface (SEI) on the electrode surface. Moreover, when tested as anode in LIBs, it showed excellent specific capacity (1230.9 mA h g^{-1} after 250 cycles at 100 mA h g^{-1}) and long-term cycling performance (500 mA h g^{-1} at 5000 mA g^{-1} after 1000 cycles). The presence of more defects and active sites at the surface of MoS_2/CCS composite electrode was responsible for enhanced Li-ion storage.

Iron sulfide (FeS) could also be a potential electrode for LIBs. For instance, Haridas et al. reported a facile and green strategy to prepare FeS NPs within graphitic carbon capsules (FeS@GCC) which are derived from sawdust biomass [37]. This nanocomposite delivered superior discharge capacity of 505 mA h g^{-1} at 1 C rate, even after 100 cycles of lithiation and delithiation. They reported that the graphitic carbon derived from sawdust biomass could serve as a viable route to encounter the volume changes and to confine the polysulfide intermediates.

Apart from the metallic sulfides, bimetallic sulfides were also anchored to biomass-derived carbon to obtain carbon nanocomposites. Thus, prepared electrodes also showed excellent electrochemical performances. For example, Dominguez et al. reported the addition of cellulose fibers to the hydrothermally synthesized bimetallic composite CoMoS followed by the carbonization in order to get CoMoS@C nanocomposite electrode [38]. The performance of carbon-incorporated electrode was compared with bare bimetallic CoMoS (Figure 8.12) where it showed enhanced anodic performance with initial discharge capacity of 1164 mA h g^{-1}. The composite electrode also retained high specific discharge capacity of 715 mA h g^{-1} after 200 cycles. They attributed the enhanced anodic performance of CoMoS@C nanocomposite electrode to the synergistic effects of CoMoS and carbonized cellulose fibers. Table 8.2 outlines the different biomasses utilized to prepare electrodes for LIBs along with the surface area and its electrochemical performance.

8.4.2 Na-Ion Batteries (SIBs)

Although LIBs have several advantages, the availability of lithium resource tends to find appropriate alternatives to LIBs. In this aspect, SIBs are emerged by simply replacing the lithium with that of sodium metal which possesses similar physiochemical properties and electrochemical behavior to that of Li. Moreover, SIBs are considered as a potential alternative to SIBs due to the high abundance of sodium

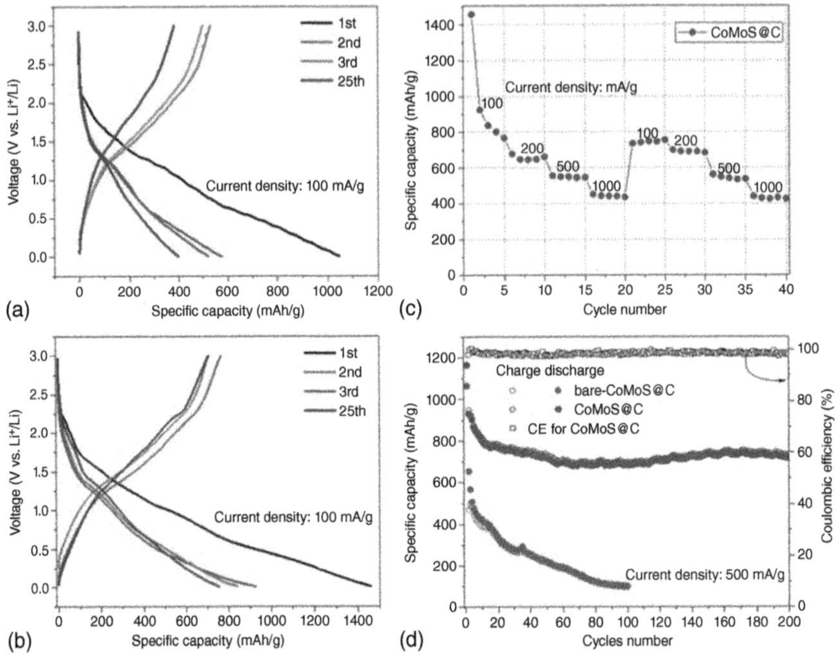

Figure 8.12 Charge/discharge voltage profiles at a current rate of $100\,mA\,g^{-1}$ for (a) bare-CoMoS and (b) CoMoS@C. (c) Rate capability of CoMoS@C from 0.1 to $1\,A\,g^{-1}$. (d) Cycling performance of CoMoS and CoMoS@C at a current rate of $500\,mA\,g^{-1}$ and Coulombic efficiency (CE) for CoMoS@C. Source: Ref. [38] / with permission of American Chemical Society.

resource. It is well noted that the working principle of SIBs are also similar to that of LIBs as discussed in Section 8.2.3. Hence, the graphite anode could also be used as anode in SIBs. However, the difference in ionic radii of Na^+ ions (1.02 Å) and Li^+ ions (0.76 Å) restricts the utilization of common electrodes of LIBs in SIBs. The energy density is also very lower when compared with LIBs due to the difference in the standard electrode potential of Na (−2.71 V vs. SHE) and Li (−3.02 V vs. SHE). But still, the charge storage capacity is well determined by the structural characteristics of the electrode materials. Therefore, variety of metal oxides, metal sulfides, and different forms of carbonaceous materials are being used as efficient anodes for SIBs. In its extension, the biomass-derived carbon materials hold great potential toward the preparation of green and sustainable anode materials for SIBs, similar to LIBs due to the several advantages as already discussed in Section 8.3. Hence, this particular section outlines the various biomass-derived carbon electrodes as well as nanocomposite electrodes for SIBs along with its electrochemical performance [39–41].

8.4.2.1 Biomass-derived Undoped Carbon Electrodes

To start with, Ding et al. in 2013 utilized peat moss as biomass precursor to obtain disordered carbon nanosheet frameworks with abundant micro-/mesopores by the pyrolysis process [42], and the corresponding interlayer distance of the attained carbon is 0.388 nm, which is larger than that of conventional graphite (0.335 nm).

Table 8.2 Various biomass-derived carbon-based electrodes for LIBs and their electrochemical performances.

Biomass	Method	End product	Surface area	Specific discharge capacity	Current density	Coulombic efficiency (%)	Year/ Reference
Spongy pomelo peels	Direct pyrolysis	Carbon powder	$114\ m^2\ g^{-1}$	$452\ mA\ h\ g^{-1}$ after 200 cycles	$90\ mA\ g^{-1}$	99.5	2013/[18]
Hazelnut shells	Hydrothermal carbonization	Nanoporous carbon	$150\ m^2\ g^{-1}$	$307\ mA\ h\ g^{-1}$ after 100 Cycles	1 C	99.35	2013/[19]
Rice husk (0.3 M NaOH)	Pyrolysis of acid as well as base-treated rice husk	Pyrolytic carbons	$371\ m^2\ g^{-1}$	$819\ mA\ h\ g^{-1}$ after 10 cycles	–	~100	2013/[20]
Rice husk	Hydrothermal carbonization followed by calcination	Porous fibrous carbon	$243\ m^2\ g^{-1}$	$403\ mA\ h\ g^{-1}$ after 100 cycles	$75\ mA\ g^{-1}$	~97	2013/[21]
Bamboo chopsticks	Controlled hydrothermal process	Uniform graphitic carbon fibers	–	$710\ mA\ h\ g^{-1}$ after 300 cycles	$0.2\ A\ g^{-1}$	93.8	2014/[23]
Rice husk	Pyrolysis followed by base treatment	Disordered carbon	$351.81\ m^2\ g^{-1}$	$502\ mA\ h\ g^{-1}$ after 100 cycles	0.2 C	99	2015/[22]
Sisal fibers	Pyrolysis followed by hydrothermal activation	Amorphous carbon	$616.4\ m^2\ g^{-1}$	$646\ mA\ h\ g^{-1}$	–	~100	2015/[24]
Loblolly pine woodchips	KOH activation followed by pyrolysis	Porous amorphous carbon	$1580\ m^2\ g^{-1}$	$650\ mA\ h\ g^{-1}$ after 250 cycles	0.2 C	99.6	2016/[25]
Coir pith	Carbonization followed by KOH activation	Microporous carbon	$2500\ m^2\ g^{-1}$	$837\ mA\ h\ g^{-1}$ after 50 cycles	$75\ mA\ g^{-1}$	–	2017/[26]
Bamboo woods	Carbonization, pulverization, and graphitization	Carbonaceous materials	$>200\ m^2\ g^{-1}$	200–260 $mA\ h\ g^{-1}$	0.1 C	40–70	2018/[27]
Corn stalks	Carbonization and $CaCl_2$ activation	Porous carbon	$370.6\ m^2\ g^{-1}$	$783.8\ mA\ h\ g^{-1}$ after 100 cycles	0.2 C	~100	2018/[28]
Degreasing cotton	In situ low-temperature catalytic graphitization	Graphitic carbon-encapsulated $\alpha-Fe_2O_3$ nanocomposite	–	$1070\ mA\ h\ g^{-1}$ after 430 cycles	0.2 C	99	2015/[31]
Seaweed biomass (sodium alginate)	Hydrogel matrix	MoO_2/carbon aerogels	$196.7\ m^2\ g^{-1}$	$490\ mA\ h\ g^{-1}$ after 120 cycles	$200\ mA\ g^{-1}$	72	2016/[29]

(Continued)

Table 8.2 (Continued)

Biomass	Method	End product	Surface area	Specific discharge capacity	Current density	Coulombic efficiency (%)	Year/ Reference
Auricularia	Biomass soaking method	Interconnected MnO_2@C nanoflake networks	$122\ m^2\ g^{-1}$	$668\ mA\ h\ g^{-1}$ after 500 cycles	$1\ A\ g^{-1}$	–	2016/[30]
Bacterial cellulose	3D carbonization and in situ thermal decomposition method	Amorphous Fe_2O_3 @ 3D-carbonized bacterial cellulose (A-Fe_2O_3 @CBC)	$327\ m^2\ g^{-1}$	$1135\ mA\ h\ g^{-1}$ after 200 cycles	$400\ mA\ g^{-1}$	~100	2016/[32]
Inner pomelo pericarp	Carbonization followed by mixing with iron source solution	Fe_3O_4 nanoparticles decorated with carbon (Fe_3O_4 NP @C)	$254.8\ m^2\ g^{-1}$	$1003.3\ mA\ h\ g^{-1}$ after 200 cycles	$100\ mA\ g^{-1}$	99	2016/[33]
Wheat straw	Flow process	Nanosized SnO_2 particles anchored on wheat straw (WS) carbon (SnO_2/C)	–	$441\ mA\ h\ g^{-1}$ after 100 cycles	0.05 C	~100	2018/[34]
Auricularia	Carbonization and reduction	Mo_2S nanosheets-incorporated hierarchical porous carbon frameworks (Mo2S@C hybrid)	–	$707.4\ mA\ h\ g^{-1}$ after 500 cycles	$200\ mA\ g^{-1}$	~100	2015/[35]
Corn stalks	Hydrothermal reaction	Mo_2S nanosheets @carbonized corn stalks	–	$1230.9\ mA\ h\ g^{-1}$ after 250 cycles	$100\ mA\ g^{-1}$	–	2018/[36]
Sawdust	Carbonization	Iron carbide-embedded graphitic carbon capsule composite (FeS@GCC)	$109.3\ m^2\ g^{-1}$	$505\ mA\ h\ g^{-1}$ after 100 cycles	1 C	>98	2019/[37]
Cellulose fibers	Hydrothermal synthesis of CoMoS followed by carbonization of cellulose fibers	Bimetallic CoMoS anchored to carbonized cellulose fibers (CoMoS@ C)	–	$715\ mA\ h\ g^{-1}$ after 200 cycles	$500\ mA\ g^{-1}$	~60	2018/[38]

Note: 1 C – $372\ mA\ g^{-1}$

When tested at the current density of 50 mA g^{-1}, its initial sodiation and desodiation capacities were 532 and 306 mA h g^{-1}, respectively, and the reversible capacity was 298 mA h g^{-1} after 10 cycles, then the current density increased to 100 mA g^{-1}, and the reversible capacity can be kept at 255 mA h g^{-1} after 200 cycles. Lotfabad et al. also prepared amorphous carbon with low specific surface area by carbonizing banana peels [43], which delivered a reversible capacity of 336 mA h g^{-1} at the current density of 100 mA g^{-1}, and only a capacity loss of 11% was observed after 300 cycles. At a higher current density of 500 mA g^{-1}, the reversible capacity is maintained at 221 mA h g^{-1}, and only 7% of the capacity was lost after 600 cycles.

Lv et al. investigated the peanut shell-derived porous hard carbons obtained by pyrolysis method as anode materials for both LIBs and SIBs [44]. The hard carbon prepared at 600 °C pyrolytic temperatures showed best electrochemical performance, and they also activated in KOH solution which also showed enhanced Li/Na storages. They observed retention capacity of 193 mA h g^{-1} after 400 cycles with improved cyclic stability where they attributed the performance to the porous structure with larger specific surface area (706.1 m^2 g^{-1}). Elizabeth et al. converted prawn shells into hierarchical (macro-, meso-, and micro-) porous carbons using an economically viable process such as carbonization followed by an activation using 0.1 M NaOH solution [45]. Thus, derived N-doped porous carbons were tested as anode materials for both LiBs and NIBs and obtained excellent electrochemical performances. It exhibited stable specific capacity of 325 mA h g^{-1} at 0.1 A g^{-1} for 200 cycles and 234 mA h g^{-1} at 0.4 A g^{-1} for 150 cycles. They also related the performance to the presence of hierarchical (macro-, meso-, and micro-) porous structure (Figure 8.13) and N-doping.

The CPC which was utilized as anode materials for LIBs by Kalaiselvi and team members were being extended to NIBs as well as capacitors [46]. The CPCs were activated at different temperatures and found an optimal temperature of 850 °C which then exhibited superior electrochemical performance, when being tested as anode for SIBs. An initial capacity of 280 mA h g^{-1} and a progressive capacity of 220 mA h g^{-1} up to 300 cycles with negligible capacity fading have been observed at 50 mA g^{-1}.

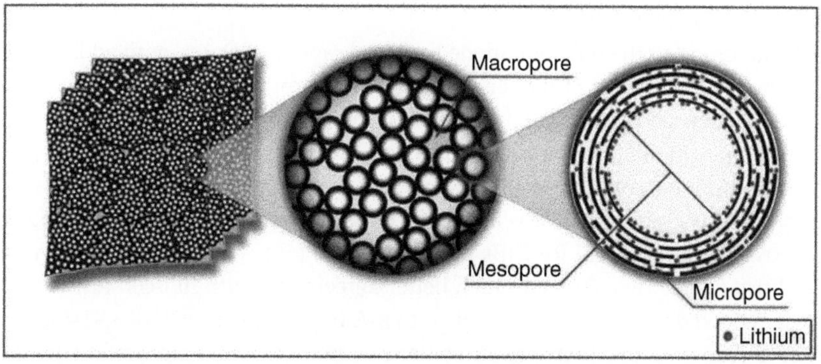

Figure 8.13 Illustration of mechanism of Na/Li insertion in hierarchical porous carbon derived from prawn shell. Source: Ref. [45] / with permission of Elsevier.

The sodium storage performances of various biomass-derived hard carbon materials were investigated by Hu and his co-workers. Hu et al. developed monodisperse hard carbon spherules (HCS) from sucrose [47]. They found that the HCS coated with a soft carbon layer provided Initial coulombic efficiency (ICE) of 83%. In an another work, they attained hard carbon microtubes (HCT) from the natural cotton pyrolyzation at different temperatures such as 1000, 1300, and 1600°C where they studied the influence of carbonization temperature on the electrochemical properties of HCT [48]. They also evaluated the sodium storage performances for the developed respective electrodes and found that the HCT electrode prepared at 1000°C showed a low reversible specific capacity of 88 mA h g^{-1} due to the lacking active sites. However, the HCT electrodes prepared at 1300°C presented the outstanding electrochemical behavior toward SIBs with the reversible capacity of 300 mA h g^{-1} at the current density of 30 mA g^{-1}, 83% ICE, and excellent cycle stability when compared to the other electrodes. Moreover, the same group was also investigated the sodium storage properties of hard carbon obtained from calcination of corn cobs at various calcination temperatures [49]. Similar to the previous work, the hard carbon prepared from corn cobs at the calcination temperature of 1300°C exhibited good reversible capacity of 298 mA h g^{-1} with good cyclic stability as well as high (97%) capacity retention after 100 cycles.

The biomass apple, coconut oil, and oatmeal were also utilized as efficient precursors to produce hard carbon material by various researchers and then examined their electrochemical behavior toward successful electrode for SIBs. Wu et al reported one such work by employing the discarded apple biomass and found that the elemental nitrogen and sulfur were also present due to the natural proteins in the apple [50]. Thus, obtained biomass-derived electrode showed capacities of 245 mA h g^{-1} at 20 mA g^{-1} and 112 mA h g^{-1} at 1000 mA g^{-1} along with long-term cycling stability up to 1000 cycles. Utilizing coconut oil as biomass precursor, Zhao et al. developed hard carbon NPs using a flame deposition method [51]. This electrode was tested as anode material for SIBs and found that it delivered a sodiation capacity of 277 mA h g^{-1} at the current density of 100 mA g^{-1}. The biomass oatmeal was also been used to produce N-doped carbon microspheres (NCSs) and tested as electrode for SIBs. The NCSs electrode exhibited 336 mA h g^{-1} capacity after 50 cycles [52].

Cao et al. reported the efficient anode based on rape seed shuck-derived lamellar hard carbon which was synthesized through the hydrothermal and pyrolysis processes [53]. They also investigated the effect of pyrolysis temperature (500–800 °C) and found that the material processed at 700 °C showed best electrochemical performance with long cycle life (143 mA h g^{-1} at current density of 100 mA g^{-1} after 200 cycles). Macadamia shell was also used to derive hard carbon and reported as an effective anode material for SIBs by Zheng et al. [54]. They performed revised half-cell test (RHT) that showed much better agreements with full-cell test results where it delivered a specific capacity of ~314 mA h g^{-1}, with an ICE of ~91.4%.

Yu et al. reported the preparation of hard carbon from the old loofah sponge through carbonization process at 800 °C for 1 hour and tested as anode for SIBs [55]. They obtained the initial discharge specific capacity of about 695 mA h g^{-1}

at 25 mA g^{-1} along with the reversible discharge specific capability of about 171 mA h g^{-1} after 1000 cycles. The observed long cycle stability indicated the promising feasibility of the old-loofah-derived hard carbon anode where the disordered microstructure and large interlayer distance are responsible for the improved performance. Zhu et al. [56] reported an effective route to obtain hard carbons from waste apricot shell. The typical processes include pyrolysis and reductive strategy, and the resulted hard carbons delivered a large interlayer spacing with well-connected structure which are beneficial for Na^+ intercalation and transport. The respective electrode exhibited reversible capacity of ca. 400 mA h g^{-1} with ICE of 79%. Moreover, certain other biomass sources such as Indonesian snake fruit peel [57], sepals of Palmyra palm fruit calyx [58], cashewnut sheath [59], and caltrop shell [60] were also used to prepare hard carbons and being tested as anode materials for SIBs. Table 8.3 shows the various biomass sources utilized to prepare electrode materials for SIBs along with the surface area and its electrochemical performances.

8.4.3 Li-S batteries

Li-S batteries are also considered as potential energy storage devices due to their high theoretical energy density of 2600 Wh kg^{-1} and high theoretical storage capacity of 1675 mA h g^{-1}. However, the practical applications of these kind of batteries are also inadequate because of the inherent problems such as low electrical conductivity of sulfur (5 × 10 g–30 S cm^{-1} at 25 °C), discharge products (Li_2S/Li_2S_2), dissolution of polysulfides in the electrolyte leading to the shuttling effect, and large volume expansion of sulfur electrode during cycling [61, 62]. Hence, intensive researches to develop novel strategies are in urge to address these challenges. One such strategy to suppress polysulfide shuttling is to incorporate porous carbon matrices with different textures and morphologies which are able to physically trap the polysulfides in the porous matrix [63]. Moreover, this strategy improves the electronic conductivity of sulfur electrode and also accommodates the volume expansion during cycling. Porous carbon materials possessing high specific surface area and excellent chemical stability are ideal host matrices for elemental sulfur. The other strategies to increase the electrochemical stability are the development of multifunctional separators, introduction of solid state electrolytes, and biopolymer binders for sulfur electrodes.

Although many of the active materials have been immensely studied, bio-based materials are more attractive toward the development of greener and sustainable devices. Hence, the exploration of biomass materials to construct key components for Li–S batteries would be worthwhile considering the potential scale of electrochemical applications. Biomass-derived carbons could serve as promising porous host materials for elemental sulfur in Li-S batteries due to their special microstructure, elemental composition, and low cost (Figure 8.14). Hence, this section provides a comprehensive summary of the carbon hosts derived from biomass employed in Li-S batteries in order to boost the electrochemical behavior of S electrode.

Table 8.3 Various biomass-derived carbon-based electrodes for SIBs and its electrochemical performance.

Biomass	Method	End product	Surface area	Specific discharge capacity	Current density	Coulombic efficiency (%)	Year/ Reference
Peat moss	Carbonization followed by air activation	3D macroporous-interconnected networks of carbon nanosheets	$196.6\ m^2\ g^{-1}$	$298\ mA\ h\ g^{-1}$ after 10 cycles	$50\ mA\ g^{-1}$	~100	2013/[42]
Banana peels	Carbonization followed by air activation	Banana peel pseudo-graphite (BPPG)	$19–217\ m^2\ g^{-1}$	$355\ mA\ h\ g^{-1}$ after 10 cycles	$50\ mA\ g^{-1}$	~100	2014/[43]
Peanut shells	Pyrolysis	Porous hard carbon	$706.1\ m^2\ g^{-1}$	$193\ mA\ h\ g^{-1}$ after 400 cycles	$0.25\ A\ g^{-1}$	~100	2015/[44]
Prawn shells	Carbonization followed by NaOH activation	N-doped hierarchical (macro-meso-micro) porous carbon	$336\ m^2\ g^{-1}$	$325\ mA\ h\ g^{-1}$ after 200 cycles	$0.1\ A\ g^{-1}$	99.9	2016/[45]
Coir pith	Carbonization and KOH activation	Coir pith-derived carbon (CPC)	$2500\ m^2\ g^{-1}$	$220\ mA\ h\ g^{-1}$ after 300 cycles	$50\ mA\ g^{-1}$	>95	2017/[46]
Sucrose	Carbonization	Amorphous monodispersed hard carbon microspheres	–	$290\ mA\ h\ g^{-1}$ after 100 cycles	$30\ mA\ g^{-1}$	83 (ICE)	2014/[47]
Natural cotton	Carbonization	Hard carbon microtubes	$14–538\ m^2\ g^{-1}$	$305\ mA\ h\ g^{-1}$ after 100 cycles	$30\ mA\ g^{-1}$	83 (ICE)	2016/[48]
Corn cobs	Carbonization	Hard carbon	$1.431–5.485\ m^2\ g^{-1}$	$275\ mA\ h\ g^{-1}$ after 100 cycles	0.2 C	99.69	2016/[49]
Apple biowaste	Two-step dehydration followed by heat treatment	Hard carbon	$196.3\ m^2\ g^{-1}$	$245\ mA\ h\ g^{-1}$ after 80 cycles	0.1 C	99.1%	2016/[50]

Coconut oil	Flame deposition method	Hard carbon nanoparticles and carboxylated carbon nanoparticles	$133\ m^2\ g^{-1}$	$206\ mA\ h\ g^{-1}$ after 20 cycles	$100\ mA\ g^{-1}$	96%	2016/[51]
Oatmeal	Hydrothermal process	N-doped carbon microspheres (NCS)	$110.7\ m^2\ g^{-1}$	$336\ mA\ h\ g^{-1}$ after 50 cycles	$50\ mA\ g^{-1}$	–	2016/[52]
Rape seed shuck	Hydrothermal and pyrolysis	Lamellar hard carbon	$11.09–360.62\ m^2\ g^{-1}$	$143\ mA\ h\ g^{-1}$ after 200 cycles	$100\ mA\ g^{-1}$	~100%	2017/[53]
Macadamia nut shell	Direct pyrolysis	Hard carbon	$0.838 – 469.0\ m^2\ g^{-1}$	$252\ mA\ h\ g^{-1}$ after 1300 cycles	0.1 C	91.4 (ICE)	2017/[54]
Old loofah sponge	Carbonization	Disordered hard carbon	-	$171\ mA\ h\ g^{-1}$ after 1000 cycles	$1000\ mA\ g^{-1}$	–	2017/[55]
Waste apricot shell	Pyrolysis and reductive strategy	Hard carbon	$2.7–56.7\ m^2\ g^{-1}$	$250\ mA\ h\ g^{-1}$ after 500 cycles	1 C	79 (ICE)	2018/[56]
Indonesian snake fruit peel	Pre-carbonization followed by KOH activation	Activated porous carbons	$845–1849\ m^2\ g^{-1}$	$255\ mA\ h\ g^{-1}$ after 100 cycles	$100\ mA\ g^{-1}$	~100	2018/[57]
Sepals of Palmyra palm	Carbonization	Hard carbon	$20–532\ m^2\ g^{-1}$	$173\ mA\ h\ g^{-1}$ after 100 cycles	$200\ mA\ g^{-1}$	99	2019/[58]
Cashewnut sheath	Carbonization followed by KOH activation	Mesoporous carbon	$1967\ m^2\ g^{-1}$	$200\ mA\ h\ g^{-1}$ after 100 cycles	$100\ mA\ g^{-1}$	99	2019/[59]
Caltrop shell	Pyrolysis followed by acid treatment	Hard carbon	$7.39–48.09\ m^2\ g^{-1}$	$288.1\ mA\ h\ g^{-1}$ after 200 cycles	0.4 C	84.09 (ICE)	2019/[60]

Note: 1 C – $372\ mA\ g^{-1}$
ICE – Initial Coulombic Efficiency.

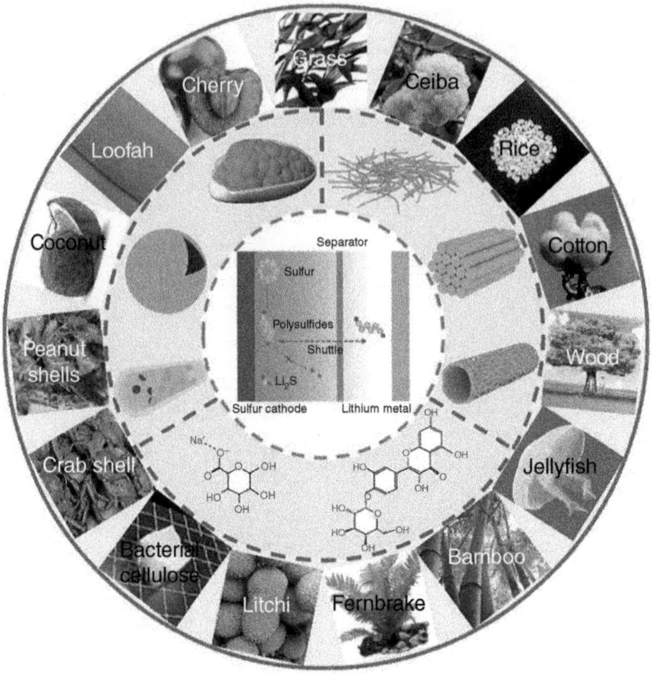

Figure 8.14 Various biomass-derived carbon utilized for Li-S batteries. Source: Ref. [61] / with permission of Royal Society of Chemistry.

8.4.3.1 Biomass-derived Carbon Hosts

Porous carbon with large surface area, excellent electronic conductivity, and hierarchical structures holds great potential to host sulfur. Hence, the hierarchical carbon–sulfur composites are the most favorable cathodes for Li-S batteries. Biomass-derived hierarchical carbon–sulfur composites are more advantageous because of the improved electrochemical behavior of the electrodes. The hierarchical porous carbon was obtained from one of the promising biomass material, pig bone via KOH activation method, and the effects of activation temperature on the pore size and volume were also investigated by Wei and his co-workers [64]. They prepared four different porous carbon by varying the activation temperatures such as 650, 750, 850, and 950 °C. Among these four carbons, the pig bone-derived hierarchical porous carbon activated at 850 °C got higher BET surface area and pore volume (2157 $m^2\ g^{-1}$ and 2.26 $cm^3\ g^{-1}$), and hence the composite electrode was made out of this carbon using gelatin as binder. Thus, prepared electrode when tested as cathode in Li-S batteries displayed a high initial capacity of 1265 $mA\ h\ g^{-1}$ and retained 643 $mA\ h\ g^{-1}$ after 50 cycles. They attributed the enhancement to the automatic adsorption of polysulfides/sulfur onto the porous carbon during the cycling process because of the high specific surface area and facile transfer of Li ions.

Zhao et al. obtained porous carbon derived from fish scales and compared the performance of sulfur cathode with porous carbon/sulfur composite cathode [65]. They ascribed the incorporation of sulfur into micropores through the capillary action. Moreover, they also demonstrated that the presence of void in the nanocomposite could accommodate the volume expansion of sulfur generated during the electrochemical reaction, reduce lithium polysulfide dissolution, and tremendously alleviate the shuttle effect. The nanocomposite with 58.8% sulfur achieved a higher initial capacity (1039 mA h g^{-1}) than that with 66.7% sulfur. And even after 70 cycles, the discharge capacity of the nanocomposite with 58.8% remained 1012 mA h g^{-1}, which is much higher than that of the nanocomposite with 66.7% (605 mA h g^{-1}).

Wang et al. derived hierarchical (micro-macro) porous carbon from one of the common biomasses, cotton, and reported the facile approach for the development of improved confinement matrix to host sulfur for Li-S batteries [66]. The chemically activated porous carbon exhibited specific surface area of 1286 m^2 g^{-1}, and as cathode, it showed good cycling stability (reversible capacity of 60 mA h g^{-1} after 200 cycles at the 0.2 C current rate). The macroporous structure not only provided channels for sulfur to be loaded into the micropores but also enhanced the rapid transfer of electrons/ions. In another work reported by Tao et al., carbon nanotiles were obtained from kapok fibers (KFCNTs) and were utilized as an efficient host to accommodate sulfur [67]. The resulted KFCNTs/S cathode exhibited high gravimetric capacity of 524 mA h g^{-1} with superior capacity retention of up to 95.4% after 90 cycles. They proposed that KFCNTs with scale-like microstructure could render sulfur particles electrically conducting, increase the kinetic inhibition of polysulfides to diffusion within the framework, and accommodate the volume expansion of sulfur during discharging.

Kalaiselvi and team members were working on different biomass materials to derive undoped carbon and doped carbon nanostructures for its effective utilization in RBs. Although some of them are already discussed in LIBs and SIBs, they also concentrated on the improvement of Li S batteries by employing the porous carbon matrices derived from various biomasses and also reported in different timelines. For instance, Balakumar et al. obtained porous carbon derived from coir pith (CPC) which was then utilized as scaffold for sulfur cathode [68]. This scaffold could accommodate 70 wt% of sulfur within the pores of biocarbon since the surface area is very high (1952 m^2 g^{-1}). However, the 50 wt% sulfur-loaded CPC exhibited a progressive capacity of 695 mA h g^{-1} with 95% Coulombic efficiency. The same team was utilized jamun seed as biomass precursor in the year 2018 in order to get porous carbon [69]. The jamun seed-derived carbon (JSC) possessed appreciable conductivity thereby enhancing the lithium polysulfide adsorption sites (Figure 8.15). The various sulfur-loaded JSC electrodes (40, 50, and 60 wt%) were prepared to act as an interlayer and scaffold for sulfur.

8.4.4 Zn-Air Batteries

In the future, metal-based ORR electrocatalysts must face the environmental problem from an ecological point of view once the deactivation is complete [70]. Thus, massive toxic wastes in the forms of oxides, hydroxides, or ions that occur in the widespread application that does not comply with the basic

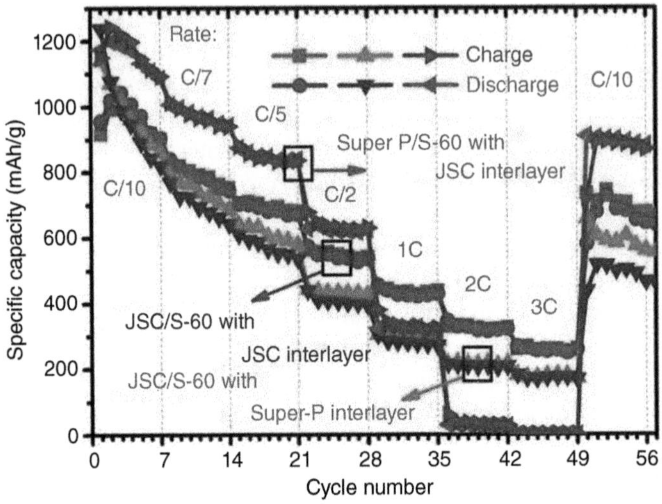

Figure 8.15 Rate capability of JSC/S-60 cathode with JSC/Super-P carbon interlayer and super-P carbon/S-60 cathode with JSC interlayer. Source: Ref. [69] / with permission of Elsevier.

principle of green chemistry. Using safe, abundant, low-cost, and quick-generation heteroatomic-doped organic carbon can be used as an alternative to ORR electrocatalysts based on graphene. Recently Li-Yuan Zhang et al. [71] developed a fast catalytic carbonization at a relatively low temperature for the obtainment of a microporous carbon as metal-free ORR catalyst precursor with biomass waste as a source of carbon. For current syntheses and experiments, the banana peel was widely used as the model biomass waste and was chosen as carbon source. The new approach consists of carbonization for 10 minutes and cooling for 15 minutes. This new approach consists of carbonization for 10 minutes and cooling for 15 minutes. It is fast, cost-effective, and easy to use with high C conversion efficiency on a large scale.

In 2016, Na Ma et al. [72] reported a facile, inexpensive, and scalable approach to prepare a new class of three-dimensional (3D) hybrid nanoaerogels composed of Ni/NiO/$NiCo_2O_4$ NPs and N-doped carbon nanotube aerogels (N-CNT-As) using alginate-derived biomass conversion strategy. In 2015, Shuyan Gao et al. presented a simple and economic strategy for the preparation of porous heteroatoms-rich carbon (PHC) derived from honeysuckles [73]. Comparing with the commercial Pt/C catalyst, the products demonstrate excellent catalytic efficiency, good stability, and high methanol tolerance. By using fermented rice as a starting material, Shuyan Gao in 2014 [74] demonstrated a simple, green, and scalable approach for the synthesis of porous N-doped carbon spheres characterized by high specific surfaces (2105.9 $m^2\ g^{-1}$) and high porosity (114 $cm^3\ g^{-1}$) that not only have an excellent electrocatalytic activity in relation to the four-electron reduction reaction with long-term fuel cell stability but also have excellent resistance to crossover effects and CO poisoning superior to that of the commercial Pt/C catalyst. In 2013, Shuyan Gao et al [75] reported the one-stone-two-bird strategy for the N-doped-carbon synthesizing via hydrothermal treatment instead of common N-doping techniques for high-energy carbon materials and expensiveness, toxic or flammable chemical substances, which requires expensive hardware and multiple-stage processes.

A modern waste-to-resource approach to convert the fallen ginkgo leaves are provided in this new form of ORR electrocatalyst (Shuyan Gao et al. [76]), and a nitrogen-doped fullerene-like carbon shell (NDCS) is presented. N is derived from fallen ginkgo leaves, where protein is 10.9–15.5% weight size. The NDCS obtained has 100% catalytic selectivity on the four-electron pathway, with the most important known carbon-dependent catalysts outperforming ORR activities. In 2015, Shuyan Gao et al [77] reported that NDCs with a high surface area were synthesized from amaranths waste, which is a natural, abundantly available, and constantly renewable source and acts as a single precursor for nitrogen as well as carbon. The technique is simple, but it is extremely effective and offers highly porous conductive doped carbon, leading to superior reduction activity in electrocatalytic oxygen and a valuable material for fuel cell and metal-air battery technologies. Chengzhou Zhu in 2012 [78] synthesized bifunctional fluorescent carbon nanodots with soy milk (simple and green), not only with favorable photoluminescence but also with strong electrocatalytic activity in order to minimize the oxygen content in the process. The biomass-derived, self-doped porous carbon nitrogen was synthesized with an easy procedure based on simple hyacinth pyrolysis (eichhornia crassipes) with $ZnCl_2$ as an activation reagent at regulated temperatures (600–800 °C) (Xiaojun Liu in 2015) [79]. Electrochemical experiments have shown that the nitrogen self-doped carbons have a high electrocatalytic activity in alkaline media for ORR, very close to the 20% marketable Pt/C. Chaozhong Guo et al. [80] proposed a new approach to develop a nitrogen-doped carbon nanomaterial to be used as a metal-free ORR catalyst based on simple pyrolysis of enoki mushroom (Flammulina velutipes) at biomass 900 °C with CNTs as conductive agent and matrix of insertion.

8.5 Biomass-derived Heteroatom-Doped Carbon Electrodes for RBs

As mentioned earlier, LIBs and NIBs play a vital role in electronic industries and electric vehicles because they are abundant in earth. In these RBs, carbon plays significant achievements as electrode materials. The different forms of carbon materials derived from biomass were used as electrode materials in commercialized RBs due to the astonishing advantages such as high safety, excellent cycling stability, low cost, high abundance, nontoxicity, stability, and durability. However, it is very imperative to do modification in the electrode materials via doping, composite, or compositional gradient films in order to enhance the battery performances. Although, the emerging electrode materials with special structures will enhance the electrochemical behavior of the batteries with high capacities and energy densities, some materials resulted in very poor cycle stability. Hence, in addition to carbonaceous, biomass-derived carbon, doped carbon materials are also very promising candidate for electrode materials in batteries [81, 82].

In 1980, carbonaceous materials were initially investigated and considered as promising candidates for the anode materials in LIBs, because intercalation of graphite compound (LiC_6) has high capacity (372 mA h g^{-1}) [40]. Numerous materials were used as electrode for batteries such as metal/alloys, oxides, chalcogenides, phosphides and carbonaceous materials. But in other metal ions, storage performance of conventional graphite is quite poor. In light of the development of LIBs electrode materials, a number of micro-/nanostructured hard carbon materials have been exploited due to the kinetically favorable transportation of ion and electron, such as, hollow nanostructured carbon materials [83], carbon nanomaterials [84], CNT, porous carbon materials [85–87], carbon nanosheets [88], and so on. However, the heteroatom doping strategy could help in boosting the electrochemical performances. Hence, the heteroatoms such as nitrogen, sulfur, and other heteroatoms could be doped into the carbon matrices for preparing various doped electrodes which could easily intercalate with ions and low potential region (flat potential plateau). Via doping or heteroatom doping, the following properties could be tailored: the nanostructures, crystalline nature, thermal, chemical, and electronic properties of pristine materials. Heteroatom doping is mostly used to improve the storage properties (Li, Na, K, etc.,). Moreover, multiple heteroatoms (dual or ternary doping) are also incorporated into a host matrix to enhance the electrochemical performance for electrode materials.

In this section, we are specifically focusing on biomass-derived carbon and its doped systems for batteries and more looked into electrode materials. The various types (source, size, and morphology) of biomass-derived carbons are beneficial for charge and ion transfer, but complete understanding of their structure, properties, and type of doping is still lacking. Tuning the hierarchical porous nature of biomass-derived carbon by simply tuning the synthetic conditions could help in achieving the excellent electrochemical performances of the respective electrode materials. Porous structures can reduce the ion diffusion length and provide large electrode–electrolyte interface to improve the electrochemical performance. Moreover, doping enhances the chemically active sites for ion adsorption to enhance the ion storage capacity.

8.5.1 Single-Heteroatom-Doped Carbon Electrodes

N-doping is an effective way to improve the electrochemical behavior of carbon materials [89]. The major concepts involving via doping are as follows: (i) enhancement of the electronic conductivity, creation of holes, and minimal agglomeration; (ii) high electronegativity difference (N-3.04 C-2.55). Therefore, it will induce polarizations of carbon chain, and this will enhance the activation cites to vary the electrochemical performance; and (iii) N atoms are covalently bonded to the carbon network leading to the stable structure. Numerous methods are involved to incorporate nitrogen in the carbon materials or lattice such as Chemical vapor deposition (CVD), hydrothermal, pyrolysis, and ball milling. Thangaian Kesavan et al. [90] developed a nitrogen-doped carbon nanosheet for high-performance electrode in batteries. The developed electrode delivered an inspiring specific capacitance of 268 and 218 $F \cdot g^{-1}$ at an applied current of 1 and 5 $A \cdot g^{-1}$, respectively. In 2015, Jianhua Hou et al. [91] developed 2D hierarchical porous carbon nanosheets (HPNC-NS) with the thickness of 15–30 nm and used as the promising electrode for LIBs. The initial reversible capacity (1913 $mA\,h\,g^{-1}$) is five times than the theoretical capacity of graphite. They achieved more than ~95% Coulombic efficiency for HPNC-NS and the reversible capacity at various cycles, indicating that HPNC-NS delivered high durability. N-doped carbon nanosheet provides more active sites for Li ion adsorption and insertion/extraction on surfaces/defects.

Lu Guan et al. [92] innovatively developed N-doped carbon nanosheet using pine nut shells because it is an abundant source. Maria K. Rybarczyk et al. [87] synthesized a micro-/mesoporous structure of carbon materials from natural RH and evaluated its properties for batteries. They achieved high capacity of 1032 $mA\,h\,g^{-1}$ RHC/S at a current density of 0.1 C. It is yet another interesting work reported by Saravanan et al. where they explored an economically viable and an environmentally benign approach to prepare nitrogen-containing porous carbon derived from the human hair (HHC) [93]. The mesoporous HHC was demonstrated as a potential anode for LIBs that delivered a steady state reversible capacity of 700 and 610 $mA\,h\,g^{-1}$ at 50 (0.13 C) and 100 $mA\,g^{-1}$ (0.26 C), respectively. The superior performance was ascribed to the high surface area 1617 m^2g^{-1}, mesoporous structure with interconnected graphene sheets, and the presence of heteroatom (N) of HHC electrode (Figures 8.16 and 8.17).

Elemental sulfur (S) is one of the crystalline materials at room temperature with high abundance. S-doped carbons also play a vital role, since they have stronger electron donor ability and more effective reduction. S-doping is performed in the normal ways, pyrolysis process, heating of sulfur and carbon source (bio-derived carbon), and also sulfurization. Zhao-Hui Chen et al. [95] reported that hierarchical

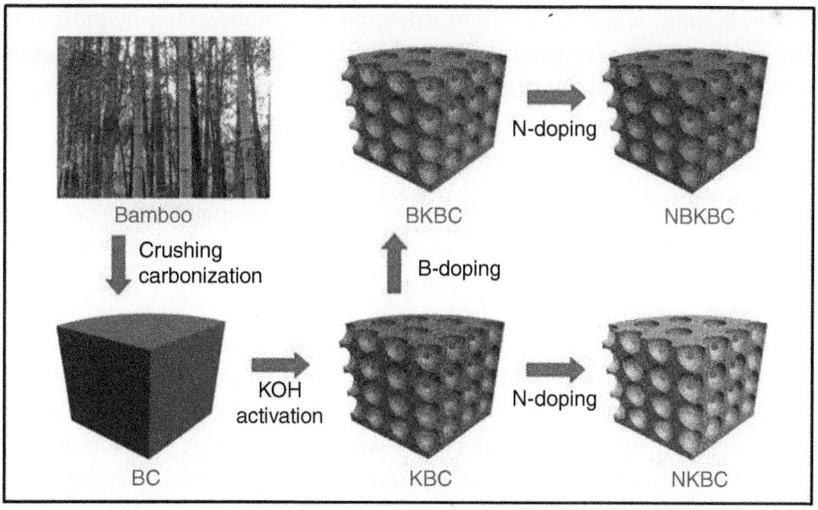

Figure 8.16 The overview of bamboo-derived carbon (BC) and its doping process. Source: Ref. [83] / with permission of Elsevier.

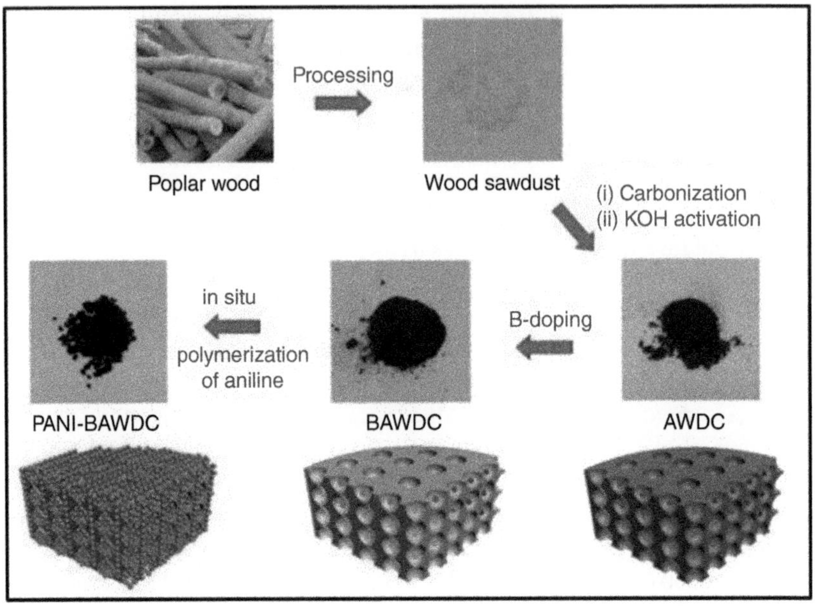

Figure 8.17 Schematic view of B-doped carbon. Source: Ref. [94] / with permission of American Chemical Society.

micro-/mesoporous carbon derived from coconut shell (CSC) could readily act as a potential sulfur host for Li-S batteries due to its high surface area. The sulfur-infiltrated CSC materials show superior discharge–charge capacity, cycling stability, and high-rate capability. High discharge capacities of 1599 and 1500 mA h g^{-1} were achieved at current rates of 0.5 and 2.0 C, respectively (Figure 8.18). The various researchers are discussed about phosphorus (P) doping in carbon, because it will cause more structural distortion, strong hybridization, electronegativity, and also good bonding nature (P-C). The phosphorus doping could be done via pyrolysis process, hydrothermal process, and thermal process [94].

8.5.2 Dual-Heteroatom-Doped Carbon Electrodes

Bongu et al. reported the development of a new nanocomposite electrode from Mn_2O_3 NPs which are wrapped in N and S containing randomly oriented graphene sheet-like carbon derived from the human hair (HHC) [96]. The Mn_2O_3/HHC nanocomposite electrode exhibited high and stable energy storage performance when tested as anode in LIBs. A reversible lithium storage capacity of 990 mA h g^{-1} over 350 cycles at 50 mA g^{-1} and 440 mA h g^{-1} at 2000 mA g^{-1} (2 C) have been exhibited this composite electrode. The enhancement was attributed to the efficient electrical conductivity that facilitates the easy and fast charge transfer, and the presence of heteroatom in HHC could increase chemical reactivity, defect sites, and electronic conductivity.

Similarly, researchers developed boron-doped carbon system to improve its physical and chemical properties. Boron-carbon iteration is very strong so it enhances the

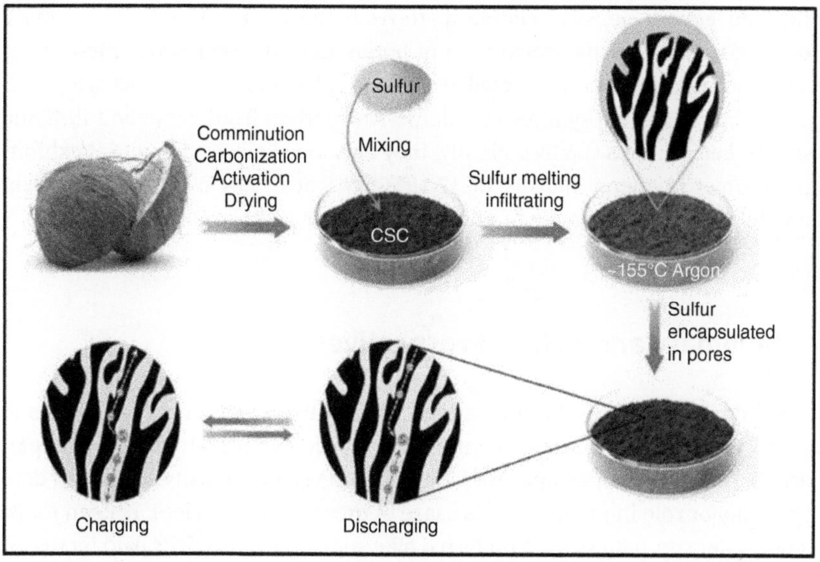

Figure 8.18 Overview of the preparation of S-doped carbon from coconut shell. Source: Ref. [95] / with permission of American Chemical Society.

electrical conductivity, surface area, and specific capacity. The doping concentration can tune the optoelectronic and transport properties of the materials. In 2015, Hao Chen et al. [83] synthesized nitrogen and boron-co-doped KOH-activated bamboo-derived carbon as a porous biomass carbon with utility as a supercapacitor electrode material. These materials exhibit enhanced electrochemical cycling

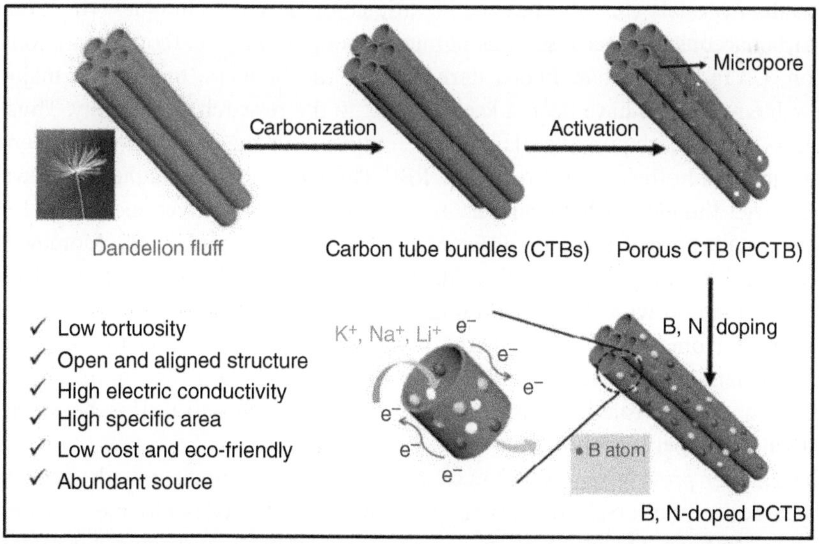

Figure 8.19 Illustration of N and B co-doping in carbon. Source: Ref. [97] / with permission of Royal Society of Chemistry.

stability and energy density relative to those based on most similar materials. The detailed scheme of the preparation of boron- and nitrogen-doped bio-derived carbon is shown as follows. The detailed synthesis procedures were discussed in ref [83]. Jing Zhao et al. [97] synthesized bio-derived carbon from dandelion fluff and doped with heteroatoms (B& N). Finally, they developed electrodes, and it exhibits a high volumetric energy density of 12.15 Wh L^{-1} at 699.84 W L^{-1} and excellent cyclic stability with capacitance retention of 92% after 10 000 cycles (Figure 8.19).

8.6 Summary and Future Prospectives

The rapid development in the utilization of renewable energy resources tends to move forward toward the advancements in the electrical energy storage devices, particularly, batteries and supercapacitors. However, the energy density and power density play a major role in predicting the usage of those storage devices. Known for its high energy density, batteries are preferred over supercapacitors. RBs paid increased attention in the past few years which could be the great support for the production and supply of energy. However, the performance of the batteries mainly depends on the electrodes which are used in the electrochemical cells. Hence, the major focus has been turned into the development of efficient and sustainable electrodes. Although, a variety of materials have been tested as electrodes in RBs, carbonaceous materials hold great potential in augmenting the electrochemical behavior of the electrodes which in turn improves the battery performance due to the high surface area and good electronic conductivity.

A variety of carbonaceous materials have been tested as efficient electrodes, in which the materials were synthesized in various routes and techniques. However, the carbonaceous materials, such as porous carbon, graphitic carbon, amorphous carbon, CNTs, and carbon fibers, derived from the biomass, one of the major renewable resources, have been a keen interest to the research community. Thus, biomass-derived carbon materials have been utilized for the preparation of green, sustainable, and efficient electrodes for RBs. Therefore, in this chapter, we have summarized the electrochemical performances of biomass-derived electrodes for batteries, with special focus on LIBs, SIBs, Li-S batteries, and ZABs. Moreover, the synthetic strategy for obtaining biomass-derived carbon electrodes for RBs and its correlation with the structural aspects are also discussed. An appropriate selection of biomass will result in the production of low-cost and high-quality carbon materials with the required porous structure, wherein the pretreatments would play a major role in removing the impurities which will further boost the electrochemical performance.

This chapter provides an overview of the recent developments and advances in biomass-derived carbon electrodes and their composite electrodes with metal oxides and metal sulfides. The electrochemical performances of the respective electrodes are discussed and summarized which clearly depicts the high specific capacities and excellent rate performances as compared to the conventional graphite used in LIBs.

Most of the works related this improved electrochemical performance to the synergistic effect of hierarchical porous carbons obtained from biomass materials as well as increased electrical conductivity. However, there are more preambles to prepare biomass-derived electrodes by optimizing different techniques where the waste to energy conversion is very essential in the current scenario.

For secondary batteries, various biomass-derived carbon electrodes were utilized in order to improve the electrochemical behavior of the respective electrodes which are also discussed in this chapter. Since the mechanism of SIBs is similar to that of LIBs, the electrodes which have been tested for LIBs could also be used. However, certain other biomasses were also converted into efficient electrodes particularly for SIBs which meant for the promising alternatives to the conventional and substantial electrode materials. However, it is very imperative to develop more efficient and sustainable electrodes from various biowaste materials that could help the cost reduction in the near future.

Although the low cost and high energy density of Li-S batteries hold great potential in the energy storage systems, the poor conductivity of sulfur, shuttling effect originating from the dissolved polysulfides, and high-volume expansion restrict the real-time applications. This problem could easily be addressed by the development of sulfur/carbon nanocomposite cathodes where the conductivity of the sulfur electrode is found to be enhanced due to the sulfur embedded into the conductive carbon frameworks. In addition, the excellent sorption properties of carbon are helpful in controlling the solubility losses of the intermediate sulfur species in the liquid electrolyte. This has been termed as one of the most promising strategies to improve the performance of Li-S batteries. The utilization of biomass-derived carbon materials as hosts for sulfur still improves the performance of Li-S batteries which are also covered in this chapter. Although the applications of biomass-derived materials in Li–S batteries have obtained great progress, there are still several challenges for their commercialization in practical battery systems which need to be addressed. One such challenge is the microscopic size of fiber-like materials which are weak in trapping polysulfides which could be overcome by the reduction of size.

Overall, the biomass-derived carbon materials and their composites as electrodes for RBs open up a new avenue for the electrical energy storage system for the production of clean energy which is very essential to meet the global energy demand. An in-depth research can significantly improve the added value of biomaterials, particularly for secondary batteries.

References

1 Gao, Y.P., Zhai, Z.B., Huang, K.J., and Zhang, Y.Y. (2017). Energy storage applications of biomass-derived carbon materials: batteries and supercapacitors. *New J. Chem.* 41: 11456–11470.

2 Liang, Y., Zhao, C., Yuan, H. et al. (2019). A review of rechargeable batteries for portable electronic devices. *InfoMat.* 1: 6–32.

3 Owens, B.B. (1986). *Batteries for Implantable Biomedical Devices*. Springer US.

4 Endo, C.A. (2001). *Anode Performance of the Li-Ion Secondary Battery*, 255–275. Springer.

5 Liang, J., Sun, Z.H., Li, F., and Cheng, H.M. (2016). Carbon materials for Li-S batteries: functional evolution and performance improvement. *Energy Storage Mater.* 2: 76–106.

6 Bruce, P.G., Freunberger, S.A., Hardwick, L.J., and Tarascon, J.M. (2012). LiO_2 and LigS batteries with high energy storage. *Nat. Mater.* 11: 19–29.

7 Peters, J.F., Cruz, A.P., and Weil, M. (2019). Exploring the economic potential of sodium-ion batteries. *Batteries* 5: 10.

8 Lu, Y., Li, L., Zhang, Q. et al. (2018). Electrolyte and interface engineering for solid-state sodium batteries. *Joule* 2: 1747–1770.

9 Moriwake, H., Kuwabara, A., Fisher, C.A.J., and Ikuhara, Y. (2017). Why is sodium-intercalated graphite unstable? *RSC Adv.* 7: 36550–36554.

10 Jache, B. and Adelhelm, P. (2014). Use of graphite as a highly reversible electrode with superior cycle life for sodium-ion batteries by making use of co-intercalation phenomena. *Angew. Chem. Int. Ed.* 53: 10169–10173.

11 Kim, H., Hong, J., Park, Y.U. et al. (2015). Sodium storage behavior in natural graphite using ether-based electrolyte systems. *Adv. Funct. Mater.* 25: 534–541.

12 Wei, Q., Fu, Y., Zhang, G., and Sun, S. (2017). Rational design of carbon-based oxygen electrocatalysts for zinc–air batteries. *Curr. Opin. Electrochem.* 4: 45–59.

13 Titirici, M.M., White, R.J., Brun, N. et al. (2015). Sustainable carbon materials. *Chem. Soc. Rev.* 44: 250–290.

14 Abioye, A.M. and Ani, F.N. (2015). Recent development in the production of activated carbon electrodes from agricultural waste biomass for supercapacitors: a review. *Renew. Sustain. Energy Rev.* 52: 1282–1293.

15 Gao, Z., Zhang, Y., Song, N., and Li, X. (2017). Biomass-derived renewable carbon materials for electrochemical energy storage. *Mater. Res. Lett.* 5: 69–88.

16 Chen, Q., Tan, X., Liu, Y. et al. (2020). Biomass-derived porous graphitic carbon materials for energy and environmental applications. *J. Mater. Chem. A* 8: 5773–5811.

17 Long, W., Fang, B., Ignaszak, A. et al. (2017). Biomass-derived nanostructured carbons and their composites as anode materials for lithium ion batteries. *Chem. Soc. Rev.* 46: 7176–7190.

18 Sun, X., Wang, X., Feng, N. et al. (2013). A new carbonaceous material derived from biomass source peels as an improved anode for lithium ion batteries. *J. Anal. Appl. Pyrolysis* 100: 181–185.

19 Unur, E., Brutti, S., Panero, S., and Scrosati, B. (2013). Nanoporous carbons from hydrothermally treated biomass as anode materials for lithium ion batteries. *Microporous Mesoporous Mater.* 174: 25–33.

20 Fey, G.T.K., Da Cho, Y., Chen, C.L. et al. (2010). Pyrolytic carbons from acid/base-treated rice husk as lithium-insertion anode materials. *Pure Appl. Chem.* 82: 2157–2165.

21 Wang, L., Schnepp, Z., and Titirici, M.M. (2013). Rice husk-derived carbon anodes for lithium ion batteries. *J. Mater. Chem. A* 1: 5269–5273.

22 Li, Y., Wang, F., Liang, J. et al. (2016). Preparation of disordered carbon from rice husks for lithium-ion batteries. *New J. Chem.* 40: 325–329.

23 Jiang, J., Zhu, J., Ai, W. et al. (2014). Evolution of disposable bamboo chopsticks into uniform carbon fibers: a smart strategy to fabricate sustainable anodes for Li-ion batteries. *Energ. Environ. Sci.* 7: 2670–2679.

24 Yu, X., Zhang, K., Tian, N. et al. (2015). Biomass carbon derived from sisal fiber as anode material for lithium-ion batteries. *Mater. Lett.* 142: 193–196.

25 Adams, R.A., Dysart, A.D., Esparza, R. et al. (2016). Superior lithium-ion storage at room and elevated temperature in an industrial woodchip derived porous carbon. *Ind. Eng. Chem. Res.* 55: 8706–8712.

26 Mullaivananathan, V., Sathish, R., and Kalaiselvi, N. (2017). Coir pith derived bio-carbon: demonstration of potential anode behavior in lithium-ion batteries. *Electrochim. Acta* 225: 143–150.

27 Fromm, O., Heckmann, A., Rodehorst, U.C. et al. (2018). Carbons from biomass precursors as anode materials for lithium ion batteries: new insights into carbonization and graphitization behavior and into their correlation to electrochemical performance. *Carbon N. Y.* 128: 147–163.

28 Li, Y., Li, C., Qi, H. et al. (2018). Formation mechanism and characterization of porous biomass carbon for excellent performance lithium-ion batteries. *RSC Adv.* 8: 12666–12671.

29 Che, Y., Zhu, X., Li, J. et al. (2016). Simple synthesis of MoO_2/carbon aerogel anodes for high performance lithium ion batteries from seaweed biomass. *RSC Adv.* 6: 106230–106236.

30 Cheng, F., Li, W.C., and Lu, A.H. (2016). Interconnected nanoflake network derived from a natural resource for high-performance lithium-ion batteries. *ACS Appl. Mater. Interfaces* 8: 27843–27849.

31 Wu, F., Huang, R., Mu, D. et al. (2016). Controlled synthesis of graphitic carbon-encapsulated α-Fe_2O_3 nanocomposite via low-temperature catalytic graphitization of biomass and its lithium storage property. *Electrochim. Acta* 187: 508–516.

32 Huang, Y., Lin, Z., Zheng, M. et al. (2016). Amorphous Fe_2O_3 nanoshells coated on carbonized bacterial cellulose nanofibers as a flexible anode for high-performance lithium ion batteries. *J. Power Sources* 307: 649–656.

33 Li, T., Bai, X., Qi, Y.X. et al. (2016). Fe_3O_4 nanoparticles decorated on the biochar derived from pomelo pericarp as excellent anode materials for Li-ion batteries. *Electrochim. Acta* 222: 1562–1568.

34 Kong, Z., Ma, Y., Ye, G. et al. (2018). Nanosized amorphous SnO_2 particles anchored in the wheat straw carbon substrate as the stabilized anode material of lithium-ion batteries. *ACS Appl. Energy Mater.* 1: 7065–7075.

35 Wang, H., Ren, D., Zhu, Z. et al. (2016). Few-layer MoS_2 nanosheets incorporated into hierarchical porous carbon for lithium-ion batteries. *Chem. Eng. J.* 288: 179–184.

36 Ma, L., Zhao, B., Wang, X. et al. (2018). MoS_2 nanosheets vertically grown on carbonized corn stalks as lithium-ion battery anode. *ACS Appl. Mater. Interfaces* 10: 22067–22073.

37 Haridas, A.K., Jeon, J., Heo, J. et al. (2019). In-situ construction of iron sulfide nanoparticle loaded graphitic carbon capsules from waste biomass for sustainable lithium-ion storage. *ACS Sustain. Chem. Eng.* 7: 6870–6879.

38 Dominguez, N., Torres, B., Barrera, L.A. et al. (2018). Bimetallic CoMoS composite anchored to biocarbon fibers as a high-capacity anode for Li-ion bsatteries. *ACS Omega.* 3: 10243–10249.

39 Zhu, J., Roscow, J., Chandrasekaran, S. et al. (2020). Biomass-derived carbons for sodium-ion batteries and sodium-ion capacitors. *ChemSusChem* 13: 1275–1295.

40 Hou, H., Qiu, X., Wei, W. et al. (2017). Carbon anode materials for advanced sodium-ion batteries. *Adv. Energy Mater.* 7: 1–30.

41 Górka, J., Vix-Guterl, C., and Matei Ghimbeu, C. (2016). Recent progress in design of biomass-derived hard carbons for sodium ion batteries. *C-J. Carbon Res.* 2: 24.

42 Ding, J., Wang, H., Li, Z. et al. (2013). Carbon nanosheet frameworks derived from peat moss as high performance sodium ion battery anodes. *ACS Nano* 7: 11004–11015.

43 Lotfabad, E.M., Ding, J., Cui, K. et al. (2014). High-density sodium and lithium ion battery anodes from banana peels. *ACS Nano* 8: 7115–7129.

44 Lv, W., Wen, F., Xiang, J. et al. (2015). Peanut shell derived hard carbon as ultra-long cycling anodes for lithium and sodium batteries. *Electrochim. Acta* 176: 533–541.

45 Elizabeth, I., Singh, B.P., Trikha, S., and Gopukumar, S. (2016). Bio-derived hierarchically macro-meso-micro porous carbon anode for lithium/sodium ion batteries. *J. Power Sources* 329: 412–421.

46 Mullaivananathan, V., Packiyalakshmi, P., and Kalaiselvi, N. (2017). Multifunctional bio carbon: a coir pith waste derived electrode for extensive energy storage device applications. *RSC Adv.* 7: 23663–23670.

47 Li, Y., Xu, S., Wu, X. et al. (2015). Amorphous monodispersed hard carbon micro-spherules derived from biomass as a high performance negative electrode material for sodium-ion batteries. *J. Mater. Chem. A* 3: 71–77.

48 Li, Y., Hu, Y.-S., Titirici, M.-M. et al. (2016). Hard carbon microtubes made from renewable cotton as high-performance anode material for sodium-ion batteries. *Adv. Energy Mater.* 6: 1600659.

49 Liu, P., Li, Y., Hu, Y.S. et al. (2016). A waste biomass derived hard carbon as a high-performance anode material for sodium-ion batteries. *J. Mater. Chem. A* 4: 13046–13052.

50 Wu, L., Buchholz, D., Vaalma, C. et al. (2016). Apple-biowaste-derived hard carbon as a powerful anode material for Na-ion batteries. *ChemElectroChem* 3: 292–298.

51 Gaddam, R.R., Yang, D., Narayan, R. et al. (2016). Biomass derived carbon nanoparticle as anodes for high performance sodium and lithium ion batteries. *Nano Energy* 26: 346–352.

52 Yan, D., Yu, C., Zhang, X. et al. (2016). Nitrogen-doped carbon microspheres derived from oatmeal as high capacity and superior long life anode material for sodium ion battery. *Electrochim. Acta* 191: 385–391.

53 Cao, L., Hui, W., Xu, Z. et al. (2017). Rape seed shuck derived-lamellar hard carbon as anodes for sodium-ion batteries. *J. Alloys Compd.* 695: 632–637.

54 Zheng, Y., Wang, Y., Lu, Y. et al. (2017). A high-performance sodium-ion battery enhanced by macadamia shell derived hard carbon anode. *Nano Energy* 39: 489–498.

55 Yu, C., Hou, H., Liu, X. et al. (2018). Old-loofah-derived hard carbon for long cyclicity anode in sodium ion battery. *Int. J. Hydrogen Energy* 43: 3253–3260.

56 Zhu, Y., Chen, M., Li, Q. et al. (2018). A porous biomass-derived anode for high-performance sodium-ion batteries. *Carbon N. Y.* 129: 695–701.

57 Arie, A.A., Kristianto, H., Demir, E., and Cakan, R.D. (2018). Activated porous carbons derived from the Indonesian snake fruit peel as anode materials for sodium ion batteries. *Mater. Chem. Phys.* 217: 254–261.

58 Damodar, D., Ghosh, S., Usha Rani, M. et al. (2019). Hard carbon derived from sepals of Palmyra palm fruit calyx as an anode for sodium-ion batteries. *J. Power Sources* 438: 227008.

59 Nagalakshmi, M. and Kalaiselvi, N. (2019). Mesoporous dominant cashewnut sheath derived bio-carbon anode for LIBs and SIBs. *Electrochim. Acta* 304: 175–183.

60 Wang, P., Fan, L., Yan, L., and Shi, Z. (2019). Low-cost water caltrop shell-derived hard carbons with high initial coulombic efficiency for sodium-ion battery anodes. *J. Alloys Compd.* 775: 1028–1035.

61 Yuan, H., Liu, T., Liu, Y. et al. (2019). A review of biomass materials for advanced lithium-sulfur batteries. *Chem. Sci.* 10: 7484–7495.

62 Liu, P., Wang, Y., and Liu, J. (2019). Biomass-derived porous carbon materials for advanced lithium sulfur batteries. *J. Energy Chem.* 34: 171–185.

63 Imtiaz, S., Zhang, J., Zafar, Z.A. et al. (2016). Biomass-derived nanostructured porous carbons for lithium-sulfur batteries. *Sci. China Mater.* 59: 389–407.

64 Wei, S., Zhang, H., Huang, Y. et al. (2011). Pig bone derived hierarchical porous carbon and its enhanced cycling performance of lithium-sulfur batteries. *Energ. Environ. Sci.* 4: 736–740.

65 Zhao, S., Li, C., Wang, W. et al. (2013). A novel porous nanocomposite of sulfur/carbon obtained from fish scales for lithium-sulfur batteries. *J. Mater. Chem. A* 1: 3334–3339.

66 Wang, H., Chen, Z., Liu, H.K., and Guo, Z. (2014). A facile synthesis approach to micro-macroporous carbon from cotton and its application in the lithium-sulfur battery. *RSC Adv.* 4: 65074–65080.

67 Tao, X., Zhang, J., Xia, Y. et al. (2014). Bio-inspired fabrication of carbon nanotiles for high performance cathode of Li-S batteries. *J. Mater. Chem. A* 2: 2290–2296.

68 Balakumar, K., Sathish, R., and Kalaiselvi, N. (2016). Exploration of microporous bio-carbon scaffold for efficient utilization of sulfur in lithium-sulfur system. *Electrochim. Acta* 209: 171–182.

69 Balakumar, K., Packiyalakshmi, P., and Kalaiselvi, N. (2018). Bio-waste derived carbon as interlayer and scaffold for Li-S batteries. *ChemistrySelect.* 3: 8901–8911.

70 Borghei, M., Lehtonen, J., Liu, L., and Rojas, O.J. (2018). Advanced biomass-derived electrocatalysts for the oxygen reduction reaction. *Adv. Mater.* 30: 1–27.

71 Zhang, L.Y., Wang, M.R., Lai, Y.Q., and Li, X.Y. (2017). Nitrogen-doped microporous carbon: an efficient oxygen reduction catalyst for Zn-air batteries. *J. Power Sources* 359: 71–79.

72 Ma, N., Jia, Y., Yang, X. et al. (2016). Seaweed biomass derived (Ni,Co)/CNT nanoaerogels: efficient bifunctional electrocatalysts for oxygen evolution and reduction reactions. *J. Mater. Chem. A* 4: 6376–6384.

73 Gao, S., Liu, H., Geng, K., and Wei, X. (2015). Honeysuckles-derived porous nitrogen, sulfur, dual-doped carbon as high-performance metal-free oxygen electroreduction catalyst. *Nano Energy* 12: 785–793.

74 Gao, S., Chen, Y., Fan, H. et al. (2014). Large scale production of biomass-derived n-doped porous carbon spheres for oxygen reduction and supercapacitors. *J. Mater. Chem. A* 2: 3317–3324.

75 Gao, S., Fan, H., Chen, Y. et al. (2013). One stone, two birds: gastrodia elata-derived heteroatom-doped carbon materials for efficient oxygen reduction electrocatalyst and as fluorescent decorative materials. *Nano Energy* 2: 1261–1270.

76 Gao, S., Wei, X., Fan, H. et al. (2015). Nitrogen-doped carbon shell structure derived from natural leaves as a potential catalyst for oxygen reduction reaction. *Nano Energy* 13: 518–526.

77 Gao, S., Geng, K., Liu, H. et al. (2015). Transforming organic-rich amaranthus waste into nitrogen-doped carbon with superior performance of the oxygen reduction reaction. *Energ. Environ. Sci.* 8: 221–229.

78 Zhu, C., Zhai, J., and Dong, S. (2012). Bifunctional fluorescent carbon nanodots: green synthesis via soy milk and application as metal-free electrocatalysts for oxygen reduction. *Chem. Commun.* 48: 9367–9369.

79 Liu, X., Zhou, Y., Zhou, W. et al. (2015). Biomass-derived nitrogen self-doped porous carbon as effective metal-free catalysts for oxygen reduction reaction. *Nanoscale* 7: 6136–6142.

80 Guo, C., Liao, W., Li, Z. et al. (2015). Easy conversion of protein-rich enoki mushroom biomass to a nitrogen-doped carbon nanomaterial as a promising metal-free catalyst for oxygen reduction reaction. *Nanoscale* 7: 15990–15998.

81 Baldinelli, A., Dou, X., Buchholz, D. et al. (2018). Addressing the energy sustainability of biowaste-derived hard carbon materials for battery electrodes. *Green Chem.* 20: 1527–1537.

82 Wang, J., Nie, P., Ding, B. et al. (2017). Biomass derived carbon for energy storage devices. *J. Mater. Chem. A* 5: 2411–2428.

83 Chen, H., Liu, D., Shen, Z. et al. (2015). Functional biomass carbons with hierarchical porous structure for supercapacitor electrode materials. *Electrochim. Acta* 180: 241–251.

84 Wang, Z., Shen, D., Wu, C., and Gu, S. (2018). State-of-the-art on the production and application of carbon nanomaterials from biomass. *Green Chem.* 20: 5031–5057.

85 Hou, H., Banks, C.E., Jing, M. et al. (2015). Carbon quantum dots and their derivative 3D porous carbon frameworks for sodium-ion batteries with ultralong cycle life. *Adv. Mater.* 27: 7861–7866.

86 Ma, B., Huang, Y., Nie, Z. et al. (2019). Facile synthesis of Camellia oleifera shell-derived hard carbon as an anode material for lithium-ion batteries. *RSC Adv.* 9: 20424–20431.

87 Rybarczyk, M.K., Peng, H.-J., Tang, C. et al. (2016). Porous carbon derived from rice husks as sustainable bioresources: insights into the role of micro-/mesoporous hierarchy in hosting active species for lithium–sulphur batteries. *Green Chem.* 18: 5169–5179.

88 Wang, H.G., Wu, Z., Meng, F.L. et al. (2013). Nitrogen-doped porous carbon nanosheets as low-cost, high-performance anode material for sodium-ion batteries. *ChemSusChem* 6: 56–60.

89 Yan, L., Yu, J., Houston, J. et al. (2017). Biomass derived porous nitrogen doped carbon for electrochemical devices. *Green Energy Environ.* 2: 84–99.

90 Kesavan, T. and Sasidharan, M. (2019). *Palm Spathe Derived N-Doped Carbon Nanosheets as a High Performance Electrode for Li-Ion Batteries and Supercapacitors*. Eng: ACS Sustain. Chem.

91 Hou, J., Cao, C., Idrees, F., and Ma, X. (2015). Hierarchical porous nitrogen-doped carbon nanosheets derived from silk for ultrahigh-capacity battery anodes and supercapacitors. *ACS Nano* 9: 2556–2564.

92 Guan, L., Pan, L., Peng, T. et al. (2019). Synthesis of biomass-derived nitrogen-doped porous carbon nanosheests for high-performance supercapacitors. *ACS Sustain. Chem. Eng.* 7: 8405–8412.

93 Saravanan, K.R. and Kalaiselvi, N. (2015). Nitrogen containing bio-carbon as a potential anode for lithium batteries. *Carbon N. Y.* 81: 43–53.

94 Liu, D., Yu, S., Shen, Y. et al. (2015). Polyaniline coated boron doped biomass derived porous carbon composites for supercapacitor electrode materials. *Ind. Eng. Chem. Res.* 54: 12570–12579.

95 Chen, Z.H., Du, X.L., He, J.B. et al. (2017). Porous coconut shell carbon offering high retention and deep lithiation of sulfur for lithium-sulfur batteries. *ACS Appl. Mater. Interfaces* 9: 33855–33862.

96 Bongu, C.S., Karuppiah, S., and Nallathamby, K. (2015). Validation of green composite containing nanocrystalline Mn_2O_3 and biocarbon derived from human hair as a potential anode for lithium-ion batteries. *J. Mater. Chem. A* 3: 23981–23989.

97 Zhao, J., Li, Y., Wang, G. et al. (2017). Enabling high-volumetric-energy-density supercapacitors: designing open, low-tortuosity heteroatom-doped porous carbon-tube bundle electrodes. *J. Mater. Chem. A* 5: 23085–23093.

9

Recent Advances in Bio-derived Nanostructured Carbon-based Materials for Electrochemical Sensor Applications

Akshat Mathur[1], Jayashankar Das[2], and Sushma Dave[3]

[1]*Jodhpur Institute of Engineering and Technology, Department of Electronics and Communication Engineering, Mogra, Jodhpur RJ 342048, India*
[2]*Valnizen Health Care, Vile Parle West, Mumbai 400056, India*
[3]*Jodhpur Institute of Engineering and Technology, Department of Applied Science, Mogra, Jodhpur RJ 342048, India*

9.1 Introduction

Sensors are devices which allow us to observe phenomena of interest by translating its state and variations into an analytical signal. The most important part of a sensor is its "sensing element" which is what "senses" or observes the desired phenomenon. The terms, however, are used interchangeably, and the term sensor usually refers to the sensing element, unless it is used to refer to a product/device.

Sensors are usually manufactured for, and classified on the basis of, their specific application. The nature of the sensing element may, however, differ by varying degrees for different sensor techniques and technologies for similar applications. In this chapter, we will discuss about sensing elements fabricated from carbon-based materials for applications in electrochemical sensing.

Electrochemical sensors, which are a variety of chemical sensors, are further categorized as voltammetric, potentiometric, chemically sensitized field effect transistor, and potentiometric solid electrolyte gas sensors [1].

An electrochemical sensor works by allowing an electrochemical reaction to occur between the substance/phenomena to be observed and the electrode, i.e. the sensing element, which produces an electric response (current) corresponding to the chemical excitation, thereby producing the analytical signal. It is an easy-to-use, low-cost tool for quantitative and qualitative determination of electroactive species.

Due to the abundant porosity and surface area provided by nanomaterials for chemical reactions to occur, they are very popular as catalysts. The same physical properties also make them ideal materials for electrochemical sensing elements.

Using a biological precursor, there are two common ways of synthesizing carbon nanomaterials: biomass pyrolysis and hydrothermal carbonization. Many studies tend to prefer pyrolysis as it offers wider flexibility in the selection of biomass precursor and requires only mild reaction conditions in an overall low-cost procedure. Due

Biomass-Derived Carbon Materials: Production and Applications, First Edition.
Edited by Alagarsamy Pandikumar, Perumal Rameshkumar, and Pitchaimani Veerakumar.

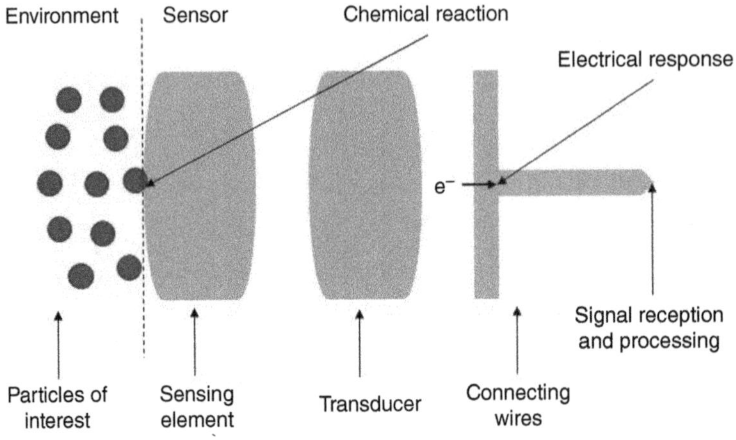

Figure 9.1 Basic structure of an electrochemical sensor.

to their nature, electrochemical sensors, in general, have found widespread applications in environmental monitoring and biosensing applications. A point of particular attraction is their low cost which enables their widespread use in large-scale environmental monitoring efforts and other industrial applications (Figure 9.1) [2–9].

In addition to their eco-friendly synthesis approach, numerous disposable sensors have also been fabricated which are based on carbon nanostructures. These sensors further encourage the use of bio-derived nanostructured carbon materials in environmental monitoring and biosensing applications, including monitoring the quality and edibility of packaged food items [10]. These nanomaterials have many applications in dye degradation and pesticide analysis [11–13].

Accidentally discovered in 2004 [14], carbon quantum dots (CQDs) have attracted significant research interest due to their many interesting properties, particularly fluorescence. CQDs are zero-dimensional structures i.e., each of their physical dimension is typically <10 nm. CQDs exhibit favorable synthesis properties such as chemical stability, water solubility, and facile synthesis methods. In addition to their optical properties, these structures exhibit robust and attractive chemical properties including low toxicity and high stability. CQDs are also relatively good conductors, and their tuneable optical properties encourage their application as sensing elements in various contexts [15, 16].

A variety of biological sources have yielded CQDs of appreciable electrochemical properties to be utilized as sensing elements. These structures exhibit varying levels of sensitivity toward various chemical compounds [7, 17–21].

Recently, carbon dots synthesized from kiwi, white sesame, and black sesame seeds have shown significant abilities as electrochemical sensors for nitrite [14]. These CQDs were synthesized by the means of pyrolysis of seeds. The resulting power was separated, purified, dried, and coated on glassy carbon electrodes for electrochemical sensing of NO^{2-} and NO^{3-}. The average size of these CQDs was reported to be 3.13, 2.97, and 2.68 nm, respectively, and they displayed a detection limit of 0.23 μM (S/N = 3) for nitrite.

Furthermore, in 2018, a CQD-based highly sensitive and selective electrochemical sensor was reported by Sha et al. for the detection of hydrazine. In the study, chia seeds were used as a biomass precursor and CQDs were synthesized by pyrolysis. The CQDs so synthesized has oxygen-rich functional groups on their surfaces and were used to modify glassy carbon electrodes. These quasi-spherical particles, each ranging from 2 to 6 nm, were able to substantially enhance the interfacial and electrocatalytic properties of the glassy carbon electrode. The CQD-enhanced sensor, thus, showed a remarkable sensitivity of 151.5 μA mM^{-1} cm^{-2} which resulted in a successful detection of hydrazine in a sample of drinking water [22].

Carbon dots synthesized from reed have also shown a significant ability to detect hydrazine [17]. The CQDs derived from reed via hydrothermal method were found to be reducing agents. For synthesis, reed powder and de-ionized water were heated at 180 °C for 3 hours and were later centrifuged to obtain the mixture which was dried for 48 hours at 50 °C to obtain CQDs. These CQDs were then formulated into a 2 mg mL^{-1} solution. This mixture was further processed to obtain a CDs-Cu_2O, and later, CDs-Cu_2O/CuO composite material.

To design the electrode, a glassy carbon electrode was decorated with the material. Along with good sensitivity, the electrode exhibited a range of 0.99–5903 μM, a sensitivity of 4.45 μA mM^{-1}, and a detection limit of 0.024 μM.

Along with being low cost, a requirement for large-scale environmental monitoring, industrial application, and manufacturing efforts to succeed is the reproducibility of sensors. The sensor reported, when examined with 0.1 M PBS solution containing 250 mM of hydrazine, showed a relative standard deviation of 3.46% only, which indicates good reproducibility of the device.

For the electrochemical detection of nitrite, synthesis of carbon nanospheres was reported by Yallappa et al. [23]. Using areca nut as the biological source of carbon, the authors were able to synthesize carbon nanospheres with tunable diameters of 30–40 nm. These structures also had a controlled morphology with a pore size of ≈2.0 nm. Without doping or activation, these carbon nanospheres exhibited good electrocatalytic activity, sensitivity, and low detection limit toward nitrites. Good reusability, stability, and repeatability were also demonstrated by the particles with the retention of satisfactory efficiency in later usage which contributes toward further reduction in the cost of sensors.

Ali et al. pyrolyzed purpureus seeds at different temperatures to produce carbon nanospheres which were fabricated into electrodes for supercapacitors. The specific capacitances of the supercapacitors so made rose with the temperature at which the carbon nanospheres were synthesized, with a capacitance of 300 F g^{-1} being demonstrated at 800 °C, the highest synthesis temperature in the study. In comparison with other approaches, Ali et al. reported that their supercapacitors had a substantially longer lifespan with the devices retaining 94% of their capacitance even after 5200 charge–discharge cycles. It is believed that the use of activated carbon can further enhance these properties [7].

Due to the availability of a large number of low-volume pores, activated carbon, also known as activated charcoal, typically offers better efficiency due to the availability of a larger surface area for chemical reactions than conventional electrodes.

This also improves the sensitivity of an electrochemical sensor. Thus, activated carbon, biologically derived from mango leaves, was used by Madhu et al. for the electrochemical detection of 4-nitrophenol based on biomass-derived activated carbons [8].

This synthesis technique, reported in a previous article [24], synthesized activated carbon decorated with spherical carbon nanoparticles by using dead mango leaves. This activated charcoal, decorated with carbon nanoparticles, exhibited a high sensitivity toward the detection of heavy metal ions. The temperatures of synthesis and surface area of these particles were 900 °C and 1555 $m^2 g^{-1}$, respectively.

While developing the electrochemical sensor for the detection of 4-nitrophenol, [8] found the presence of oxygen functional groups and heteroatoms on the large surface area of the particles to be primarily responsible for the catalytic activities and reversible redox behaviors which were observed during the detection of 4-nitrophenol. In a 0.05-M acetate buffer solution with a pH of 5.0, these particles exhibited a detection limit of 0.16 μm and a sensitivity of 5.810 $\mu A\ \mu M^{-1}\ cm^{-2}$. Furthermore, these particles also exhibited advantages in simplicity, stability, durability, reliability, and cost which are advantageous in any practical application.

Bio-derived carbon-based nanomaterials have shown significant flexibility as electrochemical sensing elements and have, thus, found wide applications, particularly in environmental monitoring and biological sensing applications.

A CQD-modified glassy carbon electrode was recently reported which offered a quick and highly selective detection of mercury. These electrodes were synthesized by using a facile one-pot synthesis technique with coconut shells as the biomass precursor. These electrodes exhibited a wide range and excellent mercury detection ability in various water samples which were also contaminated by other elements [24].

The synthesis of carbon nanoballs aggregation networks-based aerogels was reported by [25]. The structures were synthesized by using the peel of shaddock as the biomass precursor. The peels were hydrothermally treated, freeze-dried, and carbonized before being powdered. This powder was dissolved in a *N,N*-dimethylformamide solution and was drop-casted on glassy carbon electrodes. The resulting electrochemical sensing element showed a high sensitivity toward the probe molecule of potassium ferricyanide, the oxidase-related molecule of hydrogen peroxide, and the neurotransmitters of dopamine, ascorbic acid, and uric acid (Figure 9.2). These electrodes showed a wide liner range of 5–1760 $\mu mol\ L^{-1}$, a higher sensitivity of 42.4 $\mu A\ mmol\ L^{-1}\ cm^{-2}$, and a lower detection limit of 3.53 $\mu mol\ L^{-1}$, which are all desirable qualities and are an improvement of conventional carbon nanotubes and glassy carbon electrodes (Figures 9.3 and 9.4).

In a similar approach, onion peels were utilized to develop an electrochemical sensor for the nonenzymatic detection of progesterone. The onion peels were utilized to synthesize carbon nanospheres which were coated on a carbon filter paper [26]. The increase in electroactive sites due to the porosity of carbon nanostructures promotes electro-oxidation of progesterone at a negative potential. This has been attributed to oxidation of progesterone at the C-6 position which results in the formation of a

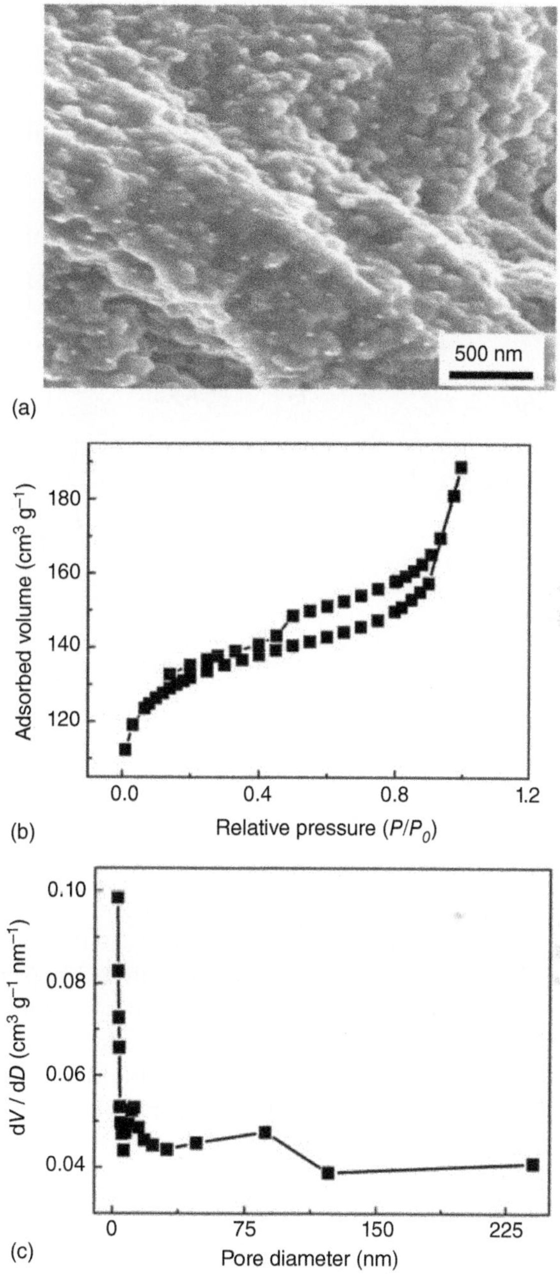

Figure 9.2 (a) Scanning tunnelling microscope (SEM) images of carbon nanoballs aggregation networks-based aerogels (CNANAs). (b) Nitrogen isotherms for CNANAs. (c) The pore size distribution for CNANAs. Source: Ref. [25] / with permission of Elsevier.

cationic radical intermediate which further undergoes oxidation in the presence of water which results in the formation of oxidized progesterone.

Activated carbon has been observed to significantly enhance the properties of electrodes. Thus, efficient processes for the activation of carbon are also a necessity and have attracted significant research scrutiny in recent years.

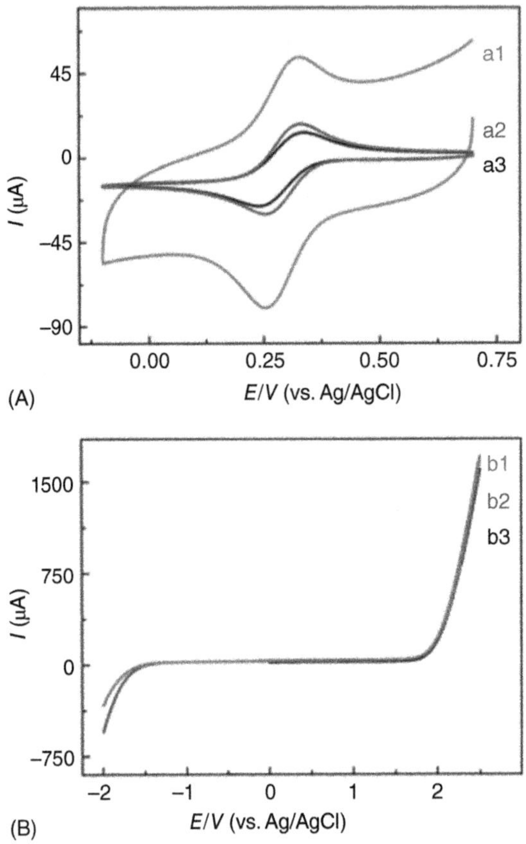

Figure 9.3 (a) Cyclic voltammetry (CV) curves recorded at CNANAs/Glassy carbon electrode (GCE) (a1), carbon nano tubes (CNTs)/GCE (a2), and GCE (a3) in 0.1 mol L^{-1} KCL containing 5 mmol L^{-1} $K_3Fe(CN)$ −3/−4. Scan rate: 10 mV s. (b) linear sweep voltammetry (LSV) curves of CNANAs/GCE (b1), CNTs/GCE (b2), and GCE (b3) in N2-saturated 0.1 mol L^{-1} pH 7.0 PBS. Scan rate: 10 mV s^{-1}. Source: Ref. [25] / with permission of Elsevier.

Kim et al. reported a two-step activation process for biomass-derived carbon. This process involves the use of $ZnCl_2$, KOH, and $ZnCl_2$-KOH for the activation of carbon. The resulting zinc-KOH activated kelp carbon resulted in the modified glassy carbon electrode showing high sensitivity and selectivity with a lower detection limit of 0.004 μm for the detection of acetaminophen. The improvement in the sensing abilities was attributed to the increase in surface area due to the addition of zinc-KOH activated kelp carbon. Although the electrode displayed promising characteristics, the two-step activation process is a resource-consuming task to perform and is generally inefficient in nature [18].

Graphene is a sp^2-hybridized sheet of carbon atoms (see Figure 9.5). The material was first studied, albeit theoretically, in 1947 by P. R. Wallace [27]. Although the interest in graphene has existed from before that time and continued to attract researchers after Wallace's 1947 paper, a single-atom-thick layer of graphene was extracted from bulk graphite in 2004 by Novoselov et al. at the University of Manchester in the United Kingdom. Later, the observation of quantum hall effect provided evidence for Barry's phase of 2D gas of massless Dirac fermions in graphene [28, 29].

One of the most common graphene nanocomposities is silver-decorated graphene which has been used in numerous electrochemical sensing applications as well.

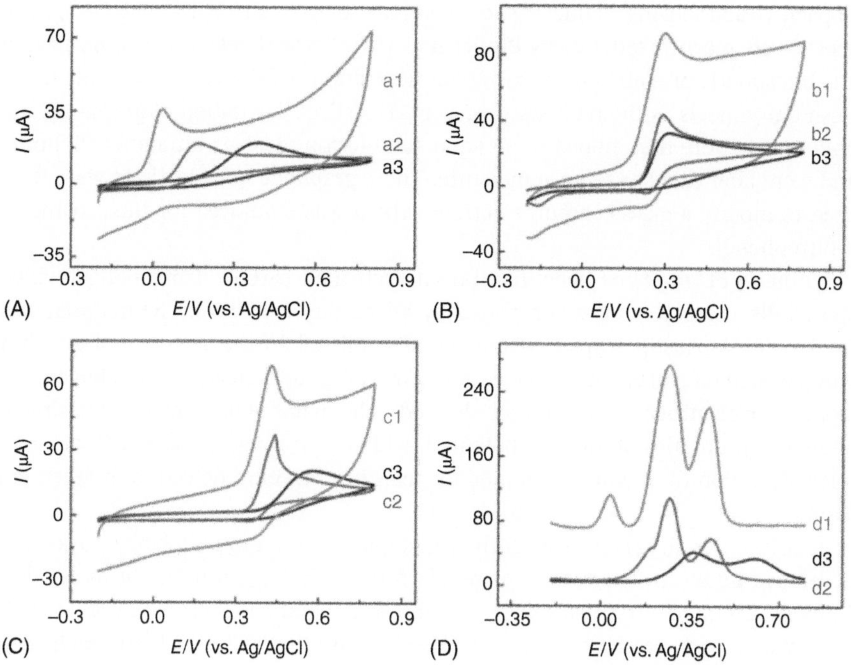

Figure 9.4 (a) CV curves for 2.5 mmol L^{-1} AA at CNANAs/GCE (a1), CNTs/GCE (a2), and GCE (a3). (b) CV curves for 2.5 mmol L^{-1} DA at CNANAs/GCE (b1), CNTs/GCE (b2), and GCE (b3). (c) CV curves for 2.5 mmol L^{-1} UA at CNANAs/GCE (c1), CNTs/GCE (c2), and GCE (c3). (d) Differential pulse voltammetry (DPV) curves for 0.1 mol L^{-1} pH 7.0 PBS containing 2.5 mmol L^{-1} AA, 2.5 mmol L^{-1} DA, and 2.5 mmol L^{-1} UA at CNANAs/GCE (d1), CNTs/GCE (d2), and GCE (d3). Scan rate: 10 mV s^{-1}. Electrolyte: N2-saturated 0.1 mol L^{-1} pH 7.0 PBS. Source: Ref. [25] / with permission of Elsevier.

Figure 9.5 Graphene.

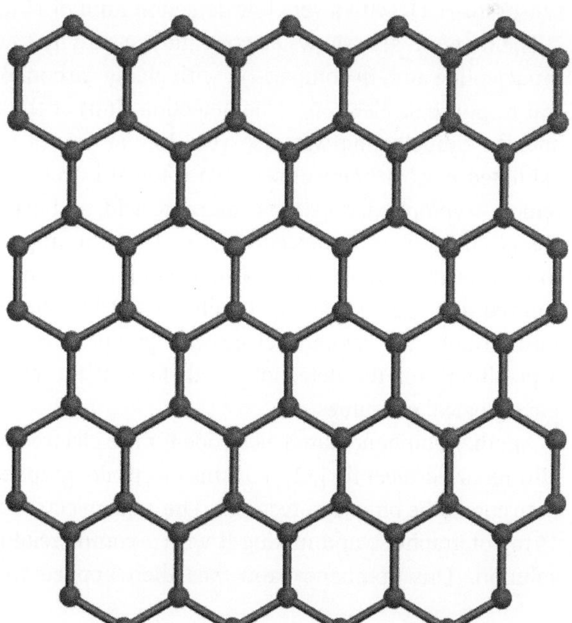

Capacitive and sensing characteristics of biomass-derived silver-decorated graphene sheets were recently reported by Bhujel et al. [30]. These sheets were synthesized in an atmospheric pressure chemical vapour deposition (APCVD) reactor while using fresh onion peels as the biomass precursor. For silver decoration of graphene, the onion peels were first dipped in an aqueous solution of silver nitrate for 12 hours and were later dried to remove moisture. These graphene nanoparticles were then used to modify a glassy carbon electrode which was employed for the sensing of *p*-nitrophenol.

Synthesis of graphene sheet-like porous activated carbon using Bougainvillea spectabilis as a precursor was reported by Veeramani et al. [31]. The material was used to develop a supercapacitor and was also utilized for the detection of catechin when coated on a glassy carbon electrode. With Ag/AgCl electrode as referece and a platinum electrode as the counter electrode, the material demonstrated desirable analytical properties of low overpotential, wide linear range, good sensitivity, and a low detection limit when examined for real-time catechin detection in green tea leaves.

In 2018, Li et al. [21] derived carbon nanoplates from ground cherry husks and developed an electrochemical sensing platform for a large number of molecules. Initially, ground cherry husks were carbonized to obtain a black solid mass which was washed with 5 M HCl and 5 M KOH solutions at 60 °C for 12 hours each. The carbonized samples were then collected by centrifugation and were subsequently washed and dried. A glossy carbon electrode was decorated with these samples to prepare the electrochemical sensor.

These carbon nanoplates were shown to be electrochemically sensitive to a large number of molecules. These electrodes show sensitivity and high selectivity toward hydrogen peroxide. A low overpotential which develops at the electrode during the electroreduction of hydrogen peroxide enables its utilization as an electrochemical sensor for H_2O_2 with a very low detection limit of 16 μM.

The material also shows electrocatalytic activity toward β-nicotinamide adenine dinucleotide and, in comparison with glassy carbon, offers enhanced electrochemical response as electrode. The detection limit of the electrode for β-nicotinamide adenine dinucleotide was also very low at 1.6 μM (S/N = 3). The material also exhibited electrochemical sensitivity toward acetaminophenol, epinephrine, cysteine, L-tryptophan, dopamine, ascorbic acid, and uric acid making it a flexible and useful choice for the detection of biologically important molecules. In response to each of these molecules, carbon nanoplate-decorated glassy carbon electrodes showed significant sensitivity for efficient utilization of the material as an electrode. Furthermore, the favorable electrocatalytic properties of this electrode also present a possibility for the detection of all four of the DNA bases (G, T, C, and A) at physiological pH values.

Another graphene-based electrode for the electrochemical detection of DNA was also reported recently [32]. For this electrode, graphene was synthesized by using corn cub as the biomass precursor. The paper electrode was fabricated by exfoliating 15 mg of graphene and mixing it with a commercial transparent polyvinyl acetate solution. This graphene paint was then applied to paper to produce low-cost, disposable, and biodegradable electrodes which were then dried for 2 hours at 150 °C to avoid peeling off of graphene paint. These electrodes exhibited remarkable mechanical properties, and the paint was not damaged even after repeated folding, bending, and twisting of the paper, thus keeping the circuit intact.

Another crucial parameter to monitor in the body is glucose. Many electrochemical sensors have thus been previously designed for the purpose, with nickel being a popular choice due to its attractive electrocatalytic properties. However, nickel structures tend to form close-packed structures after being deposited on the electrode which results in a reduction in the surface area, and, thus, the performance of the electrode [33]. To resolve this challenge, a foam-like 3D graphene porous carbon-nickel structure was recently developed and was shown to exhibit useful electrochemical sensing characteristics for glucose [17]. Dried pamelo peels were used as a biomass precursor for the synthesis of these structures which enabled nonenzymatic detection of glucose.

Recently, reduced graphene oxide nanocomposites have also been demonstrated to exhibit electrochemical sensing abilities of interest [34]. The biochar was obtained by pyrolysis of the biomass precursor: Eichhornia crassipes at 400 °C and graphene oxide was obtained by a modified Hummer's method. Then the modified carbon paste electrode was prepared by mixing 30.0% (w/w) mineral oil, 43.0% (w/w) graphite powder, and 13.5% (w/w) of biochar and 13.5% (w/w) of reduced graphene oxide. The resulting sensor was used for the detection of carbendazim with a detection limit of 2.3 nmol L^{-1}.

Nanoporous graphene-like sheets with nitrogen doping have shown a particular sensitivity toward vitamin C. Furthermore, it has been observed that the overpotential showed by glassy carbon electrodes decorated with these sheets was much lower than that shown by either undecorated glassy carbon electrodes or carbon nanotube-decorated glassy carbon electrodes [15]. These graphene-like sheets were synthesized by using dried soy pulp, also known as okara, as the biomass precursor for the carbonization process. These electrodes exhibited a wide linear range of 10–5640 μmol L^{-1}, a lower detection range of 0.51 μmol L^{-1}, and a sensitivity of 144.65 μA mmol L^{-1} cm^{-2}.

Recently, a one-pot hydrothermal synthesis route for the synthesis of nitrogen-doped carbon sheets was reported by Zhao et al. [6]. These nitrogen-doped carbon sheets with embedded zirconia particles were synthesized using orange juice as the biomass precursor. Subsequently, 5 mg of the material was evenly distributed on *N,N*-dimethylformaldihyde and 10 μL of the resulting suspension was drop-coated on a glassy carbon electrode to produce the sensor. This sensor exhibited significant sensing abilities for methyl parathion with a linear detection range of 0.01–15 μg mL^{-1} and a lower detection limit of 0.115 ng mL^{-1}.

A graphene quantum dot is another carbon-based nanomaterial which has been generating substantial research interest due to its electrochemical properties. Nanochannel-confined graphene quantum dots have been reported to be employed for the electrochemical detection of complex samples of heavy metal ions [15]. These graphene quantum dots (GQDs) were highly sensitive to Hg^{2+}, Cu^{2+}, and Cd^{2+} and were able to detect without the requirement of pretreatment of samples.

Thus, the scope of biosynthesis of GQDs have also been explored in the context of their electrochemical sensing applications. A reduced graphene oxide-prussian blue-based biosensor for the detection of dimetridazole was developed by Keerthi et al. [35]. Extract of paddy straw mushroom was used as a biomass precursor, and the resulting electrodes were screen-printed by modifying screen-printed carbon electrode (SPCE) method for economical and efficient sensors.

The synthesis of a graphene oxide–silver nanocomposite via hydrothermal route was demonstrated by using the extracts of Plectranthus amboinicus leaves as biomass precursors [5]. After simultaneous reduction of graphene oxide and silver ions which was achieved due to the reducibility of Plectranthus amboinicus leaves, the material was evaluated for its electrochemical properties and a nonenzymatic H_2O_2 electrochemical sensor was fabricated which showed high selectivity and lower overpotential toward the reduction of H_2O_2.

Carbon nanotubes, which can be thought of as rolled graphene sheets, have also proven to be of significant use in electrochemical sensing. Together with carbon nanorods, they have been shown to be sensitive to a variety of chemical species and have been widely used in electrochemical sensing applications. Furthermore, biological synthesis routes for CNTs have developed the groundwork for large-scale production of the material which has a variety of applications in many fields.

One such biological synthesis route was reported by Li et al. by using honey as the biomass precursor and crab shells as the template [16]. The prepared carbon nanorods were coated on a glassy carbon electrode for electrochemical sensing applications. As compared to glassy carbon or CNT-coated glassy carbon electrodes, the nanorod coated glassy carbon electrodes exhibited enhanced electrocatalytic activity such as low peak potential, wide linear range, high electrochemical sensitivity, and a low detection limit. The electrodes also had the unique nanostructure of MCNRs with a large surface area for electrochemical reactions, uniform mesoporous structures, and an abundance of defective sites.

Another biological route, utilized by [36], used apples as the biomass precursor for synthesis of nanorods. The apples were first heated at 80 °C for 10 hours, and the resulting hydrogel was washed, dried, and calcinated to obtain carbon nanorods-assembled hierarchical meso-macroporous carbons network aerogels. When coated on glassy carbon electrodes, these electrochemical sensors were able to detect ascorbic acid and hydrogen peroxide due to greater electrochemical sensitivity than bare glassy carbon electrodes. Electrochemical performance of carbon materials generated by biological synthesis of silver nanoparticles was also identified by Dave et al. [37].

9.2 Conclusion and Future Perspectives

To sum up, novel carbon material from biological resources can be efficiently used for the low-level detection of various compounds compared with CNTs/GCE and GCE. In addition, the prepared sensor will be able to detect the compounds in real samples (human urine, serum) with satisfactory recovery, which points toward its great potential for the detection in human body fluids for the clinical diagnostics. The electrochemical investigation suggests that the prepared bio-derived electrode shows more active electrocatalytic properties toward the oxidation of various compounds, and the field is very less explored. A large amount of work is yet to be done in this area.

References

1 Hulanicki, A., Glab, S., and Ingman, F. (1991). Chemical sensors definitions and classification. *Pure Appl. Chem.* 63 (9): 1247–1250.

2 Chen, D., Zhou, H., Li, H. et al. (2017). Self-template synthesis of biomass-derived 3D hierarchical N-doped porous carbon for simultaneous determination of dihydroxybenzene isomers. *Sci. Rep.* 7 (1): 1–10.

3 Kumar, A., Hegde, G., Manaf, S.A.B.A. et al. (2014). Catalyst free silica templated porous carbon nanoparticles from bio-waste materials. *Chem. Commun.* 50 (84): 12702–12705.

4 Liu, S., Han, T., Wang, Z. et al. (2019). Biomass-derived nitrogen and phosphorus Co-doped hierarchical micro/mesoporous carbon materials for high-performance non-enzymatic H_2O_2 sensing. *Electroanalysis* 31 (3): 527–534.

5 Zheng, Y., Wang, A., Cai, W. et al. (2016). Hydrothermal preparation of reduced graphene oxide–silver nanocomposite using Plectranthus amboinicus leaf extract and its electrochemical performance. *Enzyme Microb. Technol.* 95: 112–117.

6 Zhao, H., Liu, B., Li, Y. et al. (2020). One-pot green hydrothermal synthesis of bio-derived nitrogen-doped carbon sheets embedded with zirconia nanoparticles for electrochemical sensing of methyl parathion. *Ceram. Int.* 46 (12): 19713–19722.

7 Ali, G.A.M., Divyashree, A., Supriya, S. et al. (2017). Carbon nanospheres derived from: Lablab purpureus for high performance supercapacitor electrodes: a green approach. *Dalton Trans.* 46 (40): 14034–14044.

8 Madhu, R., Karuppiah, C., Chen, S.M. et al. (2014). Electrochemical detection of 4-nitrophenol based on biomass derived activated carbons. *Anal. Methods* 6 (14): 5274–5280.

9 Madhu, R., Sankar, K.V., Chen, S.M., and Selvan, R.K. (2014). Eco-friendly synthesis of activated carbon from dead mango leaves for the ultrahigh sensitive detection of toxic heavy metal ions and energy storage applications. *RSC Adv.* 4 (3): 1225–1233.

10 Dincer, C., Bruch, R., Costa-Rama, E. et al. (2019). Disposable sensors in diagnostics, food, and environmental monitoring. *Adv. Mater.* 31 (30): 1806739.

11 Dave, S., Dave, S., and Das, J. (2021). Photocatalytic degradation of dyes in textile effluent: a green approach to eradicate environmental pollution. In: *The Future of Effluent Treatment Plants*, 199–214. Elsevier.

12 Dave, S., Dave, S., and Das, J. (2021). Electrochemical monitoring as an emerging technology for detection of environmental pollutants with special reference to pesticides. In: *The Future of Effluent Treatment Plants*, 183–198. Elsevier.

13 Satapathy, S., Acharya, D., Dixit, P.K. et al. (2021). Mechanistic aspects and rate-limiting steps in green synthesis of metal and metal oxide nanoparticles and their potential in photocatalytic degradation of textile dye. In: *Photocatalytic Degradation of Dyes*, 605–630. Elsevier.

14 Li, K., Xu, J., Arsalan, M. et al. (2019). Nitrogen doped carbon dots derived from natural seeds and their application for electrochemical sensing. *J. Electrochem. Soc.* 166 (2): B56.

15 Sha, T., Liu, J., Sun, M. et al. (2019). Green and low-cost synthesis of nitrogen-doped graphene-like mesoporous nanosheets from the biomass waste of okara for the amperometric detection of vitamin C in real samples. *Talanta* 200: 300–306.

16 Li, X., Liu, J., Sun, M. et al. (2018). Amperometric sensing of ascorbic acid by using a glassy carbon electrode modified with mesoporous carbon nanorods. *Microchim. Acta* 185 (10): 1–7.

17 Wei, G., Wang, L., Huo, L., and Zhang, Y. (2020). Economical, green and rapid synthesis of CDs-Cu2O/CuO nanotube from the biomass waste reed as sensitive sensing platform for the electrochemical detection of hydrazine. *Talanta* 209: 120431.

18 Kim, D., Kim, J.M., Jeon, Y. et al. (2018). Novel two-step activation of biomass-derived carbon for highly sensitive electrochemical determination of acetaminophen. *Sensors Actuators B, Chem* 259: 50–58.

19 Jorio, A., Soares, E.A., Paniago, R. et al. (2020). Graphene. In: *Springer Handbook of Surface Science*, 1171–1198. Cham: Springer.

20 Xu, X., Ray, R., Gu, Y. et al. (2004). Electrophoretic analysis and purification of fluorescent single-walled carbon nanotube fragments. *J. Am. Chem. Soc.* 126 (40): 12736–12737.

21 Li, X., Li, H., Liu, T. et al. (2018). The biomass of ground cherry husks derived carbon nanoplates for electrochemical sensing. *Sensors Actuators B Chem.* 255: 3248–3256.

22 Sha, R., Jones, S.S., Vishnu, N. et al. (2018). A novel biomass derived carbon quantum dots for highly sensitive and selective detection of hydrazine. *Electroanalysis* 30 (10): 2228–2232.

23 Yallappa, S., Shivakumar, M., Nagashree, K.L. et al. (2018). Electrochemical determination of nitrite using catalyst free mesoporous carbon nanoparticles from bio renewable areca nut seeds. *J. Electrochem. Soc.* 165 (10): H614.

24 Barvin, R.K.B., Prakash, P., Ganesh, V., and Jeyaprabha, B. (2019). Highly selective and sensitive sensing of toxic mercury ions utilizing carbon quantum dot-modified glassy carbon electrode. *Int. J. Environ. Res.* 13 (6): 1015–1023.

25 Sha, T., Li, X., Liu, J. et al. (2018). Biomass waste derived carbon nanoballs aggregation networks-based aerogels as electrode material for electrochemical sensing. *Sensors Actuators, B Chem* 277: 195–204.

26 Akshaya, K.B., Bhat, V.S., Varghese, A. et al. (2019). Non-enzymatic electrochemical determination of progesterone using carbon nanospheres from onion peels coated on carbon fiber paper. *J. Electrochem. Soc.* 166 (13): B1097.

27 Wallace, P.R. (1947). The band theory of graphite. *Phys. Rev.* 71 (9): 622–634.

28 Novoselov, K.S., Geim, A.K., Morozov, S.V. et al. (2004). Electric field effect in atomically thin carbon films science. *Science* 306 (5696): 666–669.

29 Novoselov, K.S., Geim, A.K., Morozov, S.V. et al. (2005). Two-dimensional gas of massless Dirac fermions in graphene. *Nature* 438 (7065): 197–200.

30 Bhujel, R., Rai, S., Baruah, K. et al. (2019). Capacitive and sensing responses of biomass derived silver decorated graphene. *Sci. Rep.* 9 (1): 1–14.

31 Veeramani, V., Madhu, R., Chen, S.M. et al. (2015). Biomass-derived functional porous carbons as novel electrode material for the practical detection of biomolecules in human serum and snail hemolymph. *Sci. Rep.* 5 (1): 1–9.

32 Mohanraj, J., Durgalakshmi, D., Rakkesh, R.A. et al. (2020). Facile synthesis of paper based graphene electrodes for point of care devices: a double stranded DNA (dsDNA) biosensor. *J. Colloid Interface Sci.* 566: 463–472.

33 Safavi, A., Maleki, N., and Farjami, E. (2009). Fabrication of a glucose sensor based on a novel nanocomposite electrode. *Biosens. Bioelectron.* 24 (6): 1655–1660.

34 Sant'Anna, M.V.S., Carvalho, S.W.M.M., Gevaerd, A. et al. (2020). Electrochemical sensor based on biochar and reduced graphene oxide nanocomposite for carbendazim determination. *Talanta* 220: 121334.

35 Keerthi, M., Akilarasan, M., Chen, S.-M. et al. (2018). One-pot biosynthesis of reduced graphene oxide/prussian blue microcubes composite and its sensitive detection of prophylactic drug dimetridazole. *J. Electrochem. Soc.* 165 (2): B27.

36 Wang, N., Hei, Y., Liu, J. et al. (2019). Low-cost and environment-friendly synthesis of carbon nanorods assembled hierarchical meso-macroporous carbons networks aerogels from natural apples for the electrochemical determination of ascorbic acid and hydrogen peroxide. *Anal. Chim. Acta* 1047: 36–44.

37 Dave, S. and Kumbhat, S. (2018). Electrochemical and spectral characterization of silver nanoparticles synthesized employing root extract of curculigo orchioides. *Indian J. Chem. Technol.*

10

Porous Carbon Derived From Biomass for Fuel Cells

A. Sivakami[1], Aristatil Ganesan[2], P. Sakthivel[3], Kishore Sridharan[4], Sabarinathan Venkatachalam[5], and Sudhagar Pitchaimuthu[6]

[1] *Malla Reddy University, School of Sciences, Department of Physics, Medchal, Hyderabad 500100, Telangana, India*
[2] *Aatal Innovations Private Limited, Chennai 603112, Tamil Nadu, India*
[3] *Centre for Materials Science, Department of Physics, Faculty of Engineering, Karpagam Academy of Higher Education, Coimbatore 641021, Tamil Nadu, India*
[4] *University of Calicut, School of Physical Sciences, Department of Nanoscience and Technology, P.O. Thenhipalam 673635, Kerala, India*
[5] *Manonmaniam Sundaranar University, Department of Physics, Tirunelveli 627012, Tamil Nadu, India*
[6] *Heriot-Watt University, Research Centre for Carbon Solutions, Institute of Mechanical, Process and Energy Engineering, School of Engineering & Physical Sciences, Edinburgh EH14 4AS, United Kingdom*

10.1 Introduction

Modern civilization requires a power source and it also imparts the need for the life of humanity. The supply of fossil-fuel-based fuels such as coal, oil, and natural gas is still unbalanced because of increasing world economic expansion and population growth [1]. The potential biomass resources for deriving porous carbon is presented in Figure 10.1.

We have been relying on fossil fuels for energy production for more than five decades and even at present our dependence has not fallen short [2]. Already the greenhouse gases with majority being CO_2 accumulated in our environment by the persistent burning of fossil fuels have triggered a global climate change with catastrophic events such as wildfires, longer periods of drough,t and frequent high intensity tropical storms. However, the global energy demand continues to surge and has put a lot of pressure on the technologists for producing electricity without relying on the fossil fuels to reduce the global carbon footprint. Toward this end, the research on practically sustainable energy resources for producing electricity using nonpolluting, renewable, and eco-friendly systems is being widely explored. Fuel cells, solar cells, supercapacitors, and lithium-ion batteries [3–6] have been identified as the promising sustainable energy solutions. Among these, fuel cells play an essential role in commercial devices that are attributable to their high performance and durability, at a lower cost. Nonetheless, for higher performance, the materials used as catalysts in fuel cells should have excellent

Biomass-Derived Carbon Materials: Production and Applications, First Edition.
Edited by Alagarsamy Pandikumar, Perumal Rameshkumar, and Pitchaimani Veerakumar.

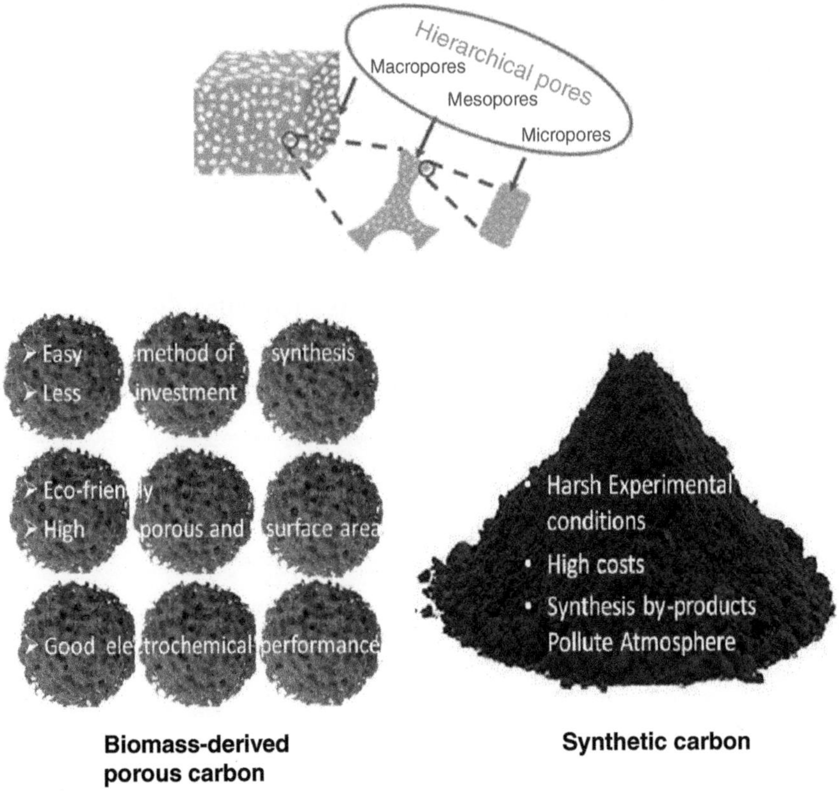

Figure 10.1 Schematic illustration of hierarchical pores and variety of biomass-derived porous carbon and synthetic carbon.

physiochemical properties and must be abundantly available, scalable, eco-friendly, and cost effective [7]. Among a wide range of materials tested as catalysts for fuel cell applications, carbon materials have drawn a greater attention due to their unique properties such as electronic conductivity, surface area, porosity, chemical stability, and durability [8, 9]. The overview of biomass-derived carbon of three important aspects for advanced catalysts for oxygen reduction reaction (ORR) as shown in Figure 10.2.

Due to excellent surface morphology and porosity, popular forms of nanostructured carbon materials such as fullerenes, carbon nanotubes (CNT), and graphene are developed that have been recognized as advanced carbon materials [10]. However, these advanced carbon materials rely on fossil fuel-based precursors (OH-based) and required harsh synthesis procedures with demanding conditions (arc melting, CVD, PVD, laser process) that are not only expensive but are also environmental unfriendly [10]. Therefore, the carbon materials processed from biomass resources are in high demand owing to their easy process ability, eco-friendliness, tailor-made surface morphology, porosity, and inexpensiveness. Generally, biomass-derived carbon materials are produced from plant and animal waste with an annual estimate of ~105 pentagrams [11]. It infers that a considerable amount of biomass waste has been taken from the burning of agricultural

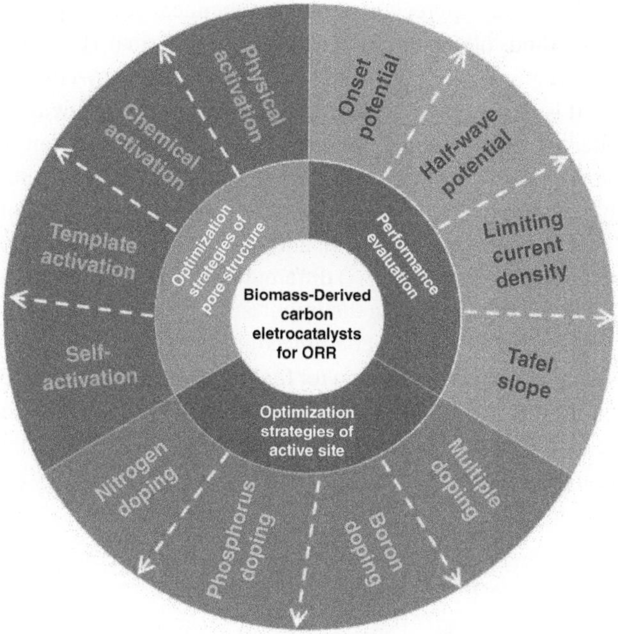

Figure 10.2 Overview of biomass-derived carbon catalysts for ORR. Source: Ref. [5] / MDPI.

residues and forest by-products. It causes severe air pollution and reducing energy utilization. Biomass resources are contributed to supplying carbon-rich precursors and carbonaceous materials in a wide range of electrochemical storage devices. The porous carbon materials are classified into three categories based on their pore sizes, i.e. microporous size (usually < 2 nm), 2 nm $<$ mesoporous < 50 nm, and macroporous > 50 nm.

The activated carbons (ACs) are typically synthesized by combining pyrolysis and other posttreatment activation methods. Researchers are investigating and examining the parameters such as pore size and internal surface area of biomass-derived carbon, which improve the performance of ORR in a fuel cell. Most ACs are restricted for fuel cell applications due to the presence of micropores, but surprisingly the ACs derived from biomass possessing nanopores are preferred. These nanopores in the biomass-derived material facilitate the diffusion of electrolytes and channelize efficient ion transfer [12]. Therefore, selecting an appropriate synthetic route and customizable procedure is essential to obtain the required pore size of a biomass-derived carbon catalyst for fuel cell applications. Using advanced synthetic methodologies, the biomass-derived carbon is anticipated to possess high surface area, high porous morphology, enhanced electrical conductivity, and good mechanical properties, as shown in Figure 10.1. It is required for all electrochemical storage devices, including fuel cells [7]. The most used carbonization method is hydrothermal carbonization (HTC), and it produces rich carbon precursors for fuel cell applications. Tuning surface morphology of biomass-derived carbons with a hierarchical porosity and high surface area ($800\ m^2\ g^{-1}$), polymer template-assisted colloidal synthesis method is highly appreciable in electrochemical devices [5, 13]. In addition to the HTC method, other methods such as ionothermal (ITC) and molten salt carbonization methods are also used to synthesize energy-efficient carbon materials [14, 15]. It is challenging for the synthesis of surface-modified carbon from biomass. So effective synthetic methodologies are required for designing efficient catalysts and catalyst support toward potential applications. Many research groups are working on biomass-derived carbon for focusing catalysis [15–17], Li-ion and sodium-ion batteries [18], supercapacitors [19], and fuel cells [20].

This chapter consists of two parts, including (i) different synthesis methods used for biomass-derived carbon and (ii) explaining the importance of ORR in fuel cells. In the first part, the mechanism of the synthetic procedure is followed to get the required characteristics of biomass-derived carbon for fuel cell applications. Second, future research directions and further development of biomass-derived carbon for fuel cell applications are outlined and discussed. The nanoporous carbon materials for fuel cell applications areshown in Figure 10.3.

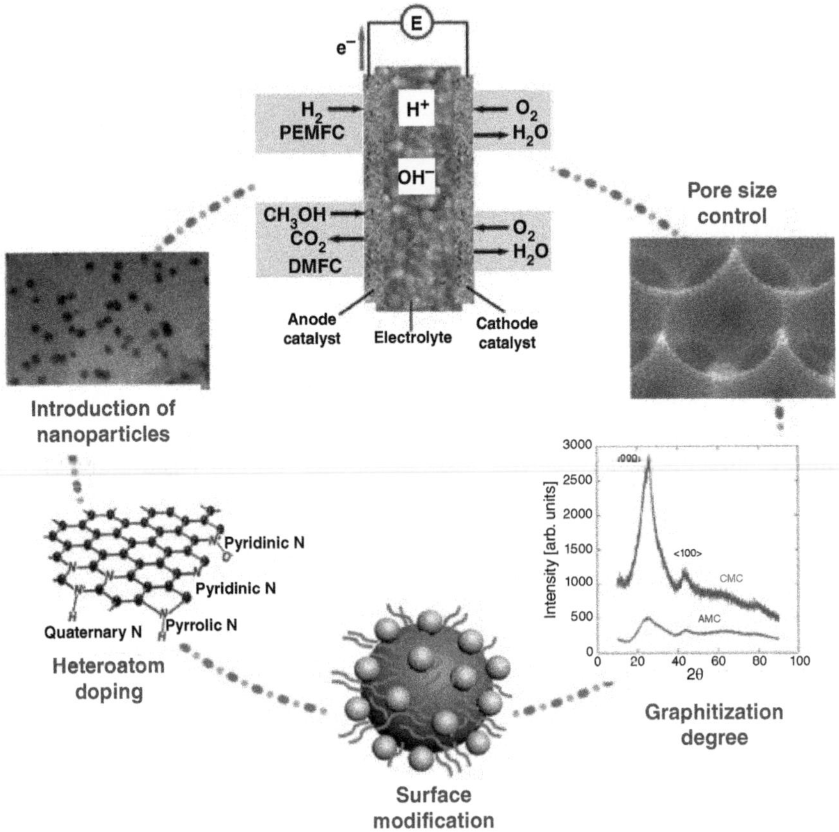

Figure 10.3 Applications of nanoporous carbon in various applications such as energy conversion and storage. Source: Ref. [21] / Reproduced with permission from Elsevier.

10.2 Fuel Cells – Theory and Fundamentals

A fuel cell is a device that converts chemical energy into electricity without combustion by combining hydrogen at the cathode and oxygen at the anode to produce water and heat. The amount of electricity produced by a single fuel cell depends on several limitations such as type, fuel cell size, temperature, and pressure supplied to the cell. Typically, a fuel cell stack consists of hundreds of fuel cells useful for larger applications. The fuel cells are mainly classified according to the type of electrolyte, fuel and oxidant used, range of temperature in which the cell operates, and other factors. Several fuel cells using biomass-derived carbon are currently under development, and each catalyst has its advantages, limitations, and potential applications. The classification of the fuel cell is presented in Figure 10.4.

The typical working principle of fuel cells is illustrated in Figure 10.5. A fuel cell is a device that uses hydrogen-rich fuel and oxygen to create electricity through the electrochemical process. As seen in Figure 10.5, the fuel cell has two electrodes, i.e. anode which supplies electrons and a cathode that absorbs electrons. The electrolyte may be liquid or solid, which carries electrically charged particles from one electrode to another. A catalyst is used for increasing the reaction speed at the electrodes. The reaction mechanism of the fuel cell is given by Equation (10.1),

$$\text{Hydrogen} + \text{Oxygen} = \text{Current} + \text{Water Vapor} \tag{10.1}$$

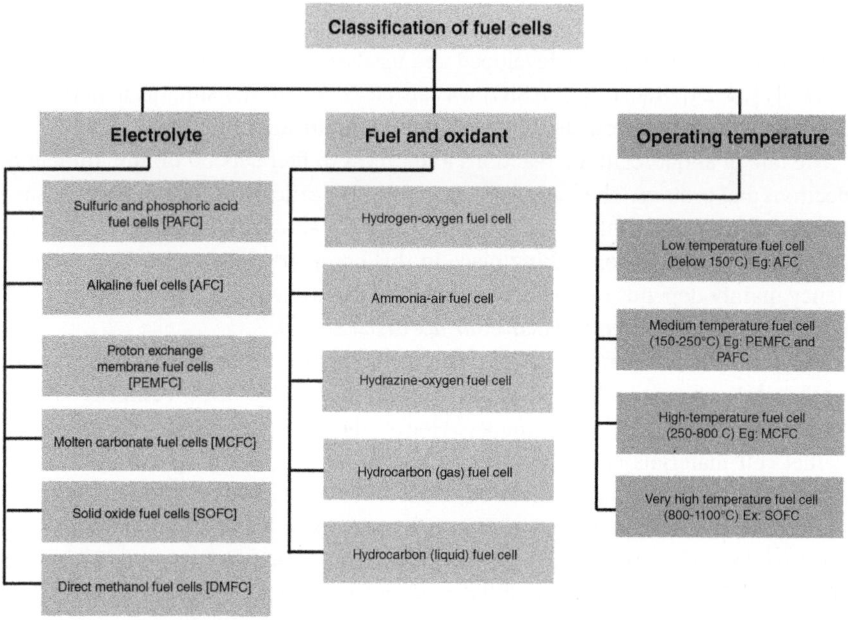

Figure 10.4 Classification of fuel cells and their varieties.

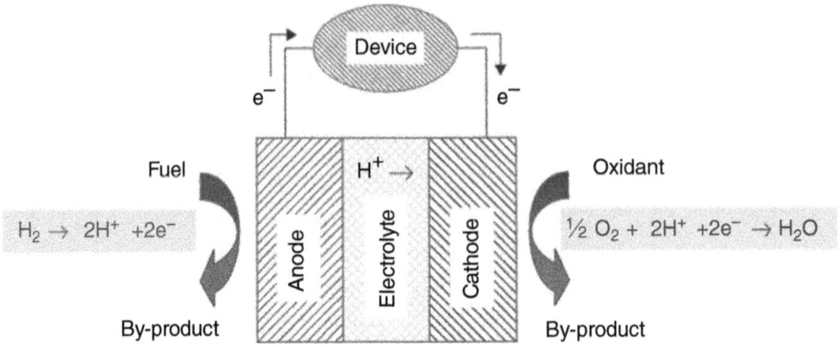

Figure 10.5 Schematic illustration of fuel cell and its operating principle.

10.3 Catalyst Support Materials

The catalyst support materials have a significant influence on the cost, performance, and durability of PEMFC. The stability and catalytic activity of the fuel cells have been improved by adding catalyst support [21]. Generally, catalyst support should have high conductivity, high surface area, good water management, high electrochemical stability, strong interaction with catalyst, strong corrosion resistance, and the ability to maximize the triple-phase boundary for the fuel cell. The catalyst support materials predominantly dictate the catalyst lifecycle by allowing the poisoning effect. Most researchers have used metal oxides [22, 23] as catalyst support for PEM fuel cells even though carbonaceous, non-carbonaceous, and ceramic-based materials are developed and used as catalyst support for PEM-based fuel cells [24, 25]. Recently, the SiO_2 was also used as catalyst support materials for the self-humidifying anode to make internal humidification of PEMFC [26, 27]

The rate of anode/cathode reactions and access to fuel depend on the number of electrons and protons. The fuel gases, nanosized Pt particles, and membrane formed one boundary called *triple-phase boundary* (TPB), as shown in Figure 10.6. The electron transfer reactions are taking place in this boundary. The overall fuel cell efficiency mainly depends on the area formed by the triple-phase boundary, and it is also called an *electrochemical active surface area (ECSA).*

The continuous operation of the fuel cell imparts severe electrochemical or mechanical stresses, which degrade catalyst durability, thus resulting in the decrease of the fuel cell performance. Hence, the degradation/failure mechanism of fuel cell materials strongly relies on the fuel cell's operating conditions. The various failure mechanisms involved, such as catalyst particle coarsening, metal dissolution, platinum bands formation in a membrane, carbon support corrosion, and membrane degradation, were investigated and reported in several kinds of literature [28]. The degradation of the catalyst layer is happening due to the failure of fuel access, a break in the proton transport pathway, and reduction of electrons conduction. The degradation rate of the material is mainly based on the selection of catalysts, catalyst support, and ionomer materials [29]. The degradation mechanism

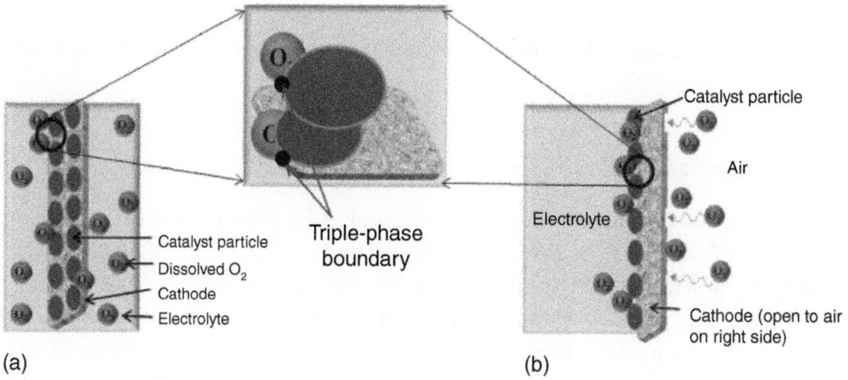

Figure 10.6 Schematic illustration of the triple-phase boundary. Source: Ref. [28] / with permission of Elsevier.

occurs due to the existence of both mechanical and electrochemical issues in the catalyst. The mechanical issues are arising due to cracks on-catalyst surface or peeling off the catalyst from the electrode. Electrochemical issues arise because of catalyst particle growth, migration, membrane degradation, and corrosion [29].

For instance, platinum (Pt) is a well-known electrocatalyst material in fuel cells. When Pt nanoparticles take part in electron transfer reaction, the surface of Pt gets oxidized into Pt ions. These Pt ions are again dissolved and deposited on the nearby existing Pt particles, thus resulting in particle growth. Finally, the agglomeration is formed and detached from the catalyst support, as shown in Figure 10.7, and the ECSA is decreased. After this detachment, the Pt ions will be dissolved into the membrane and formed as a Pt crystal as Pt bands at the middle of the membrane [31, 32]. Due to oxidation and agglomeration of platinum, the ECSA decreases,

Figure 10.7 Schematic representing Pt agglomeration and detachment from the catalyst support surface. Source: Ref. [30] / with permission of Elsevier.

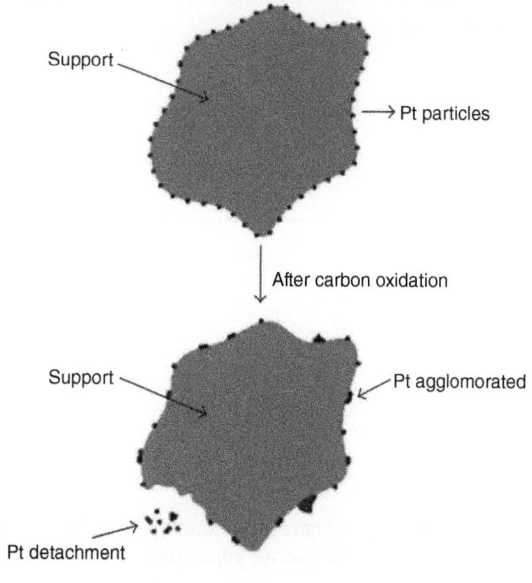

and monitoring other parameters such as catalytic stability, durability, and cyclic voltammetry characteristic studies are required to find whether they are increasing or decreasing the ECSA.

10.3.1 As a Catalyst

Biomass-derived carbon has been used as a catalyst for ORR for increasing the performance of fuel cells. Nowadays, biomass-derived carbon as an ORR catalyst is a developing field due to its low cost, high performance, and reproducibility. It plays a significant role in the development of next-generation fuel cells and batteries. The performance of a biomass-derived carbon in different environments such as neutral, alkaline, and acidic is of greater importance for finding fuel cell applications in various sectors. Many works of literature have reported and replaced the platinum catalyst for selecting the suitable nonplatinum catalysts to improve the performance of ORR electrocatalytic activity [30, 33]. At present, many biomass-derived catalysts such as active carbon [34], enzyme [35], microorganism [36, 37], transition metal porphyrins [38], $NiIn_2S_4$/CNFs [39], and phthalocyanines [40, 41] are developed as promising alternative catalysts than precious metal (Pt). Several researchers [42–44] are working on selecting the different biomass-derived catalysts for improving the ORR activity still.

10.3.2 Synthesis Methods of Porous Carbon from Biomass

So far, the maximum biomass-derived carbon content is estimated to be in the range of 45–50% [45–50]. The thermomechanical synthetic strategy removes the noncarbon elements, and the final product will be carbon alone [51–56]. The by-product nature of carbon and its characteristics mainly depends upon the type of synthesis route. The different existing synthesis methods can be classified as shown in Figure 10.8.

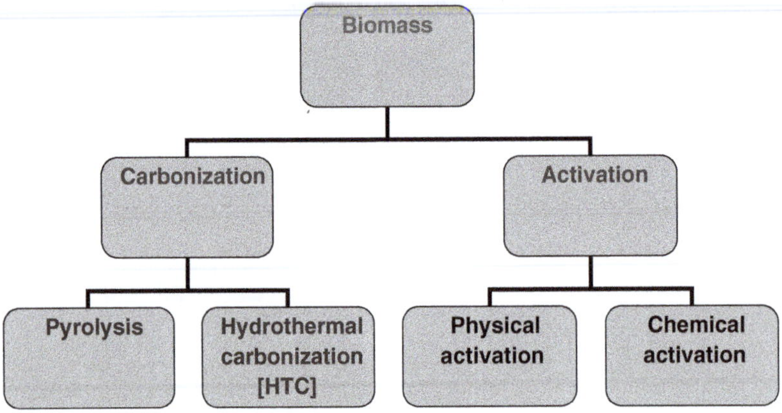

Figure 10.8 Flow chart for different synthesis methods of converting biomass to porous carbon materials for various applications.

10.4 Porous Carbon Synthesis from Different Biomass

Fuel cells require a continuous flow of hydrogen and oxygen to continue the reaction, not like lithium batteries. The hydrogen or alcohol oxidation in fuel cells occurs at the anode and the ORR at the cathode.

10.4.1 Oxygen Reduction Reaction (ORR)

Fuel cells are used as a source of electrochemical energy storage devices for electric vehicles. The overall performance of the fuel cells system depends upon ORR. For producing the best performance of fuel cells, it is necessary to increase the rate of ORR by using efficient electrocatalysts. But still, research on different ORR catalysts is developed to achieve good stability, cost, and catalytic activity for increasing the overall electrochemical system performance [57]. The ORR performance plays an essential role in the development of fuel cells. The different nanocarbons for ORR in the application of fuel cells [58]. They are

1. Metal oxides, nitrides, carbides, and sulfides
2. Non-noble metal catalysts
3. Metal-free catalysts

Still comparing all ORR catalysts, precious metal Pt showed excellent electrocatalytic activity, but it is limited to industrial because of its high cost and scarcity. So, the research is turning on finding alternative metal-free catalysts to increase the ORR performance, although it suffers from a lack of stability. To overcome the stability issue, optimizing the synthesis method for controlling the shape and structure of different electrolytes remains a challenge [58].

The fuel cell performance is greatly affected by ORR, so it is crucial to enhance the electron transfer rate for superior electrocatalytic activities than commercial Pt/C catalyst [59–61]. The electron transfer rate can be calculated from Koutecky–Levich equation, and it can be written as [62, 63],

$$\frac{1}{j} = \frac{1}{j_k} + \frac{1}{B_L \omega^{\frac{1}{2}}} \tag{10.2}$$

where j_k = limiting current density (mA cm^{-2}),
ω = electrode rotating rate (rad s^{-1}),
B_L-Levich constant which is given by
$B_L = 0.62\,nFCo_2Do_2{}^{2/3}\Upsilon^{-1/6}$
Co_2 = concentration of oxygen in 0.1 M KOH (1.2×10^{-6} mol cm^{-3}),
n = average number of electron transfer ($n = 4$),
Do_2 = oxygen diffusion coefficient in 0.1 M KOH (1.90×10^{-5} cm^2 s^{-1}),
F = Faradaic constant (96 485 C mol^{-1}),
Υ = kinematic viscosity of the electrolyte solution (0.01 cm^2 s^{-1})

The different ORR catalysts are used till now are illustrated in the Figure 10.9.

The electrolytic oxidation of alcohol in fuel cells and the addition of platinum with biomass-derived carbonaceous materials as catalysts are promising. Wang et al. [64]

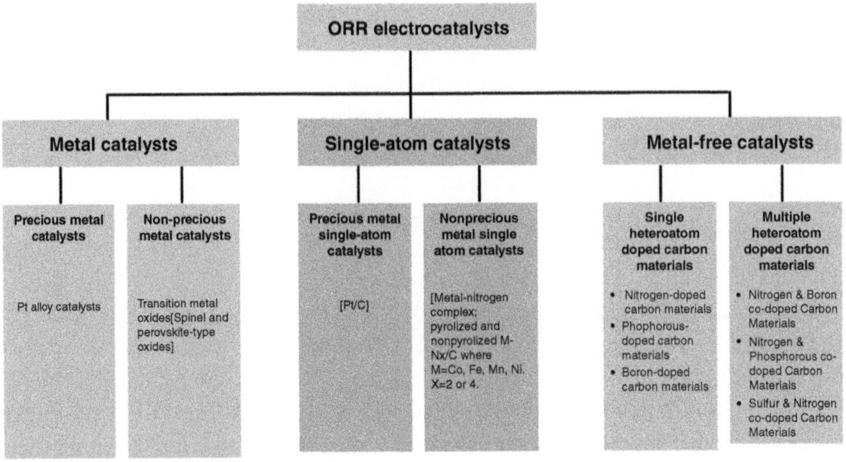

Figure 10.9 Different ORR electrocatalysts and its properties for fuel cell applications.

used okra to synthesize a composite using carbonaceous nanoparticles with platinum nanoparticles by deposition. The produced platinum okra composite as carbon catalyst used for methanol electro-oxidation and displayed superior performance compared to traditional Pt/C, with a higher peak current of $12.2\,mV\,cm^{-3}$ compared to $7\,mV\,cm^{-3}$ and an outstanding negative onset potential of 0.42 V compared to 0.49 V. In some other studies, biomass chitosan and carbon black also used as nitrogen-doped carbon precursors to produce nitrogen-doped carbon (NDPCS). The critical pathway for increasing the efficiency of the fuel cells is the transformation of energy through the ORR. Much research has been progressed to find a potential alternative material to replace the precious metal catalyst. Mainly, high surface area and more electrochemical active sites are necessary for maximizing the ORR performance. Still, there is a considerable demand for biomass-derived carbonaceous catalyst with a facile synthesis technique, high surface area, controllable pore structure, ease of heteroatom doping, good stability, and desirable morphology characteristics.

10.5 Synthesis of Biomass-Derived ORR Catalyst for Fuel Cell

Guo et al. [65] prepared the ternary-doped porous carbons from natural tea leaves treated with an iron salt by the one-step pyrolysis method. They found that the ORR activity can be enhanced due to the inserting heteroatoms in sp^2-hybridized carbon framework that can be achieved by controlling pyrolysis temperatures. It also concluded that HDPC-X catalysts are expressed as the best ORR electrocatalytic activities for direct methanol fuel cell applications because of the synergistic effect between the heteroatom doping and 3D hierarchically porous structure. Since heteroatoms such as N, B, S, P doped porous carbons have exhibited very

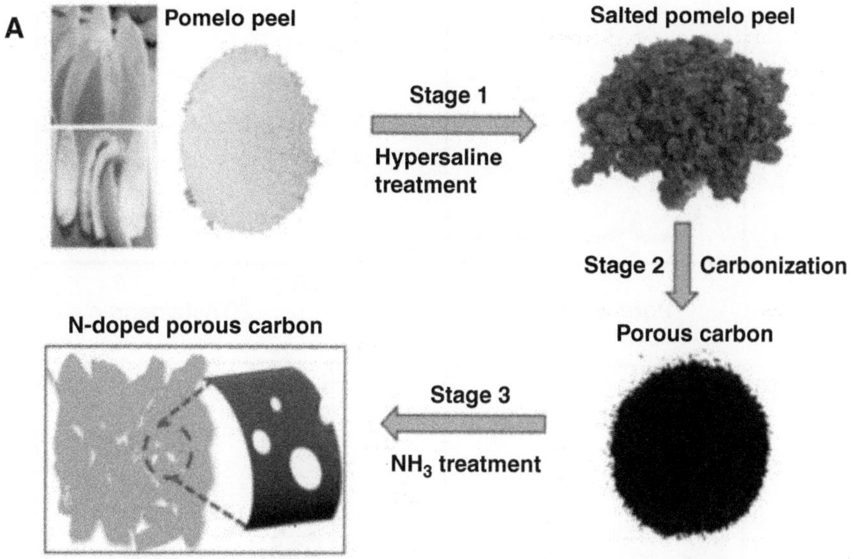

Figure 10.10 Schematic representation of the formation of porous carbon from pomelo peel by pyrolysis method. Source: Ref. [66] / Reproduced with permission from Elsevier.

high electrocatalytic activity and stability for the ORR in alkaline environments. Metal-free carbon materials can be found as one of the best alternative potential candidates to the well-known precious metal Pt for fuel cells because of their lower cost. Wang et al. [66] synthesized nitrogen-doped porous carbons from pomelo peel using the pyrolysis carbonization method. They observed that the hypersaline treatment of pomelo peel is one of the efficient ways to tune the pore structure of carbon effectively, as shown in Figure 10.10.

The ORR performance of bioderived carbon has improved significantly compared to direct pyrolytic carbon for two reasons (i) hypersaline treatment of bio-derived carbon, which increases the total pore volume and (ii) synergistic effect of total pore volume and nitrogen doping content (Figure 10.11).

Chatterjee et al. [67] prepared the rich doped nitrogen–carbon nano-onion architectures from the renewable biological resource, i.e. collagen, used as a metal-free ORR catalyst (Figure 10.12). The obtained product consists of a high percentage of nitrogen (7.5%) integrated into the carbon molecular skeleton. Because of this reason, the ORR electrocatalytic activity is increased with low onset potential, high current density, and better durability than commercial Pt/C catalyst in alkaline medium. Three different N configurations exist pyridinic-N, amino-N, and pyrrolic-N, as shown in Figure 10.12a. The linear sweep voltammetry (LSV) and cyclic voltammetry (CV) studies were used to estimate ORR's kinetics. From the LSV graph as shown in Figure 10.13b, the durability of the fuel cell sample 750-8 N-CNO was measured before and after 3000 cycles. The 750-8 N-CNO catalysts showed significant durability and stability because of decreasing half-wave potential. The results reveal that the simple method of the N-CNO catalyst derived from collagen waste is the best option for direct fuel cells in alkaline media.

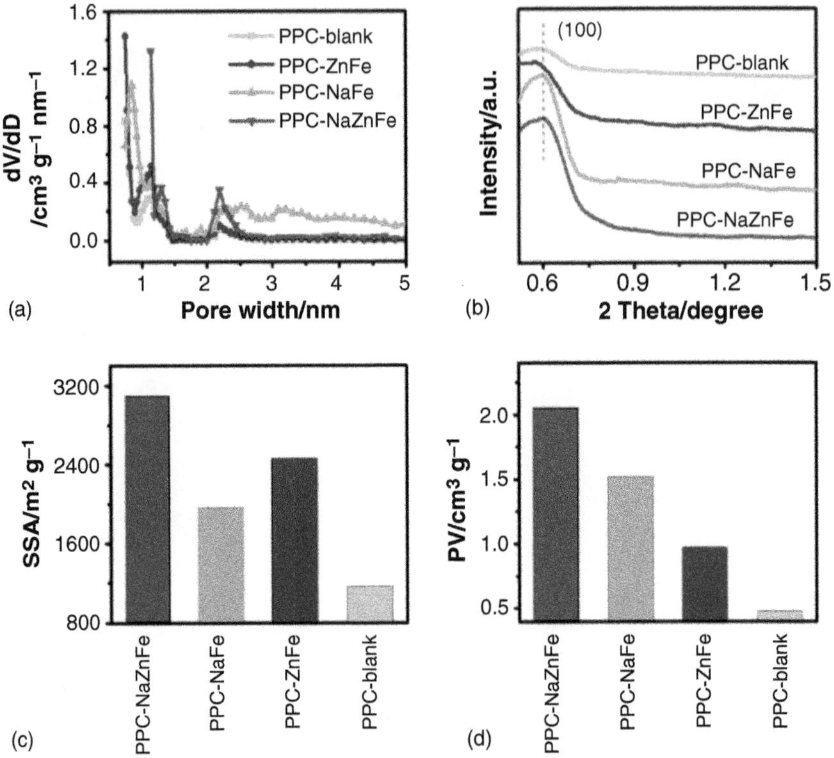

Figure 10.11 (a) PPCs pore size distribution, (b) XRD patterns of the PPCs, (c) specific surface area (SSA) of the PPCs, (d) PPCs pore volume (PV). Source: Ref. [66] / with permission of Elsevier.

Huang et al. [68] prepared self-doped nitrogen porous carbon nanosheets [NPCNS] derived from *Euonymus japonicus* leaves by a facile and green pyrolysis method (Figure 10.13a). They found that the biomass-derived NPCNS showed excellent electrocatalytic activities in both oxygen reduction and evolution reactions. The morphological studies of NPCNS (700, 800, 900, 1000) samples under different temperature conditions were characterized by scanning electron microscope (SEM), transmission electron microscope (TEM), and atomic force microscope (AFM), as shown in Figure 10.13b–g. The results indicated that the sample possesses an irregular shape and porous sponge nanostructure used for enhancing a large surface area to increase the electrocatalytic activities of ORR and OER. Figure 10.13h showed the durability results of NPCNS-900 for OER was characterized by continuous electrolysis at 1.6 V for 20 hours. It is found that the loss of current density is only 16.4%, and NPCNS-900 maintains its original morphology (insert in Figure 10.13h) after the long-term catalysis.

Li et al. [69] prepared ternarily doped (N, S and Fe) porous carbon aerogel [HDCA-X] from the cocoon by pyrolysis method and ternary doping levels are modified by adjusting the different pyrolysis temperatures. Due to the synergistic effect of different ternary doped atoms, the ORR electrocatalytic activity can be increased and used for fuel cells. They found that HDPC is one of the promising

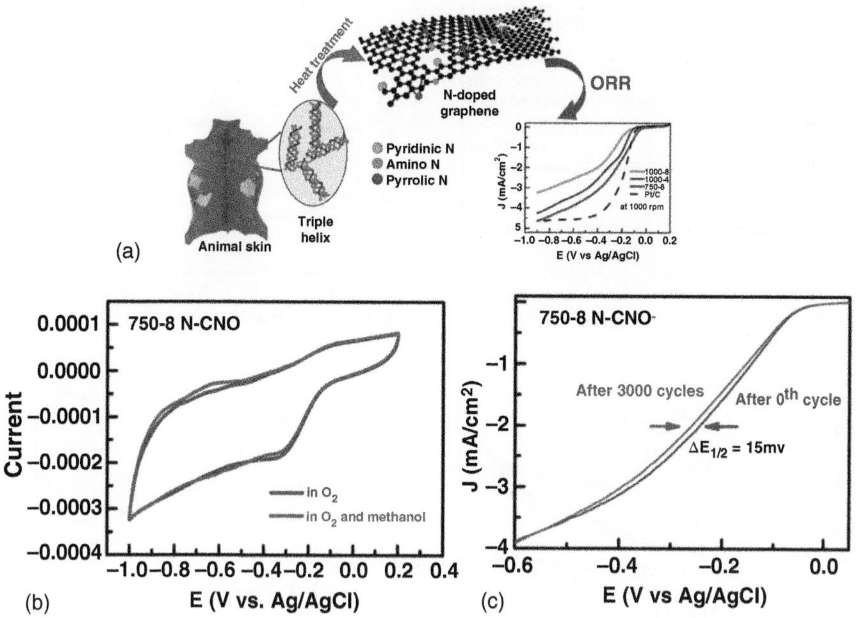

Figure 10.12 (a) Synthesis of N-doped graphene from animal waste. Tolerance and stability test for 750-8 N-CNO catalyst, (b) CVs of 750-8 N-CNO catalyst, (c) RDE polarization curves of 750-8 N-CNO catalyst with a scan rate of $10\,mV\,s^{-1}$. Source: Ref. [67] / Reproduced with permission from Elsevier.

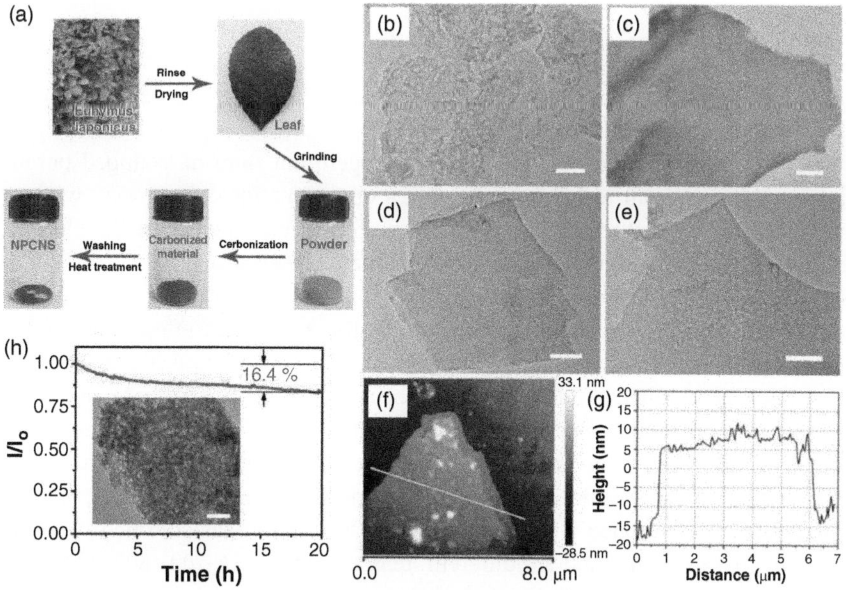

Figure 10.13 (a) Preparation of self-doped nitrogen porous carbon from *Euonymus japonicus* leaves. TEM images of (b) NPCNS-700, (c) NPCNS-800, (d) NPCNS-900, (e) NPCNS-1000. (f) AFM image and corresponding (g) height profiles of NPCNS-900. (h) Time-dependent current density curve for NPCNS-900 at 1.6 V for 20 h. Insert in (h) is the TEM image of NPCNS-900 after the durability test. Source: Ref. [68] / Reproduced with permission from Elsevier.

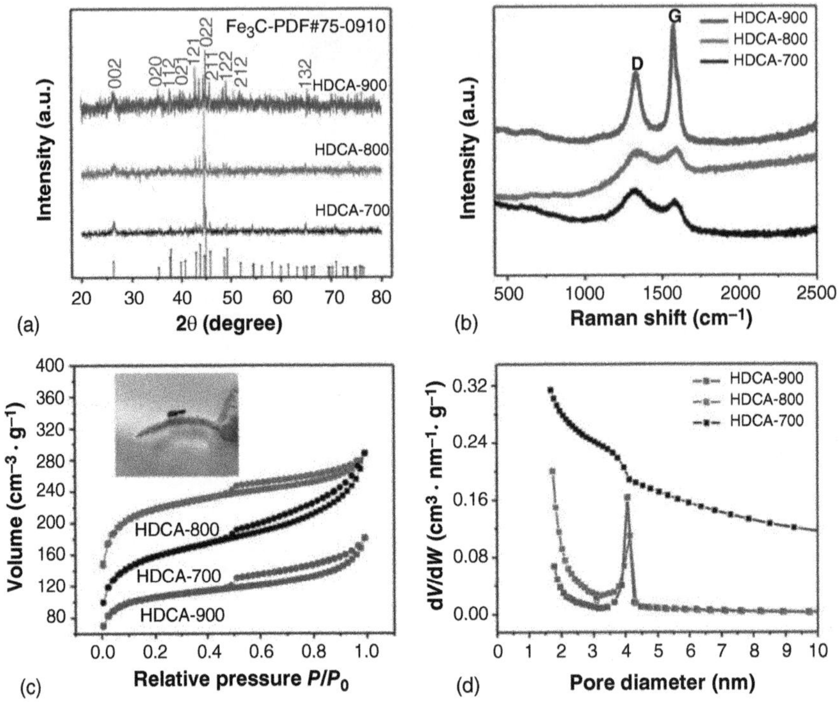

Figure 10.14 (a) XRD pattern, (b) Raman spectra, (c) N_2 adsorption/desorption isotherm, (d) pore diameter distribution of HDCA-700, HDCA-800, and HDCA-900. Source: Ref. [69] / Reproduced with permission from Elsevier.

alternative materials to commercial Pt/C catalysts. The XRD, pore size distribution and Raman spectra as shown in Figure 10.14.

Wu et al. [70] prepared the metal-free nitrogen and fluorine codoped porous carbon materials from tea residue through a one-step annealing process without any activation or posttreatment. The ORR performance depends on the electron transfer number. Due to the doping of heteroatoms, the electronegativity difference is induced, which is helpful for increasing the electron transfer rate in ORR comparable to commercial Pt/C one. Due to the synergistic effect of N and F codoping, the electrocatalyst properties are enhanced and proved an excellent alternative potential candidate for ORR because of high selectivity and high stability with the high electrocatalytic property. The codoping of heteroatoms enhances properties such as high specific surface area, high pore structure, and high contents of pyridinic and graphitic nitrogen. Huang et al. [68] prepared and studied the N-doped porous carbon materials from Malachium Aquaticum (MA) by pyrolysis method at different temperatures (MA-700, 800, 900, and 1000) as shown in Figure 10.15. Out of all samples, MA-900 having high charge transfer, which enhances the electrocatalytic activities of ORR. They found that the current density of the prepared sample ($1.3\,mA\,cm^{-2}$) is double that of Pt/C ($0.69\,mA\,cm^{-2}$).

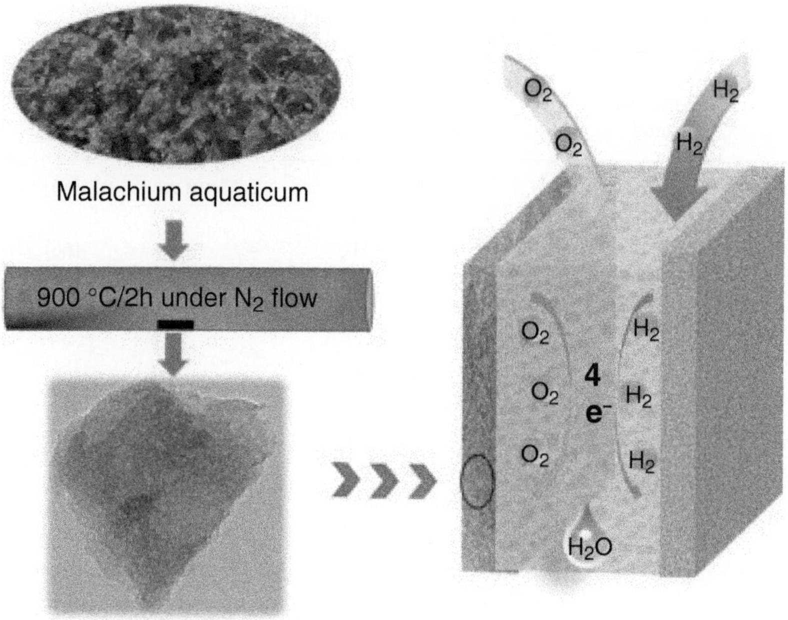

Figure 10.15 Synthesis of N-doped porous carbon materials from Malachium Aquaticum. Source: Ref. [71] / Reproduced with permission from Elsevier.

Chaudhari et al. [72] prepared highly porous carbon-containing heteroatoms such as N, S, P, Si as a carbon precursor from human urine waste [URC]. Due to N-doping, the electronegativity of nitrogen (3.04) and carbon (2.55) in the carbon matrix changing the C—C bond length of sp^2 hybridization. Because of these changes, the carbon surface became asymmetric and showed more activeness in ORR. Due to the correct combination between URC and doping content, the primary keys of ORR, such as high porosity for greater surface area, doping of heteroatom content, and electrical conductivity, can be enhanced and proved valuable electrode materials for the fuel cell. The catalysts derived from various biomass for fuel cell applications are enlisted in Table 10.1.

Khan et al. [77] synthesized the large-scale graphitic shells like carbon nano onions (GS-CNOs) of 15 nm particle size with maximum BET surface area 214 m^2/gram by direct solution method with a current density of 5.9 mA $cm^{-2,}$ which is higher than that of commercial Pt/C. The CV and LSV curves are used to find the onset potential, half-wave potential, and current density information, which are very useful for calculating the fuel cell efficiency. Chung et al. [74] synthesized N-doped porous carbon materials from coffee waste by single-step carbonizations, as shown in Figure 10.16. N-doped porous carbon materials are auspicious alternative materials than Pt/C for electrochemical energy storage systems.

Liu et al. [75] prepared the honeycomb-like Fe-N doped porous carbon materials [Fe-N-PC] from soybean straw biomass as a precursor by a one-step pyrolysis method. Due to multiple doping contents, the carbon can form more active sites with iron and nitrogen. The Fe-N-PC showed excellent ORR electrocatalytic activity

Table 10.1 List of catalysts derived from various biomass for fuel cell applications.

Biomass source	Synthesis method	brunauer–emmett–teller (BET) surface area	Current density	Reference
Pristine tea leaves	One step pyrolysis method	HDPC-800 345.76 $m^2\,g^{-1}$	5.07 $mA\,cm^{-2}$	[65]
Pomelo peel	Pyrolysis method treated with hypersaline salts	PPC-NaZnFe 3092 $m^2\,g^{-1}$	–	[66]
		PPC-1165 m^2/gram		
Animal skin	Simple synthesis	750-8 N-CNO catalyst	10 $mA\,cm^{-2}$	[67]
Euonymus japonicus leaves	Green pyrolysis	NPCNS-700 750.45 $m^2\,g^{-1}$	2.20 $mA\,cm^{-2}$	[68]
		NPCNS-800 882.67 $m^2\,g^{-1}$	2.73 $mA\,cm^{-2}$	
		NPCNS-900 842.33 $m^2\,g^{-1}$	3.37 $mA\,cm^{-2}$	
		NPCNS-1000 683.26 $m^2\,g^{-1}$	2.22 $mA\,cm^{-2}$	
Cocoon	Pyrolysis	HDCA-700 527.5 $m^2\,g^{-1}$	2.33 $mA\,cm^{-2}$	[69]
		HDCA-800 714.4 $m^2\,g^{-1}$	3.8 $mA\,cm^{-2}$	
		HDCA-900 349.3 $m^2\,g^{-1}$	1.31 $mA\,cm^{-2}$	
Tea residue	Pyrolysis without activation	T-NFC 855.6 m^2g^{-1}	5.1 $mA\,cm^{-2}$	[70]
Sargassum spp.	Pyrolysis	SDO- 133.871 $m^2\,g^{-1}$	4.78 $mA\,cm^{-2}$	[73]
Malchium Aquaticum	Pyrolysis	851.42 $m^2\,g^{-1}$	1.30 $mA\,cm^{-2}$	[71]
Human urine (URC)	Template-free route synthesis	URC-700 1080.8 $m^2\,g^{-1}$	−3.5 $mA\,cm^{-2}$ (±0.01) less or more for all samples	[72]
		URC-800 1436.8 $m^2\,g^{-1}$		
		URC-900 1064.9 $m^2\,g^{-1}$		
		URC-1000 811.4 $m^2\,g^{-1}$		
		URC-1100 602.2 $m^2\,g^{-1}$		
Coffee waste	Single-step carbonization	126.6 $m^2\,g^{-1}$	2.0 $mA\,cm^{-2}$	[74]
Soyabean straw	One-step pyrolysis	520.9 $m^2\,g^{-1}$	–	[75]
Golden shower pods	Solvent-free synthesis	N-PC 839 $m^2\,g^{-1}$	2.76 $mF\,cm^{-2}$	[76]

Figure 10.16 Synthesis of N-doped porous carbon from coffee waste. Source: Ref. [74] / with permission of American Chemical Society.

than the commercial Pt/C catalyst. All electrochemical measurements indicate that Fe-N-PC catalysts are the best promising alternative material compared with other N-PCs and Pt/C. The SEM and TEM images are confirmed the honeycomb structure of Fe-N-PC.

10.6 Future Outlook

The primary studies need to be done in terms of the properties of biomass-derived carbon concerning its various synthesis methods. The interrelation of pore morphology and surface chemistry on carbon materials influences performance characteristics needs to be improved. The cheap resources of biomass such as waste should be explored. The synthesis conditions should be tuned to get biomass-derived carbons with high surface area, amorphous region, and sufficient functional groups. Instead of doping, codoping (polyatomic doping) can enhance the properties of carbon in terms of synergic effects so that other atoms doping may be improved. The development of cost-effective, durability, and efficient biomass-derived ORR catalysts remains a significant challenge for fuel cell applications.

10.7 Summary

The high electronic conductivity, low cost, readily available, and environmentally friendly features of carbon nanomaterials influence much potential in electrochemical energy storage devices. The N-doped carbon porous materials show excellent ORR performance, high durability, large surface area, and methanol tolerance because of changing the electronic structure of carbon. The metal-free electrocatalysts are among the best promising alternative materials for ORR to replace the commercial Pt/C one. Due to the heterodoping of atoms such as N, S, P with carbon porous materials, the ORR performance significantly improved because of the synergistic effect between heteroatoms and sp^2 hybridized carbon. The ORR performance is also greatly affected by the pore size. By creating more electrochemical active sites, we need to adjust doping content for tuning the pore size. Extensive research is carrying out on different coordination doping with heteroatoms such as N, P, S to increase the performance of ORR. This chapter summarized the various synthesis methods of porous carbon from biomass feedstock and addressed ORR's importance for fuel cell applications. This chapter concluded that extensive research on different types of ORR electrocatalysts preparation to date shows more excellent selectivity, high stability, low cost, and excellent electrocatalytic properties for increasing the overall performance of the fuel cell system.

References

1 Dai, L., Chang, D.W., Baek, J.-B., and Lu, W. (2012). Carbon nanomaterials for advanced energy conversion and storage. *Small* 8: 1130–1166. https://doi.org/10.1002/small.201101594.

2 Chheda, J.N., Huber, G.W., and Dumesic, J.A. (2007). Liquid-phase catalytic processing of biomass-derived oxygenated hydrocarbons to fuels and chemicals. *Angew. Chem. Int. Ed.* 46: 7164–7183. https://doi.org/10.1002/anie.200604274.

3 Hu, B., Wang, K., Wu, L. et al. (2010). Engineering carbon materials from the hydrothermal carbonization process of biomass. *Adv. Mater.* 22: 813–828. https://doi.org/10.1002/adma.200902812.

4 Lai, X., Halpert, J.E., and Wang, D. (2012). Recent advances in micro-/nano-structured hollow spheres for energy applications: from simple to complex systems. *Energ. Environ. Sci.* 5: 5604–5618. https://doi.org/10.1039/c1ee02426d.

5 Jiang, M., Yu, X., Yang, H., and Chen, S. (2020). Optimization strategies of preparation of biomass-derived carbon electrocatalyst for boosting oxygen reduction reaction: a minireview. *Catalysts* 10: 1472. https://doi.org/10.3390/catal10121472.

6 Noked, M., Soffer, A., and Arubach, D. (2011). The electrochemistry of activated carbonaceous materials: past, present, and future. *J. Solid State Electrochem.* 15: 1563–1578. https://doi.org/10.1007/s10008-011-1411-y.

7 Titirici, M.M., White, R.J., Brun, N. et al. (2015). Sustainable carbon materials. *Chem. Soc. Rev.* 44: 250–290. https://doi.org/10.1039/c4cs00232f.

8 Upare, D.P., Yoon, S., and Lee, C.W. (2011). Nano-structured porous carbon materials for catalysis and energy storage. *Korean J. Chem. Eng.* 28: 731–743. https://doi.org/10.1007/s11814-010-0460-8.

9 Frackowiak, E. and Béguin, F. (2001). Carbon materials for the electrochemical storage of energy in capacitors. *Carbon N. Y.* 39: 937–950. https://doi.org/10.1016/S0008-6223(00)00183-4.

10 Dai, H. (2002). Carbon nanotubes: synthesis, integration, and properties. *Acc. Chem. Res.* 35: 1035–1044. https://doi.org/10.1021/ar0101640.

11 Field, C.B., Behrenfeld, M.J., Randerson, J.T., and Falkowski, P. (1998). Primary production of the biosphere: Integrating terrestrial and oceanic components. *Science* 281: 237–240. https://doi.org/10.1126/science.281.5374.237.

12 Wang, D.-W., Li, F., Liu, M. et al. (2008). 3D aperiodic hierarchical porous graphitic carbon material for high-rate electrochemical capacitive energy storage. *Angew. Chem. Int. Ed.* 47: 373–376. https://doi.org/10.1002/anie.200702721.

13 Kubo, S., White, R.J., Tauer, K., and Titirici, M.M. (2013). Flexible coral-like carbon nanoarchitectures via a dual block copolymer-latex templating approach. *Chem. Mater.* 25: 4781–4790. https://doi.org/10.1021/cm4029676.

14 Liu, X., Fechler, N., and Antonietti, M. (2013). Salt melt synthesis of ceramics, semiconductors and carbon nanostructures. *Chem. Soc. Rev.* 42: 8237–8265. https://doi.org/10.1039/c3cs60159e.

15 Xie, Z.-L. and Su, D.S. (2015). Ionic liquid based approaches to carbon materials synthesis. *Eur. J. Inorg. Chem.* 2015: 1137–1147. https://doi.org/10.1002/ejic.201402607.

16 Li, M., Xu, F., Li, H., and Wang, Y. (2016). Nitrogen-doped porous carbon materials: promising catalysts or catalyst supports for heterogeneous hydrogenation and oxidation. *Catal. Sci. Technol.* 6: 3670–3693. https://doi.org/10.1039/c6cy00544f.

17 Zhang, P., Zhu, H., and Dai, S. (2015). Porous carbon supports: recent advances with various morphologies and compositions. *ChemCatChem* 7: 2788–2805. https://doi.org/10.1002/cctc.201500368.

18 Yao, Y. and Wu, F. (2015). Naturally derived nanostructured materials from biomass for rechargeable lithium/sodium batteries. *Nano Energy* 17: 91–103. https://doi.org/10.1016/j.nanoen.2015.08.004.

19 Song, M., Zhou, Y., Ren, X. et al. (2019). Biomass-based porous carbon for supercapacitor: the influence of preparation processes on structure and performance. *J. Colloid Interface Sci.* 535: 276–286. https://doi.org/10.1016/j.jcis.2018.09.055.

20 Kaur, P., Verma, G., and Sekhon, S.S. (2019). Biomass derived hierarchical porous carbon materials as oxygen reduction reaction electrocatalysts in fuel cells. *Prog. Mater. Sci.* 102: 1–71. https://doi.org/10.1016/j.pmatsci.2018.12.002.

21 Tang, J., Liu, J., Torad, N.L. et al. (2014). Tailored design of functional nanoporous carbon materials toward fuel cell applications. *Nano Today* 9: 305–323. https://doi.org/10.1016/j.nantod.2014.05.003.

22 Ganesan, A., Narayanasamy, M., Shunmugavel, K., and Jayanthi Chinnappa, I. (2016). Ultra low loading of anode catalyst for direct methanol fuel cells with ZrO pyrolysed (PANI-melamine) as catalyst support. *Int. J. Hydrogen Energy* 41: 8963–8977. https://doi.org/10.1016/j.ijhydene.2016.03.135.

23 Ganesan, A., Narayanasamy, M., and Shunmugavel, K. (2018). Self-humidifying manganese oxide-supported Pt electrocatalysts for highly-durable PEM fuel cells. *Electrochim. Acta* 285: 47–59. https://doi.org/10.1016/j.electacta.2018.08.001.

24 Antolini, E. (2009). Carbon supports for low-temperature fuel cell catalysts. *Appl. Catal. Environ.* 88: 1–24. https://doi.org/10.1016/j.apcatb.2008.09.030.

25 Antolini, E. and Gonzalez, E.R. (2009). Ceramic materials as supports for low-temperature fuel cell catalysts. *Solid State Ion.* 180: 746–763. https://doi.org/10.1016/j.ssi.2009.03.007.

26 Han, M., Chan, S.H.H., and Jiang, S.P.P. (2007). Investigation of self-humidifying anode in polymer electrolyte fuel cells. *Int. J. Hydrogen Energy* 32: 385–391. https://doi.org/10.1016/j.ijhydene.2006.08.034.

27 Zheng, L., Zeng, Q., Liao, S., and Zeng, J. (2012). Highly performed non-humidification membrane electrode assembly prepared with binary RuO 2-SiO 2 oxide supported Pt catalysts as anode. *Int. J. Hydrogen Energy* 37: 13103–13109. https://doi.org/10.1016/j.ijhydene.2012.04.110.

28 Butti, S.K., Velvizhi, G., Sulenen, M.L.K. et al. (2016). Microbial electrochemical technologies with the perspective of harnessing bioenergy: Maneuvering towards upscaling. *Renew. Sustainable Energy Resources* 53: 462–476. https://doi.org/10.1016/j.rser.2015.08.058.

29 Wang, C.H., Lai, C.M., Huang, W.H. et al. (2012). Failure mode analysis of membrane electrode assembly (MEA) for PEMFC under low humidity operation. *J. Chinese Chem. Soc.* 59: 1313–1322. https://doi.org/10.1002/jccs.201200319.

30 Shao, Y., Yin, G., and Gao, Y. (2007). Understanding and approaches for the durability issues of Pt-based catalysts for PEM fuel cell. *J. Power Sources* 171: 558–566. https://doi.org/10.1016/j.jpowsour.2007.07.004.

31 Zhang, J., Litteer, B.A., Gu, W. et al. (2007). Effect of hydrogen and oxygen partial pressure on Pt precipitation within the membrane of PEMFCs. *J. Electrochem. Soc.* 154: B1006. https://doi.org/10.1149/1.2764240.

32 Ascarelli, P., Contini, V., and Giorgi, R. (2002). Formation process of nanocrystalline materials from x-ray diffraction profile analysis: application to platinum catalysts. *J. Appl. Phys.* 91: 4556–4561. https://doi.org/10.1063/1.1453495.

33 Shao, M., Chang, Q., Dodelet, J.P., and Chenitz, R. (2016). Recent advances in electrocatalysts for oxygen reduction reaction. *Chem. Rev.* 116 (359): 4–3657. https://doi.org/10.1021/acs.chemrev.5b00462.

34 Deng, Q., Li, X., Zuo, J. et al. (2010). Power generation using an activated carbon fiber felt cathode in an upflow microbial fuel cell. *J. Power Sources* 195: 1130–1135. https://doi.org/10.1016/j.jpowsour.2009.08.092.

35 Qiao, Y., Bao, S.J., and Li, C.M. (2010). Electrocatalysis in microbial fuel cells – from electrode material to direct electrochemistry. *Energ. Environ. Sci.* 3: 544–553. https://doi.org/10.1039/b923503e.

36 Majidi, M.R., Shahbazi Farahani, F., Hosseini, M., and Ahadzadeh, I. (2019). Low-cost nanowired α-MnO2/C as an ORR catalyst in air-cathode microbial fuel cell. *Bioelectrochemistry* 125: 38–45. https://doi.org/10.1016/j.bioelechem.2018.09.004.

37 Papiya, F., Das, S., Pattanayak, P., and Kundu, P.P. (2019). The fabrication of silane modified graphene oxide supported Ni–Co bimetallic electrocatalysts: a catalytic system for superior oxygen reduction in microbial fuel cells. *Int. J. Hydrogen Energy* 44: 25874–25893. https://doi.org/10.1016/j.ijhydene.2019.08.020.

38 Zheng, Y., Yang, D.S., Kweun, J.M. et al. (2016). Rational design of common transition metal-nitrogen-carbon catalysts for oxygen reduction reaction in fuel cells. *Nano Energy* 30: 443–449. https://doi.org/10.1016/j.nanoen.2016.10.037.

39 Fu, G., Wang, Y., Tang, Y. et al. (2019). Superior oxygen electrocatalysis on Nickel Indium Thiospinels for rechargeable Zn-air batteries. *ACS Mater. Lett.* 1: 123–131. https://doi.org/10.1021/acsmaterialslett.9b00093.

40 Bhowmick, G.D., Kibena-Põldsepp, E., Matisen, L. et al. (2019). Multi-walled carbon nanotube and carbide-derived carbon supported metal phthalocyanines as cathode catalysts for microbial fuel cell applications. *Sustain. Energy Fuel* 3: 3525–3537. https://doi.org/10.1039/c9se00574a.

41 Kaare, K., Kruusenberg, I., Merisalu, M. et al. (2016). Electrocatalysis of oxygen reduction on multi-walled carbon nanotube supported copper and manganese phthalocyanines in alkaline media. *J. Solid State Electrochem.* 20: 921–929. https://doi.org/10.1007/s10008-015-2990-9.

42 Liu, L., Yang, X., Ma, N. et al. (2016). Scalable and cost-effective synthesis of highly efficient Fe_2N-based oxygen reduction catalyst derived from seaweed biomass. *Small* 12: 1295–1301. https://doi.org/10.1002/smll.201503305.

43 Sawant, S., Han, T., and Cho, M. (2016). Metal-free carbon-based materials: promising electrocatalysts for oxygen reduction reaction in microbial fuel cells. *Int. J. Mol. Sci.* 18: 25. https://doi.org/10.3390/ijms18010025.

44 Zhao, J., Liu, Y., Quan, X. et al. (2017). Nitrogen-doped carbon with a high degree of graphitisation derived from biomass as high-performance electrocatalyst for oxygen reduction reaction. *Appl. Surf. Sci.* 396: 986–993. https://doi.org/10.1016/j.apsusc.2016.11.073.

45 Magnussen, S. and David Reed Knowledge reference for national forest assessments - modeling for estimation and monitoring (n.d.). http://www.fao.org/forestry/17111/en/ ().

46 Basu, P. (2010). *Biomass Gasification and Pyrolysis*, 1–376. Academic Press. 9780080961620.

47 Yang, H., Yan, R., Chen, H. et al. (2007). Characteristics of hemicellulose, cellulose and lignin pyrolysis. *Fuel* 86: 1781–1788. https://doi.org/10.1016/j.fuel.2006.12.013.

48 Wang, G., Zhang, L., and Zhang, J. (2012). A review of electrode materials for electrochemical supercapacitors. *Chem. Soc. Rev.* 41: 797–828. https://doi.org/10.1039/c1cs15060j.

49 Kurosaki, F., Koyanaka, H., Tsujimoto, M., and Imamura, Y. (2008). Shape-controlled multi-porous carbon with hierarchical micro-meso-macro pores synthesised by flash heating of wood biomass. *Carbon N. Y.* 46: 850–857. https://doi.org/10.1016/j.carbon.2008.02.014.

50 White, R.J., Budarin, V., Luque, R. et al. (2009). Tuneable porous carbonaceous materials from renewable resources. *Chem. Soc. Rev.* 38: 3401–3418. https://doi.org/10.1039/b822668g.

51 Zhang, L.L. and Zhao, X.S. (2009). Carbon-based materials as supercapacitor electrodes. *Chem. Soc. Rev.* 38: 2520–2531. https://doi.org/10.1039/b813846j.

52 Williams, P.T. and Reed, A.R. (2006). Development of activated carbon pore structure via physical and chemical activation of biomass fibre waste. *Biomass Bioenergy* 30: 144–152. https://doi.org/10.1016/j.biombioe.2005.11.006.

53 Wang, J. and Kaskel, S. (2012). KOH activation of carbon-based materials for energy storage. *J. Mater. Chem.* 22: 23710–23725. https://doi.org/10.1039/c2jm34066f.

54 Deng, J., Xiong, T., Xu, F. et al. (2015). Inspired by bread leavening: one-pot synthesis of hierarchically porous carbon for supercapacitors. *Green Chem.* 17: 4053–4060. https://doi.org/10.1039/c5gc00523j.

55 Kang, Z., Wu, C., Dong, L. et al. (2019). 3D porous copper skeleton supported zinc anode toward high capacity and long cycle life zinc ion batteries. *ACS Sustain. Chem. Eng.* 7: 3364–3371. https://doi.org/10.1021/acssuschemeng.8b05568.

56 Luo, W., Wang, B., Heron, C.G. et al. (2014). Pyrolysis of cellulose under ammonia leads to nitrogen-doped nanoporous carbon generated through methane formation. *Nano Lett.* 14: 2225–2229. https://doi.org/10.1021/nl500859p.

57 Li, Y., Li, Q., Wang, H. et al. (2019). Recent progresses in oxygen reduction reaction electrocatalysts for electrochemical energy applications. *Electroche. Ene. Revs.* 2: 518–538. https://doi.org/10.1007/s41918-019-00052-4.

58 Raj, C.R., Samanta, A., Noh, S.H. et al. (2016). Emerging new generation electrocatalysts for the oxygen reduction reaction. *J. Mater. Chem. A* 4: 11156–11178. https://doi.org/10.1039/C6TA03300H.

59 Choi, C.H., Lim, H.K., Chung, M.W. et al. (2018). The Achilles' heel of iron-based catalysts during oxygen reduction in an acidic medium. *Energ. Environ. Sci.* 11: 3176–3182.

60 Chiwata, M., Yano, H., Ogawa, S. et al. (2016). Oxygen reduction reaction activity of carbon-supported Pt–Fe, Pt–Co, and Pt–Ni alloys with stabilised Pt-skin layers. *Electrochemistry* 84: 133–137. https://doi.org/10.5796/electrochemistry.84.133.

61 Beermann, V., Gocyla, M., Willinger, E. et al. (2016). Rh-doped Pt–Ni octahedral nanoparticles: understanding the correlation between elemental distribution, oxygen reduction reaction, and shape stability. *Nano Lett.* 16: 1719–1725. https://doi.org/10.1021/acs.nanol ett.5b04636.

62 Zhou, R., Zheng, Y., Jaroniece, M., and Qiao, S.-Z. (2016). Determination of the electron transfer number for the oxygen reduction reaction: from theory to experiment. *ACS Catal.* 6: 4720–4728. https://doi.org/10.1021/acscatal.6b01581.

63 Bard, A.J. and Faulkner, L.R. (1980). *Electrochemical Methods*. John Wiley & Sons.

64 Wang, Z.-B., Li, C.-Z., Gu, D.-M., and Yin, G.-P. (2013). Carbon riveted PtRu/C catalyst from glucose in-situ carbonisation through hydrothermal method for

direct methanol fuel cell. *J. Power Sources* 238: 283–289. https://doi.org/10.1016/j.jpowsour.2013.03.082.

65 Guo, Z., Ren, G., Xiao, G. et al. (2016). Natural tea-leaf-derived, ternary-doped 3D porous carbon as a high-performance electrocatalyst for the oxygen reduction reaction. *Nano Res.* 9: 1244–1255. https://doi.org/10.1007/s12274-016-1020-2.

66 Wang, N., Li, T., Song, Y. et al. (2018). Metal-free nitrogen-doped porous carbons derived from pomelo peel treated by hypersaline environments for oxygen reduction reaction. *Carbon* 130: 692–700. https://doi.org/10.1016/j.carbon.2018.01.068.

67 Chatterjee, K., Ashokkumar, M., Gullapalli, H. et al. (2018). Nitrogen-rich carbon nano-onions for oxygen reduction reaction. *Carbon* 130: 645–651. https://doi.org/10.1016/j.carbon.2018.01.052.

68 Huang, Y., Feng, D., Wu, D.C., and Cheng, D. (2018). Facile preparation of biomass-derived bifunctional electrocatalysts for oxygen reduction and evolution reactions. *Int. J. Hydrogen Energy* 43: 8611–8622. https://doi.org/10.1016/j.ijhydene.2018.03.136.

69 Li, C., Sun, F., and Lin, Y. (2018). Refining cocoon to prepare (N, S, and Fe) ternary-doped porous carbon aerogel as efficient catalyst for the oxygen reduction reaction in alkaline medium. *J. Power Sources* 384: 48–57. https://doi.org/10.1016/j.jpowsour.2018.01.020.

70 Wu, D., Shi, Y., Jing, H. et al. (2018). Tea-leaf-residual derived electrocatalyst: hierarchical pore structure and self nitrogen and fluorine co-doping for efficient oxygen reduction reaction. *Int. J. Hydrogen Energy* 43: 19492–19499. https://doi.org/10.1016/j.ijhydene.2018.08.201.

71 Xianjun, H.H. and Wei, S.G. (2016). Nitrogen-doped porous carbon derived from malachium aquaticum biomass as a highly efficient electrocatalyst for oxygen reduction reaction. *Electrochim. Acta* 220: 427–435. https://doi.org/10.1016/j.electacta.2016.10.108.

72 Chaudhari, N.K., Young Song, M., and Yu, J.-S. (2014). Heteroatom-doped highly porous carbon from human urine. *Sci. Rep.* 4: 5221–5230. https://doi.org/10.1038/srep05221.

73 Escobar, B., Perez-Salcedo, K.Y., Alonso-Lemus, I.L. et al. (2017). N-doped porous Carbon from Sargassum spp. as metal-free electrocatalysts for oxygen reduction reaction in alkaline media. *Int. J. Hydrogen Energy* 42: 30274–30283. http://dx.doi.org/10.1016/j.ijhydene.2017.06.240.

74 Chung, D.Y., Son, Y.J., Yoo, J.M. et al. (2017). Coffee waste-derived hierarchical porous carbon as a highly active and durable electrocatalyst for electrochemical energy applications. *ACS Appl. Mater. Interfaces* 9: 413033–441313. https://doi.org/10.1021/acsami.7b13799.

75 Liu, Y., Su, M., Li, D. et al. (2020). Soybean straw biomass-derived Fe–N co-doped porous carbon as an efficient electrocatalyst for oxygen reduction in both alkaline and acidic media. *RSC Adv.* 10: 6763–6771. https://doi.org/10.1039/C9RA07539A.

76 Sathiskumar, C., Ramakrishnan, S., Vinothkannan, M. et al. (2020). Nitrogen-doped porous carbon derived from biomass used as trifunctional electrocatalyst toward oxygen reduction, oxygen evolution and hydrogen evolution reactions. *Nanomaterials* 10: 76–92. https://doi.org/10.3390/nano10010076.

77 Khan, K., Tareen, A.K., Aslam, M. et al. (2020). Facile synthesis of mayenite electride nanoparticles encapsulated in graphitic shells like carbon nano onions: non-noble-metal electrocatalysts for oxygen reduction reaction (ORR). *Front. Chem.* 7 (934): 1–10. https://doi.org/10.3389/fchem.2019.00934.

11

Biomass-Derived Carbon-Based Materials for Supercapacitor Applications

G. Murugadoss[1], M. Rajaboopathi[2], M. Rajesh Kumar[3], and A. M. Kamalan Kirubaharan[1,4]

[1] *Centre for Nanoscience and Nanotechnology, Sathyabama Institute of Science and Technology, Chennai 600119, Tamil Nadu, India*
[2] *Aalto University, School of Chemical Engineering, Department of Chemical and Metallurgical Engineering, P.O. Box 16100, FI-00076, Aalto, Espoo, Finland*
[3] *Ural Federal University, Institute of Natural Science and Mathematics, Yekaterinburg 620002, Russia*
[4] *Alexander Dubcek University of Trencin, FunGlass - Centre for Functional and Surface Functionalised Glass, Coating Department, 91150 Trencin, Slovakia*

11.1 Introduction

Almost every single person in the globe consumes energy in some way for the survival of day-to-day life. The energy consumption is rapidly increasing and in future the energy can be considered importantly as equivalent to oxygen for the life. The rapid increase of consumption can be realized everyone easily by smart phone charge inadequacy. A decade before, the smart phone battery had power with a minimum capacity of 700 mAh. Nowadays, it has been increased more than seven times, and the normal battery comes with capacity of 5000 mAh. This increase can impact the entire sector associated with battery production and usage including hot topic of global warming and climate change [1].

Before to know what is supercapacitor or ultracapacitor or electrochemical capacitor, the following few paragraphs explain the basics of capacitor and battery. Capacitor and battery are devices used to store electric energy. Based on the device modeling, the way in which energy stored is different. The battery stores energy in the form of chemical, whereas capacitor stores energy as electric field.

11.1.1 Capacitor

The capacitor consists of two conducting plates separated by a distance as shown in Figure 11.1a. The capacitance of the capacitor is determined by ability of energy stored between the plates in the form of electric field [2]. The electric field is defined as $E = (\sigma/\varepsilon_0)$, where σ refers surface charge density of the plate and ε_0 is the vacuum permittivity. The surface charge density is defined as the amount of charges stored

Biomass-Derived Carbon Materials: Production and Applications, First Edition.
Edited by Alagarsamy Pandikumar, Perumal Rameshkumar, and Pitchaimani Veerakumar.

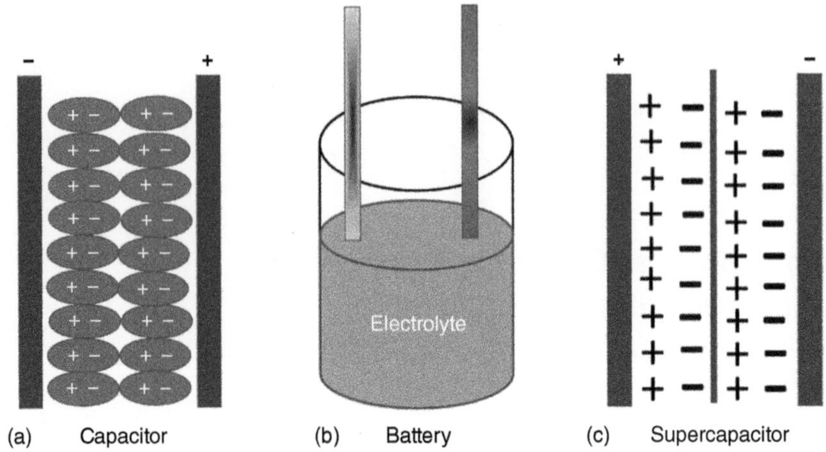

Figure 11.1 Simple diagrammatic representation of capacitor, battery, and supercapacitor.

per unit area (A) and potential difference between the plate is $V = Ed$. Therefore, the capacitance of the capacitor is given by

$$C = \frac{Q}{V} = \frac{Q}{\left(\frac{Qd}{A\varepsilon_0}\right)} = \frac{\varepsilon_0 A}{d} \tag{11.1}$$

The capacitance value will be tuned by varying the values in right side of the Eq. (11.1). More charges can be stored, if larger the surface of each conductor plate used. Furthermore, the distance between the plates and the type of materials filled between the gap also determines the capacitance. The gap may be empty or filled by dielectric medium. The gap does not pass the electron from one plate to another plate and thus it creates electric field between the plates.

11.1.2 Battery

The component in the battery is similar to the capacitor, but the gap is filled by electrolytes, and anode and cathodes are made by different materials, as shown in Figure 11.1b. The electrolytes will vary based on the battery type. The use of electrolyte is to conduct charged atoms, molecules, or ions without allowing electrons to pass through it. The cathode contains materials that release electron easily such as lithium and graphite. The anode and cathode materials are specially prepared to hold more charges so that it can store more energy [3].

The energy density of the battery is high compared to capacitor. It helps to develop modern electronic devices lighter and portable. A battery can store energy thousands folds more than a capacitor. Batteries can also supply energy in a steady and reliable stream. However, the power density of capacitor is larger than battery and provide energy as quickly based on the demand (e.g., flashbulb in a camera).

11.2 Supercapacitor

Supercapacitor holds the advantages of both battery and capacitor. It has two identical conducting electrodes electrically connected with ionic liquid called electrolyte. A membrane is kept between the electrodes. When a voltage is applied to electrode, the positive and negative ions present in the electrolyte travel opposite side toward electrodes. The supercapacitor stores energy on the electrodes in the form of electric field but not in chemical form as battery. Therefore, supercapacitor can store more energy than normal capacitor. Though increasing surface area of the electrodes by coating coarse materials hold more electric charges, the supercapacitor cannot meet the energy density of the battery. A capacitor with huge capacitance values and energy density can be called as supercapacitor [4].

Energy Density: The energy density is defined as the sum of energy stored per unit volume. The unit for energy density is Wh kg^{-1}. The energy stored in the normal capacitor in the form of potential energy is given by equation $E = (½)CV^2$ and for the supercapacitor is $(½)C_{sp}(dV)^2$. The specific capacitance is $C_{sp} = (Idt)/(mdV)$ for a three-electrode system and $C_{sp} = (2Idt)/(mdV)$ for two-electrode system where I referred as discharge current, dt refers discharge time, m is the mass of the active electrode materials, and dV is the potential window in the cyclic voltammetry measurement.

Power Density: The power density is defined as the amount of power stored per unit volume. The unit is W kg^{-1}. The power density is $P = 3.6E/dt$ where E is the electric field developed for the applied voltage per unit time.

In general, batteries having higher energy density but lesser power density than capacitors. The capacitor storing the electrical energy directly between the conducting plates; therefore, the discharging rate for capacitors is higher than batteries. The capacitor can charge faster than a battery due to its energy storage mechanism.

11.2.1 Types of Supercapacitors

The capacitor can be classified based on its working principle as displayed in Figure 11.2. The electrode materials used in the supercapacitors are important and can be selected based on certain criteria. (i) High specific area that needs to hold more charges. (ii) Long-term stability to show good performance without losing its efficiency. (iii) High cyclability for multiple run charge and discharge. (iv) High electrical conductivity to speedy charge and discharge. (v) Resistance to electrolyte solution without corrosion and (vi) high temperature in case of high voltage is applied. The electrolyte materials used in the supercapacitor should have cost effective, high chemical stability, availability, high ionic conductivity, and low viscosity.

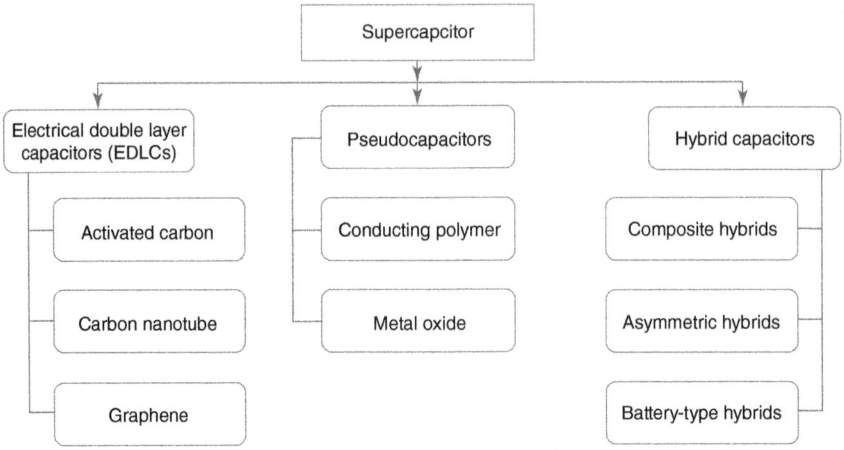

Figure 11.2 Different types of supercapacitors.

11.2.2 Electrical Double-Layer Capacitors (EDLC)

The block diagram of electrical double-layer capacitor (EDLS) is presented in Figure 11.3. The two different carbon-based electrodes E1 and E2 are kept apart by a separator, and an electrolyte is filled in between the electrodes. The EDLS stores energy via non-Faradic process where the double layer formation occurs, and the adsorbed ions are presented at the surface of either positively or negatively charged electrodes. When a voltage is applied between the electrodes, the charge is accumulated on the electrode surfaces and ions present in the electrolyte diffuse across the separator and moves into the pores of the electrode of opposite charge. The formation of charge double layer coupled with high surface area carbon electrode allows EDLC to achieve high-energy density than classical capacitor.

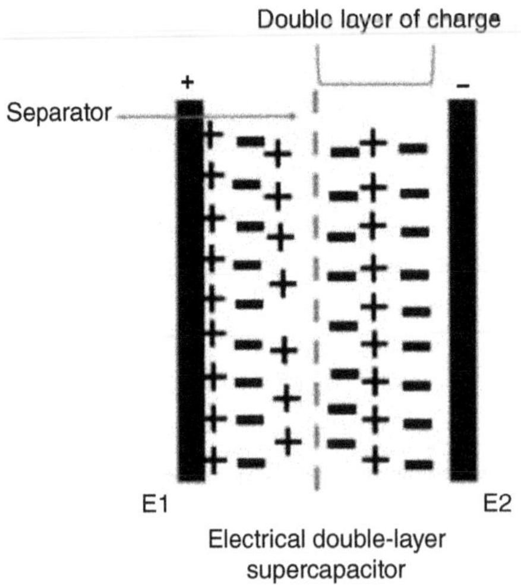

Figure 11.3 Block diagram of electrical double-layer supercapacitor.

Activated carbon electrodes: The increased specific surface area of electrode made from activated carbon (AC) increases the capacitance as it holds more electric charges. The uniform small size pores can increase the ion transport and thus increase the energy and power density compared with conventional capacitor. However, in the AC, the wide pores are common that limits the efficiency. The pore size of AC nanosphere's includes macropores (>50 nm), mesopores (2–50 nm), and micropores (<2 nm). Carbon-based electrodes, particularly those derived from AC, are extensively used as electrodes in most of the supercapacitors, which is available commercially due to its high surface area, low cost, and excellent chemical and thermal stability [5]. The electrochemical performance of ACs in supercapacitor electrodes is largely dependent on the source of carbon and method of fabrication. The desired properties such as electrical conductivity, porosity, pore size, and particle size can be obtained via choosing carefully the carbon source and processing parameters of ACs. Any commercially available ACs are produced from the precursors of fossil fuel based, such as pitch, coal, and polymers, which make them expensive and ecologically unfavorable [6, 7]. Hence, there is a demand to find and utilize sustainable and renewable sources for AC synthesis to develop supercapacitor electrodes with better performance. In this era, researchers interested to focus on the search for biomass- derived green carbon electrode materials development.

Carbon nanotubes and graphene electrodes: The high electrical conductivity and chemical stability attract the carbon nano tubes (CNT) as electrodes for supercapacitor. A sheet of carbon is rolled to cylindrical form where the two opposite side edges (not top and bottom of cylinder) act as electrodes. The ionic charge transport of graphene electrodes is higher than CNT.

11.2.3 Pseudocapacitor

In pseudocapacitor, when the battery is connected with the electrodes, fast and reverse redox reaction takes place on the electrodes and across the double layer flow of charge happened similar to charge–discharge process in battery. In this way, a Faradaic current is passing through the supercapacitor cell. The redox reaction is highly reversible process. The electrodes used are made up of several metal oxides and conducting polymers. It means the electrodes are not that much porous compared to EDLC. The conducting polymers are polypyrrole (PPy), polyaniline (PANI), polyacetylene (PA), and the metal oxides, including Fe_2O_3, MnO_2, RuO_2, Co_3O_4, and SnO_2. There are three types of Faradaic process that occur in a pseudocapacitor; (i) reversible adsorption or underpotential deposition, (ii) redox reaction, and (iii) intercalation process. The reversible adsorption is that the metal ions for, e.g. lead are adsorbed monolayer at different metal electrode, e.g. gold. The ions are electrochemically adsorbed on the electrode surface due to oxidation and reduction process (redox) that occurs at the surface of the electrodes. Intercalation happens when the ions are inserted into the layers of a reduction–oxidation active materials along with Faradaic charge transfer process.

11.2.4 Hybrid Capacitors

The two different electrodes are used in hybrid capacitors in which one electrode is made up of carbon materials and another electrode is from conducting polymers or metal oxides. The limiting factors of pseudocapacitor and EDLC are avoided in this hybrid capacitor. The hybrid capacitor shows better electrochemical behavior than pseudocapacitor and EDLC. The hybrid capacitor can be classified into three types based on configuration of electrodes: battery, asymmetric, and composite type. A single electrode in a hybrid capacitor that combines carbon-based materials either with metal oxide or polymers that are conductive in nature. In asymmetric, the carbon source is set as a negative electrode, and conducting polymer or metal oxide is used as positive electrode. The supercapacitor electrode and battery electrodes are jointly used in battery- type hybrid capacitor.

In Figure 11.4, it is noticed that supercapacitors hold a significant portion in specific power and specific energy. Because of this high-power capability and huge energy density, the need for energy storage can be met in this present scenario [9]. These supercapacitors can be operated in a wide range of operating temperatures; they provide high-power density as well as long cycle life [10]. It is obviously revealed that the supercapacitor is a good choice for future application of functional device with high-power density devices.

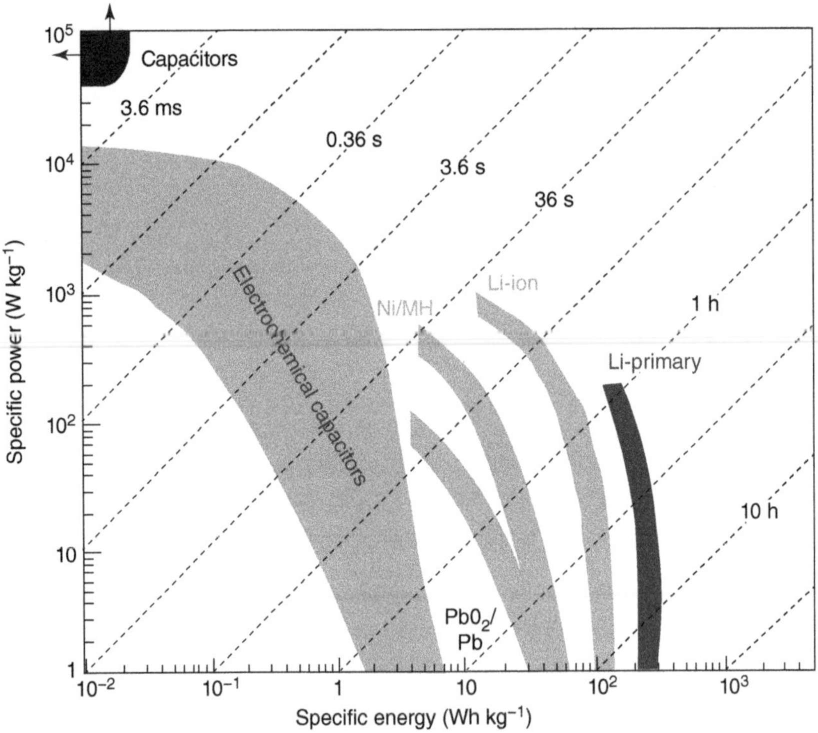

Figure 11.4 Regone plot of energy storage device. Source: Reprinted with permission from Ref. [8].

11.3 Activated Carbon Obtained from Biomass for Supercapacitor Application

Since the AC has been eco-friendly nature and low cost, it has been a main choice of electrode material to the EDLC compared to the other forms of carbon. The AC materials were produced by conventional methods from petroleum, coal and their by-products. However, because of environmental pollution and limited hold of petroleum derivatives, researchers are examining biomass squander as a potential asset for readiness of AC. The AC has been prepared from different biomasses such as tobacco stem, cow dung, animal bones, coconut shell, waste coffee beans, fungi, torreya grandis shell, human hair, nut shells, cotton stalk, coal liquefaction residue, olive pits, and dead leaves have been reported in literature (see Table 11.1).

The biomass materials obtained from natural are expected to replace the conventional fossil fuel in the near future and will play a key role in energy storage and conversion field. Still, the performance of supercapacitors obtained from biomass-derived carbons was not completely satisfactory, especially in poor cyclic performance and low storage capacity. The surface morphology of the electrodes is affected by the original morphology of the biomass waste, and the presence of impurity is greatly affected on materials property. Therefore, the most efficient way to improve the performance of biomass-based carbon materials is through surface modification of these materials. Table 11.1 represents a consolidated details of difference biomass materials and their corresponding Brunauer–Emmert–Teller (BET) surface area values for the ACs. These represented properties greatly promote charge accumulation at the electrode interface and electrolyte, with the formation of electrostatic adsorption of negative and positive charges.

Table 11.1 shows various activation methods used for the preparation of carbon materials derived from biomass and the list of biomass materials that are extensively used for the AC materials production for electrode application in supercapacitors.

11.3.1 Essential for Carbon-based Electrodes

For a decade, there is a huge interest among researchers to study on carbon-based materials owing to its superior electrochemical energy storage. Carbon portrays considered as a right candidate material for supercapacitor electrodes because of its existence in different structures and allotropes. The degree of graphitization owes to a range of micro textures and their dimensionality extends from 3D to 0D. Carbon-based electrodes prepared mainly because of its cost, process ability, and easy accessibility. Additionally, these carbon-based materials are suitable for both basic as well as in acidic conditions [53]. In addition, this can be used in a wide range of operating temperatures owing to its thermal stability. With the help of various physical activation methods, it is desirable to produce materials with large surface area and controlled porosity distribution that majorly determines the electrolyte/electrode interface for electrochemical applications [54].

Table 11.1 Supercapacitor performance results for the biomass carbon electrode materials.

Biomass resources	Activation method	Specific surface area ($cm^2\ g^{-1}$)	Pore volume ($cm^3\ g^{-1}$)	Electrolyte	Highest specific capacitance ($F\ g^{-1}$)	Reference
Pineapple crown	Physical method	684	0.364	1 M H_2SO_4	150	[11]
Coffee endocarp	Physical method	1050	0.50	1 M H_2SO_4	176	[12]
Oil palm	Physical method	1704	0.89	–	150	[13]
Rubber wood sawdust	Physical method	913	0.61	1 M H_2SO_4	138	[14]
Firwoods	Physical method	1016	0.747	1 M $NaNO_3$	105	[15]
Pistachio shell	Physical method	1009	0.667	1 M $NaNO_3$	80	[15]
Lignocellulosic	Physical method	750	0.37	6 M KOH	80	[16]
Corncob	Physical method	1210	0.67	6 M KOH	314	[17]
Shaddok peel	HCl	1272	0.49	0.1 M HCl	–	[18]
Fish scale	KOH	2273	2.74	7 M KOH	68	[19]
Sugarcane bagasse	$ZnCl_2$	2871	0.81	1 M H_2SO_4	109	[20]
Waste paper	KOH	416	0.225	6 M KOH	160	[21]
Water bamboo	KOH	2352	1.11	6 M KOH	268	[22]
Yeast cell	KOH	–	–	1 M KOH	330	[23]
Fallen leaves	KOH/K_2CO_3	2869	1.598	6 M KOH	242	[24]
Coffee beans	$ZnCl_2$	999.67	0.57	1 M H_2SO_4	286	[25]
Pine-cone	KOH	3950	2.395	Organic	198	[26]
Pig bone	KOH	2157	0.77	7 M KOH	185	[27]
Corncob	KOH	3530	1.95	1 M H_2SO_4	401	[28]
Willow catkins	KOH	1586	0.78	6 M KOH	253	[29]
Celtuce leaves	KOH	3404	1.88	2 M KOH	421	[30]

Cherry stone	KOH	1100–1300	–	2 M H_2SO_4	230	[31]
Waste tea leaves	KOH	2841	1.366	2 M KOH	330	[32]
Sunflower seed	KOH	2585	1.41	30 wt.% KOH	311	[33]
Acacia gum	KOH	1764	0.99	1 M HCl	272	[34]
Ginkgo shells	KOH	1775	–	6 M KOH	178	[35]
Cow dung	KOH	1984	0.91	Organic	124	[36]
Silk	KOH	2494	2.28	Liquid ionic	242	[37]
Carrageenan	KOH	2502	1.43	6 M KOH	230	[38]
Human hair	KOH	1306	0.9	6 M KOH	340	[39]
Sewage sludge	KOH	2839	2.65	1 M Na_2SO_4	379	[40]
Wood saw dust	HTC/KOH	2967	1.35	1 M $TEABF_4$	236	[41]
Human hair	KOH	1306	–	1 M H_2SO_4	304	[39]
Hemicellulose	HTC/KOH	2200–2300	~1	0.5 M H_2SO_4	300	[42]
Paper pulp	HTC/KOH	2980	1.75	Organic	166	[43]
D-glucosamine	HTC	598	0.34	1 M H_2SO_4	300	[44]
Rice husk	Microwave/$ZnCl_2$ activation	1442	0.71	6 M KOH	243	[45]
Peanut shell	Microwave/$ZnCl_2$ activation	1634	1.39	6 M KOH	245	[46]
Fungi	HTC	80.08	0.496	6 M KOH	196	[47]
Fungus	HTC/KOH	1103	0.54	6 M KOH	360	[48]
Watermelon	HTC	–	–	6 M KOH	333	[49]
Hemp	HTC/KOH	2287	1.45	Liquid ionic	142	[50]
Bamboo waste	HTC/KOH	1472	–	6 M KOH	301	[51]
Microalgae	HTC/KOH	2190	0.94	6 M LiCl	200	[52]

11.4 Electrochemical Measurements

Electrochemical analysis is the promising method to evaluate capacitive behavior of the all kind of nanomaterials and carbon-based materials. The electrochemical properties of the bio-wasted carbon materials can be obtained using two- and three-electrode electrochemical systems. For the three-electrode system, AC is used as a working electrode, platinum wire is used as counter electrode, and Ag/AgCl is used as a reference electrode. Activated-carbon-based working electrode is prepared by mixing of carbon material (80 wt.%) and conducting carbon black super P (10 wt.%) and 10 wt.% polyvinylidine fluoride (PVDF) dissolved in *N*-methyl-2-pyrrolidone (NMP). In general, working electrode is prepared through coating the slurry solution on nickel foam, dried at 100 °C, and pressed before the electrochemical measurements at 6 M KOH aqueous electrolyte solution. The mass loading was used between the range of 3 and 5 mg, and all the electrochemical measurements are carried out at room temperature. Galvanostatic charge and discharge, cyclic voltammetry (CV), electrochemical impedance spectroscopy (EIS), and cycling tests are used to study the capacitive properties and stability of the materials. For a two-electrode system with two symmetrical electrodes, measuring with desired active area, followed by the same experimental procedure using a separator in between two electrodes that was soaked in an electrolyte.

11.5 Structural Diversities of Biomass-Derived Carbon for Supercapacitor Applications

Specific surface area of the electrode materials used in supercapacitor plays a key role to tailor the energy and power density. AC possesses high specific surface area in the range 500–3000 m^2/g as well as good conductivity that has attracted researchers to enhance the charge storage and also be able to tune the storage power density. The AC can be produced from biomass that is spurred demands for green products, waste to wealth of circular economy, and sustainability. The widely used carbon-based materials are AC, graphene, carbon aerogels, carbon nanotubes, carbon nanofibers, and nanosized carbon particles. The natural biomass contains nitrogen and sulfur that can be used as dopants by external process to improve the active sites and electrical conductivity [55, 56]. The commercially used sources for carbon are bamboo, husk, coconut and walnut shell, bagasse, jute, wheat straw, and rice husk and activating agent such as KOH and K_2CO_3 will be used to obtain the AC. There are plenty of reports and reviews available on the development of carbon from biomass and its activations. The specific capacitance value has been achieved in the range of 150–300 F/g. The important factors influencing the carbon-based supercapacitors performances are pore size and specific surface area. In view of structural diversity, graphene and carbon nanotubes are highly attracted exhibiting specific structural and attractive physicochemical properties [57, 58]. Furthermore, inherent specific morphology such as fibrous, spherical, tubular, sheet, rod, and with adjustable pore size is an added advantage.

11.5.1 Spherical Structure

The carbon spheres obtained from sugars, oatmeal, and coconut oil possess smooth surface with size ranging from 50 nm to 2 μm. The materials can be used for sodium- and lithium-ion batteries. Compared to solid sphere, porous and hollow carbon spheres received attention as it provides good charge accommodation and withstand at high current loads. This porosity can be produced either by using biomass as carbon source or removable template [42, 59]. After treatment, AC nanospheres surface area was found to possess 2871 $m^2\ g^{-1}$, pore volume of 0.81 $cm^3\ g^{-1}$, and specific capacitance of 109 $F\,g^{-1}$ in 1 M H_2SO_4 electrolyte. It also showed the retention of 77% even after 10 000 cycles [20].

11.5.2 Fibrous Structure

Carbon fibers are one-dimensional material that has received huge attention due to high electron transport, surface area, tuneability, and tensile strength. The electrospinning is commonly used technique for fiber production. Similar to sphere, porosity can be introduced in fibers too. Different materials such as seaweed, lotus seedpods, and bacterial cellulose are used as precursors to produce fibrous structure [60, 61].

11.5.3 Tubular Structure

Tubular structure provides better electron transport than fibrous structure. Furthermore, this structure controls the ion transport from bulk electrolyte solution to interior pore thereby shorten the diffusion path. Mostly biomass materials such as cotton [62], willow catkins [63], and dandelion [64] are used as precursors to produce tubular structure. Similar to tubular structure, hollow tubes facilitate energy storage and fast electrolyte ions diffusion, demanding for the supercapacitor capability [65].

Yin et al. [66] developed hollow structure based multitubular AC from coconut fibers using KOH solution. This structure exhibited with a pore volume of 1.59 $cm^3\ g^{-1}$ (30% mesopores) and specific surface area of 2898 $m^2\ g^{-1}$. The supercapacitor electrode prepared with 6 M KOH electrolyte solution exhibited specific capacitance value of 266 $F\,g^{-1}$ at a current value of 0.1 $A\,g^{-1}$, having 76% of its capacitance at 100 $A\,g^{-1}$. They also obtained that the AC electrode with 3D hierarchical porous exhibited a high capacitance of 155 $F\,g^{-1}$ at 0.1 $A\,g^{-1}$ and 142 $F\,g^{-1}$ at 10 $A\,g^{-1}$ and showed a power density of 8224 $W\,kg^{-1}$ and high energy density of 53 $Wh\,kg^{-1}$.

11.5.4 Sheet Structure

Atgrocery stores, there are different food sources that contains cancer fighting properties attributed to the irradical scavenger and antioxidant properties. Within these, catechin belongs to the flavonoid family with polyphenol groups. It also contained in an assortment of normal food varieties and beverages, such as natural products, vegetables, wines, and a few types of tea. Recently, Veeramani et al. synthesized two-dimensional (2D) based sheet-like graphene with large surface area by utilizing *Bougainvillea spectabilis* source material with the aid of a simple chemical method [67]. The as-prepared AC was studied with different physicochemical and scientific procedures to explore their porous and structural properties. Furthermore, the biomass-derived graphene-based carbon materials were used as an anode for supercapacitor application and catechin detecting, and related electrochemical investigations were done utilizing galvanostatic charge–release, cyclic voltammetry, and pulse voltammetry. In supercapacitors, the graphene-based carbon shows energy density of 7.2 W h kg^{-1} and specific capacitance of 233 F g^{-1} at a current density of 1.6 A g^{-1} using symmetric configuration. For sensing applications, GPAC exhibits sensitive detection of catechin (CA). It also demonstrates the conversion of waste biomass into useful electrochemical devices successfully (Figures 11.5 and 11.6).

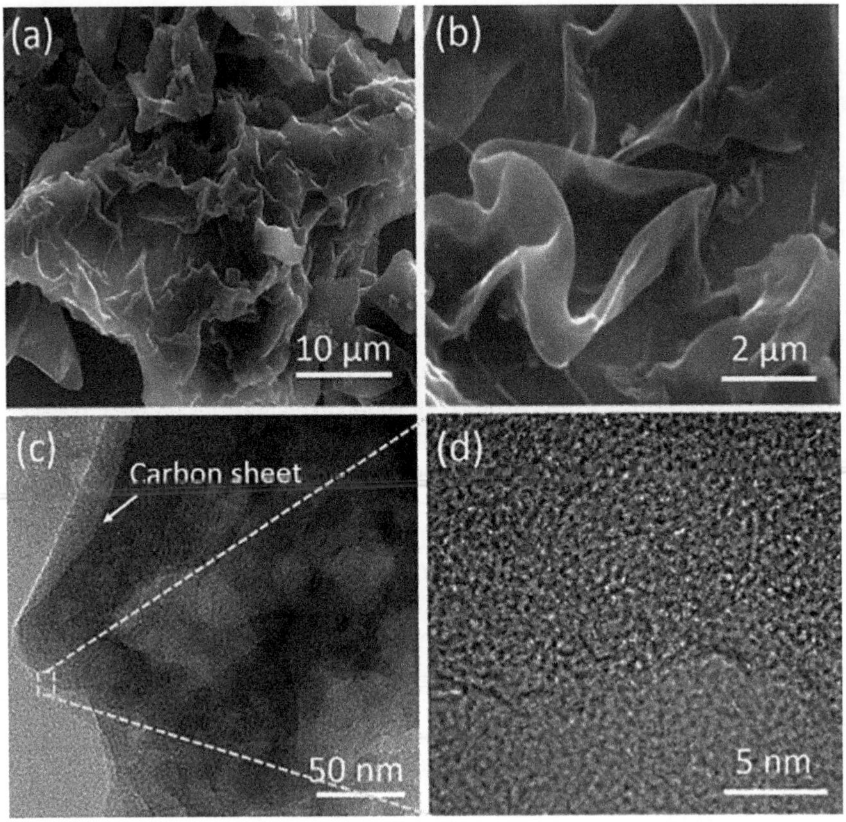

Figure 11.5 (a, b) SEM and (c, d) TEM images of the graphene sheet-like carbon material with different magnifications. Source: From Ref. [67], Royal Society of Chemistry, CC BY 3.0.

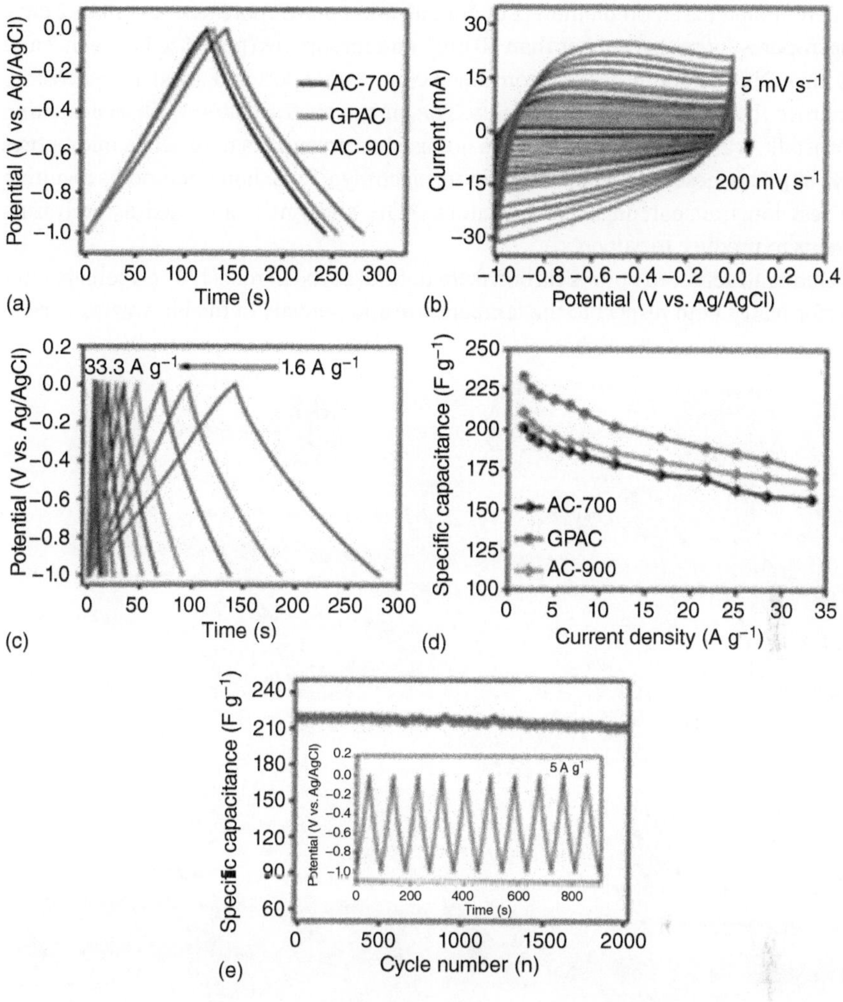

Figure 11.6 Electrochemical performance of various graphene sheet-like carbon material based modified electrodes (at 2 M KOH). (a) Charge–discharge result at 1.6 A g^{-1} for different modified electrodes. (b) Graphene sheet-like carbon electrode CV curves at different scan rate in the range of 5–200 mV s^{-1}, and (c) charge–discharge curves at different current densities. (d) The variation of specific capacitance with respect to the current density. (e) Cyclic stability of the electrode material at 5 A g^{-1} for 2000 cycle and the inset image shows charge–discharge curves at particular cycles. Source: Ref. [67]. Copyright (2016) American Chemical Society.

11.5.5 Porous Structure

To smooth transport of ion diffusion, homogeneous pore size distribution is required. Carbon materials derived from biomass are appropriate to produce controlled pores as the macromolecules in the materials mainly decompose as H_2O and CO_2 resulted in the generation of pores. The porous materials can be differentiated based on diameters of pore as micropores (pore size less than 2 nm), macropores (pore size greater than 50 nm), and mesopores (pore size between 2 and 50 nm). Different pores feature controlled charge transport and acted as reservoir to improve the supercapacitor capability. The ginkgo leaves, cotton stalk, acacia gum, cornstalk, walnut shell, and popcorn are used as precursors to produce micropores [68, 69]. The mesopores can occupy more electrolyte ions than micropores resulting in less ion transport in supercapacitors. KOH and $ZnCl_2$ are used as activating agents to produce mesopores.

Guo et al. achieved porous carbon from natural soybean roots [70]. Supercapacitor performance with respect to the temperature and porosity of the bio-wasted carbon

Figure 11.7 (a) Schematic representation of the scalable preparation of soybean root-derived carbons (SRPCs); (b) SEM surface image of the SRC carbonized at 500 °C; (c–e) SEM surface morphology of SRPC-4K, functionalized at 900 °C for 2 h under nitrogen atmosphere. Source: Reproduced with permission from Ref. [70], American Chemical Society.

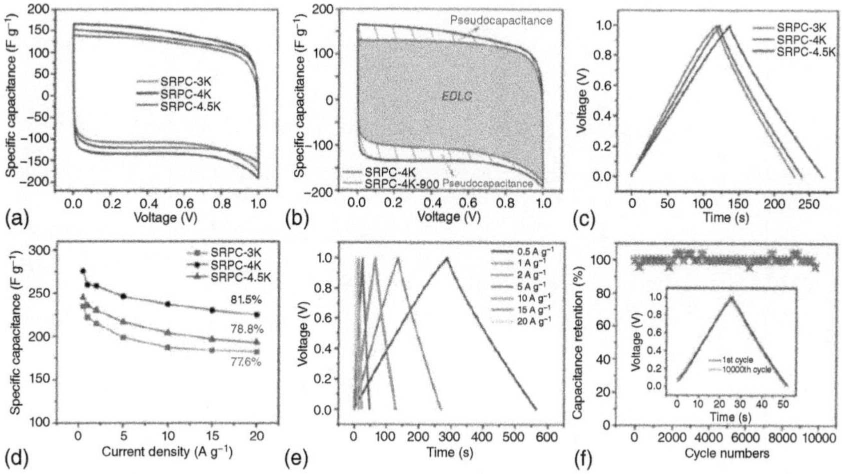

Figure 11.8 (a) Cyclic voltammetry curves at the scan rate of 5 mV s^{-1} for SRPC-3K, SRPC-4K, and SRPC-4.5Ksamples in a two-electrode symmetric supercapacitor in 6 M KOH solution; (b) CV curves of SRPC-4K and SRPC-4K-900 at a scan rate of 5 mV s; (c) Charge–discharge curves at 1 A g^{-1}; (d) GCD performance of SRPC-SRPC-and SRPC-4.5K; (e) GCD curves at different current densities; (f) cycling stability at 5 A g^{-1} for 10 000 cycles of SRPC-4K, insert curves shows GCD of the 1st and 10 000th cycles). Source: Reprinted (adapted) with permission from [70].

was systematically investigated. The collected soybeans roots were heat calcinated at 500 °C for 2 hours in nitrogen atmosphere and functionalized at 900 °C for 2 hours under nitrogen atmosphere. Figure 11.7 shows the sample collection and SEM surface micrographs. Samples were denoted as SRC (soybean root-derived carbons), and the obtained activated porous carbons were named SRPC-nK, where n denotes KOH/char weight ratio.

Using a symmetric two electrode supercapacitor setup in 6 M KOH solution, they described the presence of quasi-rectangular shapes for CV plots (Figure 11.8). The authors found that the sample SRPC-4K showed high specific capacitance of 276 F g^{-1} at a current value of 0.5 A g^{-1} with a capacitance retention of 98% even after 10 000 cycles. A supercapacitor setup is assembled using a liquid ionic electrolyte (EMIM BF4) and reported the power density and energy as 4353 W kg^{-1} and 100.5 Wh kg^{-1}, respectively. Ahmed et al. [71] successfully prepared nitrogen-doped AC from orange peels and reported. The electrochemical performance of the AC was tested with the help of two-electrode assembly using 6 M KOH solution as an electrolyte. They noticed a specific surface area of 1577 m^2 g^{-1} and obtained a specific capacitance of 167 F g^{-1} at 0.7 A g^{-1}. Xie et al. [72] prepared a composite from corn straw and soy protein to obtain AC with a surface area of 1412.9 m^2 g^{-1} with volumetric capacitance and gravimetric capacitance values of 213 F cm^{-3} and 321.1 F g^{-1}, respectively, at 20 A g^{-1} in 6 M KOH solution.

Qu et al. [17] utilized corncob residue to develop porous carbon material for supercapacitor electrodes through a green and cost-effective steam activation method. Porous structure carbon obtained at 850 °C was further treated for ash elimination,

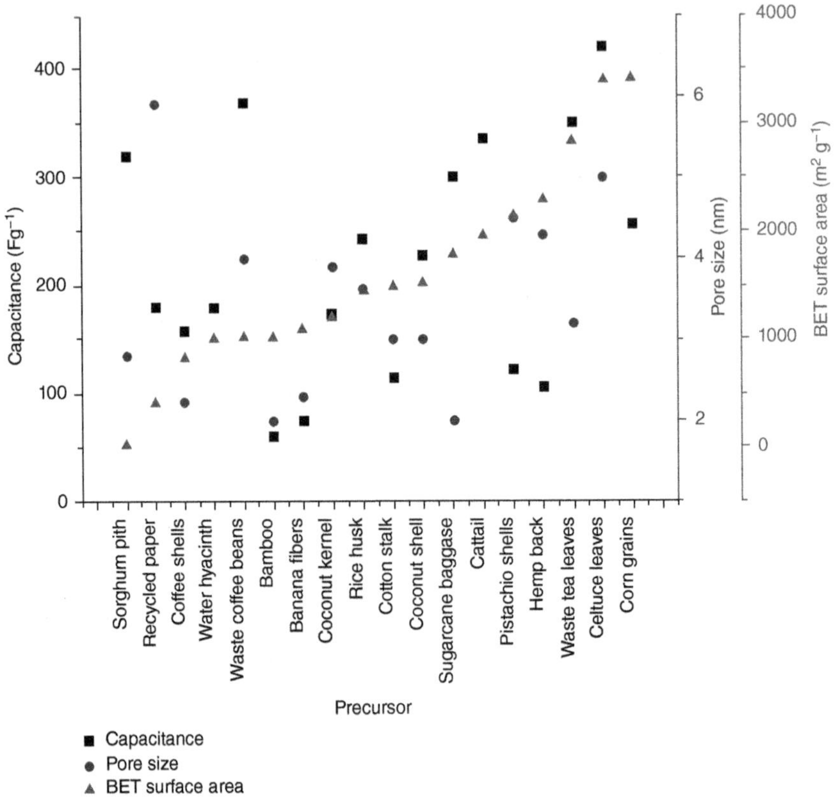

Figure 11.9 BET surface area, capacitance, and pore size of activated carbon derived from different plant sources. Source: Ref. [73].

and acid soaking showed SBET of 1210 $m^2\ g^{-1}$ with a yield of 23.2 wt.%. The authors obtained specific capacitance of 314 $F\ g^{-1}$ at the scan rate of 5 $mV\ s^{-1}$ and showed the capacitance retention up to 82%. They also claimed that the good performance of the samples was ascribed to the porosity, pore distribution, and conductivity of the prepared porous carbon.

Figure 11.9 represents that there is no direct relation between pore size and capacitance of the electrodes. From the figure, the pore size, surface area, and capacitance of the AC are independently varied. Many parameters influence a huge impact on the capacitance, method of preparing electrodes, electrode testing method, the materials surface chemistry, and the electrolyte used.

11.5.6 Mesocrystal Structure

Mesocrystals are crystallographically oriented attachment of nanoparticles that has been recently attracted among the materials researchers because to its superior properties such as high surface area, internal voids, and micron size. It has been proved itself by exhibiting better charge transport properties as photocatalyst. As charge transport is one among the requisite for the electrodes in battery and supercapacitor, this mesocrystals strategy has been also utilized to improve the electrochemical performance for energy storage. Interestingly, instead of carbon source, the marine biomass-squid ink has been used as a template to fabricate novel TiO_2 hollow sphere successfully with mesocrystal nanostructure [74]. In future, we hope that the biomass- derived mesocrystals structure will be used as electrodes for supercapacitor.

11.6 Conclusion and Future Perspectives

In summary, biomass-derived carbon materials produced from various sources and different methods have been discussed. It is also indicated that the energy density and capacitance of biomass-derived supercapacitor electrodes depend on carbon nanostructures created from singular carbon antecedent. In this chapter, we have collected information of ACs obtained from biomass for electrochemical energy storage systems, deliberating the capacitor parameters and storage mechanisms of different types of supercapacitors. It was found that the supercapacitor performance varied with respect to the morphology and porous size of the carbon materials. Most of the AC materials showed excellent capacitance behavior and high cyclic capability. The development of alternate supercapacitor cathode material obtained from biomass sources serves two main purposes: (i) it assists with waste removal; changing over waste to a valuable item and (ii) it gives a financial contention to the substantiality of supercapacitor innovation. The understanding of these major exploration premise could assist us to reveal the conversion mechanism of the convoluted process and give significant direction to synthesis biomass-derived carbon material with the desired structure and properties for the supercapacitor application. Besides, biomass carbon with various dimensions can be integrated together to develop electrode materials for supercapacitors. These blends can take full benefits of each building block and as needs be bring about the composite electrode with enhanced flexibility and mechanical robustness as well as maintained their capacitance. In addition, the electrodes prepared in the form mesocrystals could improve the performance for energy storage due to its unique features such as charge transport and uniform porous structure.

References

1 Duncan, R.C. (2001). World energy production, population growth, and the road to the Olduvai Gorge. *Popul. Environ.* 22: 503–522.
2 Andelman, M. (2011). Flow through capacitor basics. *Sep. Purif. Technol.* 80: 262–269.
3 Nelson, J.P. and Bolin, W.D. (1995). Basics and advances in battery systems. *IEEE Trans.Ind.Appl.* 31: 419–428.
4 Libich, J., Máca, J., Vondrák, J. et al. (2018). Supercapacitors: properties and applications. *J. Energy Storage* 17: 224–227.

5 Inagaki, M., Konno, H., and Tanaike, O. (2010). Carbon materials for electrochemical capacitors. *J. Power Sources* 195: 7880–7903.

6 Abioye, A.M. and Ani, F.N. (2015). Recent development in the production of activated carbon electrodes from agricultural waste biomass for supercapacitors: a review. *Renew. Sustain. Energy Rev.* 52: 1282–1293.

7 Zhang, J., Jin, L., Cheng, J., and Hu, H. (2013). Hierarchical porous carbons prepared from direct coal liquefaction residue and coal for supercapacitor electrodes. *Carbon* 55: 221–232.

8 Divyashree, A. and Hegde, G. (2015). Activated carbon nanospheres derived from biowaste materials for supercapacitor applications–a review. *RSC Adv.* 5: 88339–88352.

9 Conway, B.E., Birss, V.I., and Wojtowicz, J. (1997). The role and utilization of pseudocapacitance for energy storage by supercapacitors. *J. Power Sources* 66: 1–14.

10 Hongsheng, Y., Xiao, Z., Cuiling, J., and Dequan, W. (2003). Recent advances in the study on electrochemical capacitors II. Redox capacitors. *Electron. Compon. Mater.* 22: 38–42.

11 Taer, E., Apriwandi, A., Ningsih, Y.S. et al. (2019). Preparation of activated carbon electrode from pineapple crown waste for supercapacitor application. *Int. J. Electrochem. Sci.* 14: 2462–2475.

12 Nabais, J.M.V., Teixeira, J.G., and Almeida, I. (2011). Development of easy made low cost bindless monolithic electrodes from biomass with controlled properties to be used as electrochemical capacitors. *Bioresour. Technol.* 102: 2781–2787.

13 Farma, R., Deraman, M., Awitdrus, A. et al. (2013). Preparation of highly porous binderless activated carbon electrodes from fibres of oil palm empty fruit bunches for application in supercapacitors. *Bioresour. Technol.* 132: 254–261.

14 Taer, E., Deraman, M., Talib, I.A. et al. (2011). Preparation of a highly porous binderless activated carbon monolith from rubber wood sawdust by a multi-step activation process for application in supercapacitors. *Int. J. Electrochem. Sci.* 6: 3301–3315.

15 Wu, F.C., Tseng, R.L., Hu, C.C., and Wang, C.C. (2005). Effects of pore structure and electrolyte on the capacitive characteristics of steam and KOH-activated carbons for supercapacitors. *J. Power Sources* 144: 302–309.

16 Jiang, C., Yakaboylu, G.A., Yumak, T. et al. (2020). Activated carbons prepared by indirect and direct CO_2 activation of lignocellulosic biomass for supercapacitor electrodes. *Renew. Energy* 155: 38–52.

17 Qu, W.-H., Xu, Y.-Y., Lu, A.-H. et al. (2015). Converting biowaste corncob residue into high value added porous carbon for supercapacitor electrodes. *Bioresour. Technol.* 189: 285–291.

18 Hong, K.L., Qie, L., Zeng, R. et al. (2014). Biomass derived hard carbon used as a high performance anode material for sodium ion batteries. *J. Mater. Chem. A* 2: 12733–12738.

19 Chen, W., Zhang, H., Huang, Y., and Wang, W. (2010). A fish scale based hierarchical lamellar porous carbon material obtained using a natural template for high performance electrochemical capacitors. *J. Mater. Chem.* 20: 4773–4775.

20 Rufford, T.E., Hulicova-Jurcakova, D., Khosla, K. et al. (2010). Microstructure and electrochemical double-layer capacitance of carbon electrodes prepared by zinc chloride activation of sugar cane bagasse. *J. Power Sources* 195: 912–918.

21 Kalpana, D., Cho, S.H., Lee, S.B. et al. (2009). Recycled waste paper – a new source of raw material for electric double-layer capacitors. *J. Power Sources* 190: 587–591.

22 Li, J. and Wu, Q. (2015). Water bamboo-derived porous carbons as electrode materials for supercapacitors. *New J. Chem.* 39: 3859–3864.

23 Sun, H., He, W., Zong, C., and Lu, L. (2013). Template-free synthesis of renewable macroporous carbon via yeast cells for high performance supercapacitor electrode materials. *ACS Appl. Mater. Interfaces* 5: 2261–2268.

24 Li, Y.T., Pi, Y.T., Lu, L.M. et al. (2015). Hierarchical porous active carbon from fallen leaves by synergy of K_2CO_3 and their supercapacitor performance. *J. Power Sources* 299: 519–528.

25 Kondamudi, N., Mohapatra, S.K., and Misra, M. (2008). Spent coffee grounds as a versatile source of green energy. *J. Agric. Food Chem.* 56: 11757–11760.

26 Karthikeyan, K., Amaresh, S., Lee, S.N. et al. (2014). Construction of high-energy-density supercapacitors from pine-cone-derived high-surface-area carbons. *ChemSusChem* 7: 1435–1442.

27 Huang, W., Zhang, H., Huang, Y. et al. (2011). Hierarchical porous carbon obtained from animal bone and evaluation in electric double-layer capacitors. *Carbon* 49: 838–843.

28 Zhang, C., Geng, Z., Cai, M. et al. (2013). Microstructure regulation of super activated carbon from biomass source corncob with enhanced hydrogen uptake. *Int. J. Hydrogen Energy* 38: 9243–9250.

29 Wang, K., Zhao, N., Lei, S. et al. (2015). Promising biomass based activated carbons derived from willow catkins for high performance supercapacitors. *Electrochim. Acta* 166: 1–11.

30 Wang, R., Wang, P., Yan, X. et al. (2012). Promising porous carbon derived from celtuce leaves with outstanding supercapacitance and CO_2 capture performance. *ACS Appl. Mater. Interfaces* 4: 5800–5806.

31 Olivares-Marin, M., Fernandez, J.A., Lazaro, M.J. et al. (2009). Cherry stones as precursor of activated carbons for supercapacitors. *Mater. Chem. Phys.* 114: 323–327.

32 Peng, C., Yan, X.B., Wang, R.T. et al. (2013). Promising activated carbons derived from waste tea-leaves and their application in high performance supercapacitors electrodes. *Electrochim. Acta* 87: 401–408.

33 Li, X., Xing, W., Zhuo, S. et al. (2011). Preparation of capacitor's electrode from sunflower seed shell. *Bioresour. Technol.* 102: 1118–1123.

34 Fan, Y., Liu, P., Zhu, B. et al. (2015). Microporous carbon derived from acacia gum with tuned porosity for high-performance electrochemical capacitors. *Int. J. Hydrogen Energy* 40: 6188–6196.

35 Jiang, L., Yan, J., Hao, L. et al. (2013). High rate performance activated carbons prepared from ginkgo shells for electrochemical supercapacitors. *Carbon* 56: 146–154.

36 Bhattacharjya, D. and Yu, J.S. (2014). Activated carbon made from cow dung as electrode material for electrochemical double layer capacitor. *J. Power Sources* 262: 224–231.

37 Hou, J., Cao, C., Idrees, F., and Ma, X. (2015). Hierarchical porous nitrogen-doped carbon nanosheets derived from silk for ultrahigh-capacity battery anodes and supercapacitors. *ACS Nano* 9: 2556–2564.

38 Fan, Y., Yang, X., Zhu, B. et al. (2014). Micro-mesoporous carbon spheres derived from carrageenan as electrode material for supercapacitors. *J. Power Sources* 268: 584–590.

39 Qian, W., Sun, F., Xu, Y. et al. (2014). Human hair-derived carbon flakes for electrochemical supercapacitors. *Energ. Environ. Sci.* 7: 379–386.

40 Feng, H., Zheng, M., Dong, H. et al. (2015). Three-dimensional honeycomb-like hierarchically structured carbon for high-performance supercapacitors derived from high ash-content sewage sludge. *J. Mater. Chem. A* 3: 15225–15234.

41 Wei, L., Sevilla, M., Fuertes, A.B. et al. (2011). Hydrothermal carbonization of abundant renewable natural organic chemicals for high-performance supercapacitor electrodes. *Adv. Energy Mater.* 1: 356–361.

42 Falco, C., Sieben, J.M., Brun, N. et al. (2013). Hydrothermal carbons from hemicellulose-derived aqueous hydrolysis products as electrode materials for supercapacitors. *ChemSusChem* 6: 374–382.

43 Wang, H., Li, Z., Tak, J.K. et al. (2013). Supercapacitors based on carbons with tuned porosity derived from paper pulp mill sludge biowaste. *Carbon* 57: 317–328.

44 Zhao, L., Fan, L.Z., Zhou, M.Q. et al. (2010). Nitrogen-containing hydrothermal carbons with superior performance in supercapacitors. *Adv. Mater.* 22: 5202–5206.

45 He, X., Ling, P., Yu, M. et al. (2013). Rice husk-derived porous carbons with high capacitance by $ZnCl_2$ activation for supercapacitors. *Electrochim. Acta* 105: 635–641.

46 He, X., Ling, P., Qiu, J. et al. (2013). Efficient preparation of biomass-based mesoporous carbons for supercapacitors with both high energy density and high power density. *J. Power Sources* 240: 109–113.

47 Zhu, H., Wang, X., Yang, F., and Yang, X. (2011). Promising carbons for supercapacitors derived from fungi. *Adv. Mater.* 23: 2745–2748.

48 Long, C., Chen, X., Jiang, L. et al. (2015). Porous layer-stacking carbon derived from in-built template in biomass for high volumetric performance supercapacitors. *Nano Energy* 12: 141–151.

49 Wu, X.L., Wen, T., Guo, H.L. et al. (2013). Biomass-derived sponge like carbonaceous hydrogels and aerogels for supercapacitors. *ACS Nano* 7: 3589–3597.

50 Wang, H., Xu, Z., Kohandehghan, A. et al. (2013). Interconnected carbon nanosheets derived from hemp for ultrafast supercapacitors with high energy. *ACS Nano* 7: 5131–5141.

51 Tian, W., Gao, Q., Tan, Y. et al. (2015). Bio-inspired beehive-like hierarchical nanoporous carbon derived from bamboo based industrial by-product as a high performance supercapacitor electrode material. *J. Mater. Chem. A* 3: 5656–5664.

52 Sevilla, M., Gu, W., Falco, C. et al. (2014). Hydrothermal synthesis of microalgae-derived microporous carbons for electrochemical capacitors. *J. Power Sources* 267: 26–32.

53 Chunming, N., Sichel, E.K., Hoch, R. et al. (1997). High power electrochemical capacitors based on carbon nanotube electrodes. *Appl. Phys. Lett.* 70: 1480–1482.

54 Hoffman, P. and Ferrante, J. (1971). Energy storage capacitors of high energy density. *IEEE Trans. Nucl. Sci.* 18: 235–239.

55 Liu, B., Liu, Y., Chen, H. et al. (2017). Oxygen and nitrogen co-doped porous carbon nanosheets derived from Perilla frutescens for high volumetric performance supercapacitors. *J. Power Sources* 341: 309–317.

56 Ling, Z., Wang, Z., Zhang, M. et al. (2016). Sustainable synthesis and assembly of biomass-derived B/N Co-doped carbon nanosheets with ultrahigh aspect ratio for high-performance supercapacitors. *Adv. Funct. Mater.* 26: 111–119.

57 Wen, L., Li, F., and Cheng, H.M. (2016). Carbon nanotubes and graphene for flexible electrochemical energy storage: from materials to devices. *Adv. Mater.* 28: 4306–4337.

58 Wu, H., Zhang, Y., Cheng, L. et al. (2016). Graphene based architectures for electrochemical capacitors. *Energy Storage Mater.* 5: 8–32.

59 Tang, K., Fu, L., White, R.J. et al. (2012). Hollow carbon nanospheres with superior rate capability for sodium-based batteries. *Adv. Energy Mater.* 2: 873–877.

60 Li, D., Lv, C., Liu, L. et al. (2015). Egg-box structure in cobalt alginate: a new approach to multifunctional hierarchical mesoporous n-doped carbon nanofibers for efficient catalysis and energy storage. *ACS Cent. Sci.* 1: 261–269.

61 Liu, B., Zhou, X., Chen, H. et al. (2016). Promising porous carbons derived from lotus seedpods with outstanding supercapacitance performance. *Electrochim. Acta* 208: 55–63.

62 Liu, Y., Shi, Z., Gao, Y. et al. (2016). Biomass-swelling assisted synthesis of hierarchical porous carbon fibers for supercapacitor electrodes. *ACS Appl. Mater. Interfaces* 8: 28283–28290.

63 Xie, L., Sun, G., Su, F. et al. (2016). Hierarchical porous carbon microtubes derived from willow catkins for supercapacitor applications. *J. Mater. Chem. A* 4: 1637–1646.

64 Wang, C., Huang, J., Qi, H. et al. (2017). Controlling pseudographtic domain dimension of dandelion derived biomass carbon for excellent sodium-ion storage. *J. Power Sources* 358: 85–92.

65 Qu, Y., Zan, G., Wang, J., and Wu, Q. (2016). Preparation of eggplant-derived macroporous carbon tubes and composites of EDMCT/$Co(OH)(CO_3)_{0.5}$

nano-cone-arrays for high-performance supercapacitors. *J. Mater. Chem. A* 4: 4296–4304.

66 Yin, L., Chen, Y., Li, D. et al. (2016). 3-Dimensional hierarchical porous activated carbon derived from coconut fibers with high-rate performance for symmetric supercapacitors. *Mater. Design* 111: 44–50.

67 Veeramani, V., Sivakumar, M., Chen, S.-M. et al. (2017). Lignocellulosic biomass-derived, graphene sheetlike porous activated carbon for electrochemical supercapacitor and catechin sensing. *RSC Adv.* 7: 45668–45675.

68 Zhu, G., Ma, L., Lv, H. et al. (2017). Pine needle-derived microporous nitrogen-doped carbon frameworks exhibit high performances in electrocatalytic hydrogen evolution reaction and supercapacitors. *Nanoscale* 9: 1237–1243.

69 Qiu, X., Wang, L., Zhu, H. et al. (2017). Lightweight and efficient microwave absorbing materials based on walnut shell-derived nano-porous carbon. *Nanoscale* 9: 7408–7418.

70 Guo, N., Li, M., Wang, Y. et al. (2016). Soybean root-derived hierarchical porous carbon as electrode material for high-performance supercapacitors in ionic liquids. *ACS Appl. Mater. Interfaces* 8: 33626–33634.

71 Ahmed, S., Rafat, M., and Ahmed, A. (2018). Nitrogen doped activated carbon derived from orange peel for supercapacitor application. *Adv. Nat. Sci. Nanosci. Nanotechnol.* 9: 35008.

72 Xie, Q., Bao, R., Zheng, A. et al. (2016). Sustainable low-cost green electrodes with high volumetric capacitance for aqueous symmetric supercapacitors with high energy density. *ACS Sustainable Chem. Eng.* 4: 1422–1430.

73 Enock, T.K., King'ondu, C.K., Pogrebnoi, A., and Jande, Y.A.C. (2017). Status of biomass derived carbon materials for supercapacitor application. *Int. J. Electrochem. (Hindawi)* 2017: 14.

74 Yang, R., Sun, L., Liu, W. et al. (2019). Bio-derived 3D TiO2 hollow spheres with a mesocrystal nanostructure to achieve improved electrochemical performance of Na-ion batteries in ether-based electrolytes. *J. Mater. Chem. A* 7: 3399.

12

Biomass-Derived Carbon for Dye-Sensitized and Perovskite Solar Cells

N. Santhosh[1], P. Vijayakumar[2], M. Senthil Pandian[1], and P. Ramasamy[1]

[1]*SSN Research Centre, Sri Sivasubramaniya Nadar College of Engineering, Chennai 603 110, Tamil Nadu, India*
[2]*Xi'an Jiaotong University, State Key Laboratory for Mechanical Behavior of Materials, Xi'an 710049, China*

12.1 Introduction

Increasingly, people around the world are concerned about the availability and cost of energy. Researchers have made numerous attempts to reduce energy usage by utilizing energy conversion services. The production of electricity from renewable sources has been found to be a viable option. Solar energy is the best at harvesting the sun's rays to create electricity. There have typically been three generations of solar cell technology. Most first-generation solar cells use silicon wafers, and they currently account for more than 90% of the market. Thin-film deposition techniques utilized in second generation solar cells include amorphous silicon, CIGS, and CdTe. Solar cells of this new class consume less energy than the first generation of solar cells. However, solar cells fall short of silicon wafer-based solar cells in terms of performance. The preponderance of third-generation solar cells has largely made up of organic and polymeric materials. Its simple manufacturing process and affordability as well as high performance have led to the growing popularity of this third generation [1–3]. Solar cells made of dye sensitized and perovskite technology are much more appealing for third-generation solar cells due to their high-energy conversion and low cost of manufacture.

A new photovoltaic DSSC system has been developed by Michael Graetzel and coworkers since 1991. The first successful DSSC system used nanocrystalline TiO_2 as a photoanode and platinum as a conductor [4]. The DSSC consists of four constituent parts: the nanocrystalline TiO_2 photoanode, the photosensitizer, the electrolyte, and the counter electrode. [5]. The DSSC's components all have their own benefits and drawbacks.

Biomass-Derived Carbon Materials: Production and Applications, First Edition.
Edited by Alagarsamy Pandikumar, Perumal Rameshkumar, and Pitchaimani Veerakumar.

12.2 DSSC Working Principle

A sensitizer molecule is excited by an electron transition from HOMO to LUMO upon being irradiated by sunlight (D -> D*). They take place in accordance with equations (12.1) and (12.2). When the electrons were injected, they were collected by the conductive substrate (CS), then flowed through an external load before reaching the CE.

The electrolyte's iodide molecule donates electrons to the dye molecule, regenerating it. Equations (12.5) and (12.6) explain how iodide is regenerated by reducing triiodide on the cathode. Electron recombination is linked to the electron transfer procedure. Recombination between the dyes, photoanode (CB), and redox electrolyte is demonstrated in Equations (12.7) and (12.8). The open circuit voltage of the cells varies with the work function of the carrier collector (Figure 12.1) [6, 7].

$$D \longrightarrow D^* \tag{12.1}$$

$$D^* \longrightarrow D^+ + e- \tag{12.2}$$

$$e-(CE) \longrightarrow e-(CS) \tag{12.3}$$

$$e-(CS) \longrightarrow e-(CE) \tag{12.4}$$

$$Ox + e-(CE) \longrightarrow Red- \tag{12.5}$$

$$Red - +D^+ \longrightarrow Ox + D \tag{12.6}$$

$$e-(CB)D^+ \longrightarrow D \tag{12.7}$$

$$e-(CD)Ox \longrightarrow Red- \tag{12.8}$$

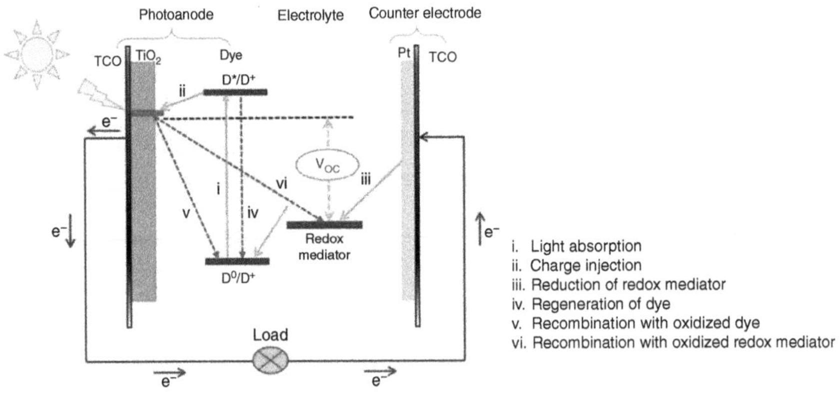

Figure 12.1 Working principle of DSSCs.

12.3 DSSC Components

12.3.1 Transparent Conducting Substrate (TCO)

DSSCs are generally made up of two transparent conductive substrates that aid in the deposition of photoanode and counter electrodes, as well as a current collector [8]. For DSSC applications, two types of transparent conducting oxide substrate are commonly used: (i) fluorine doped tin oxide (FTO) and (ii) indium-doped tin oxide (ITO). ITO and FTO transmit about 80% of their light in the visible region, while sheet resistances are $8\,\Omega\,sq^{-1}$.

12.3.2 Photoanode

The photoanode is the strength of DSSC due to its morphology, size, and material are carefully optimized with another compound. A good photoanode should have the following characteristics:

- To increase dye loading, the photoanode should have a large surface area, and smaller pores will make electrolyte transport easier.
- The photoanode serves as an ideal interface between dye molecules and the TCO substrate, acting as an electron acceptor.
- The conduction band edge should be slightly below the excited LUMO level of dye molecules for an efficient electron injection.
- It should have scattering characteristics that multiple photons, leading to an increase in carrier generation.

The photoanodes in DSSC that have been studied include SnO_2, ZnO, TiO_2, Nb_2O_5, and $SrTiO_3$ [9–14]. In the field of DSSC, anatase TiO_2 material satisfies the above-mentioned photoanode.

12.3.3 Dye Sensitizer

The following requirements should be met by a potential dye sensitizer:

- The dye sensitizer should ideally have a high absorption coefficient in the visible to near-infrared range of light.
- Anchoring groups (for example, —COOH, —CHO, H_2PO_3) that strongly bind on the TiO_2 surface should be included in the dye sensitizer.
- When exposed to sunlight, it should have high thermal and photochemical stability.
- The HOMO and LUMO energy levels should be higher than and lower than their corresponding semiconductor and redox potentials.
- When reduced by an electrolyte, it should remain stable.

On the basis of the above requirements, a different type of dye molecules have been created and effectively tested in the DSSC, namely, metal-free organic sensitizers, natural sensitizers, and metal complex sensitizers [15].

12.3.4 Electrolyte

Electrolytes aid in DSSC by donating electrons to the dye molecules and serving as carriers for charge transfer between the photoanode and CE. The following characteristic should be present in an effective electrolyte.

- ✓ In order for the oxidized dye to be revived efficiently, the redox electrolyte must regenerate it.
- ✓ The electrolyte should not deteriorate the dye molecules, and it should refuse to absorb visible light.
- ✓ Electrolytes with high thermal conductivity, conductivity, and viscosity may improve device stability.
- ✓ The electrolyte's redox potential should be greater than the dye sensitizer's LUMO.

Iodide and cobalt-based electrolytes have been widely utilized in DSSC during the previous two decades [16]. To facilitate hole transport, the electrolytes or hole conductors act as a mediator. Most important function of an electrolyte is to contribute electrons to sensitizer and the DSSC system as a whole. The regeneration of sensitizers using a standard iodide/triiodide redox electrolyte may be explained as follows:

$$D^+ + I^- \longrightarrow D + I_3^- \tag{12.9}$$

12.3.5 Counter Electrode

In DSSC, a counter electrode, is among the most important components. The counter electrode's primary function is as follows:

- It acts as a catalyst by reducing the redox electrolyte.
- It should have good thermal and electrical conductivity.
- It should have corrosion resistance with the iodide/triiodide electrolyte.
- The transmitted light from the dye molecule reaches the counter electrode, which should act as a reflector or scattered to multiple the photons as back.
- Platinum is commonly utilized mostly as counter electrode for DSSC. Because of its strong catalytic activity, conductivity, surface area, and chemical resistance [17].

12.4 Perovskite Solar Cells

The term "perovskite" was first used to describe the crystal structure of calcium titanate, which was discovered in 1839 by German mineralogist Gustav Rose at Ural Mountain, Russia and named after Russian mineralogist Lev Perovski. The halide perovskite absorber has a general chemical formula of ABX_3.

Where, A = Mono cation (organic or inorganic cation such as $CH_3NH_3^+$ or Cs^+)
B = Divalent metal cation (e.g. Pb^{2+} or Sn^{2+})
X = Halide anions (I^-, Br^- and Cl^-)

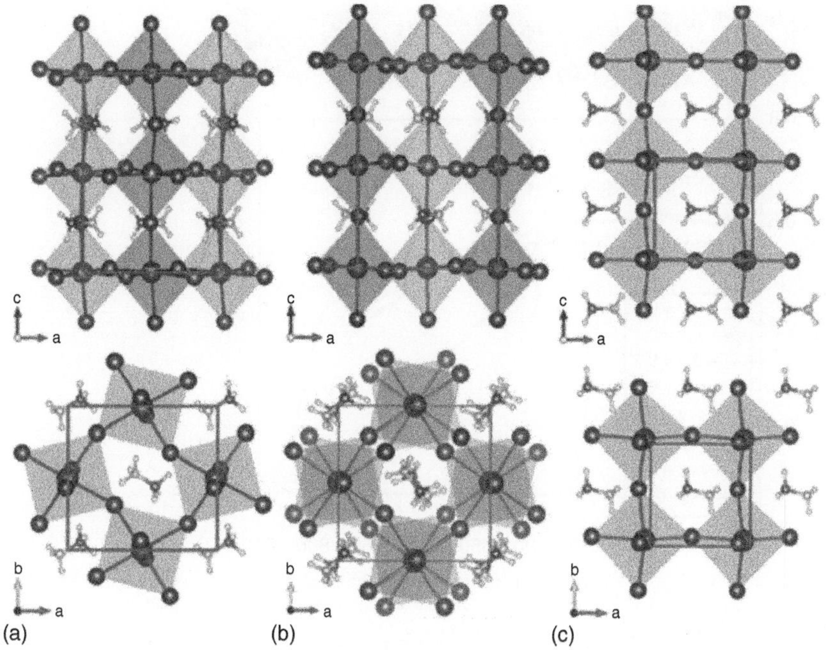

Figure 12.2 Comparison of (a) orthorhombic, (b) tetragonal, and (c) cubic perovskite phase obtained from structural optimization of $MAPbI_3$. Top row: a–c-plane and bottom row: a–b-plane. Source: Ref. [18] / with permission of John Wiley & Sons.

The A site cation located at eight corner of the unit cell, B site metal cation faced at the body center, and X site anion at the six-face center (see Figure 12.2). Nevertheless, perovskites containing halogen anions permit monovalent and divalent cations to occupy the A and B sites, respectively. Here, $MAPbI_3$ or $FAPbI_3$ perovskite light absorber are typically used in PSC device. Perovskites exhibit different phase structures depending on their temperature as follows:

1. Orthorhombic phase (less than 100 K)
2. Tetragonal phase (starts at 160 K)
3. Cubic phase (at 330 K)

The halide perovskites phase stability can be found using Goldschmidt tolerance factor (t) as follows [19]:

$$t = \frac{r_A + r_X}{\sqrt{2}\left(r_A + r_X\right)} \tag{12.10}$$

where r_A, r_B, and r_X are the ionic radii of A, B, and X, respectively, and the ratio of r_B/r_X is defined as octahedral (μ). By specifying the tolerance (t) and octahedral (μ) parameters, one can determine the crystal structure and crystallographic stability of a material. The tolerance factor value is influencing the perovskite structure. If it is close to 1, it is perfectly fit perovskite structure and t value is more than 1 the ionic radii of A site cation are too big, which forbids the formation of perovskite. Further, if t value is less than 0.8, the ionic radii of A site cation is too small and hinders the formation of perovskite [20, 21].

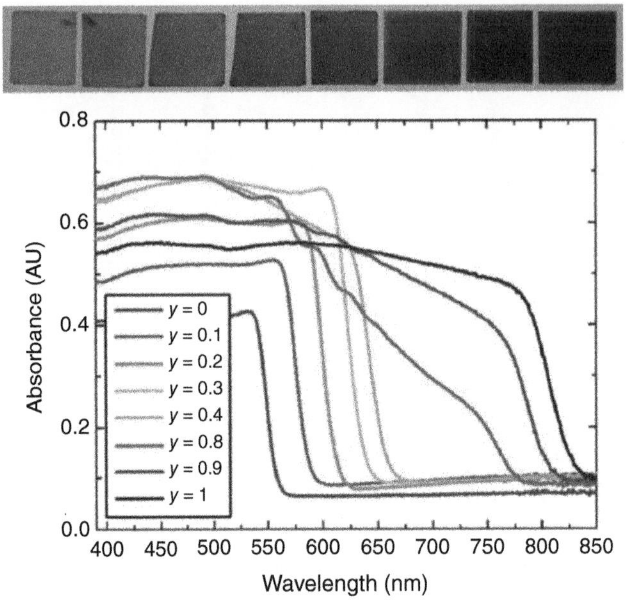

Figure 12.3 UV-Vis absorbance of the $FAPbIyBr_{3y}$ perovskites with varying "y." Source: Ref. [22] / with permission of Royal Society of Chemistry.

12.5 Tunability of Bandgap Energy

One of the most interesting factors for halide perovskite has tunable bandgap material. The bandgap engineering can be accomplished by altering or mixing the ratio of halide anions, which allows continuous bandgap tuning. For example, $FAPbI_{3-y}Br_y$ halide perovskite absorber, by increasing the iodine content, the absorbing behavior changes from lower wavelength to longer wavelength ($y = 0$ to 1 see Figure 12.3) [22]. This tunability benefits to the development of multifunctional perovskite solar cells and tandem solar cells [23, 24].

Halide perovskites exhibit optoelectronic properties mainly as a function of the valence band maximum (VBM) and the conduction band minimum (CBM). VB originates when halide p-state orbital and metal s-state orbital are combined, while lower CB arises from the p-orbital of metal atom along with s-orbital of halide atom [25]. Here, the A site cation does not contribute to the electronic bandgap of perovskite structure. However, the A site cation with different ionic radii had an effect on the perovskite structure's bandgap.

12.6 Development of Perovskite Solar Cells from Dye-Sensitized Solar Cells

Halide perovskite absorbers were used in solar cells for the first time by Miyasaka et al. in 2009. It was used as a sensitizer in DSSC and reported PCE of 3.8% [26].

A couple of years later, Park et al. fabricated perovskite $(CH_3NH_3)PbI_3$ nanocrystals with a thick TiO_2 layer with redox electrodes based on iodide/triiodide, achieved a PCE of 6.5% [27]. Perovskite is degraded or bleached when using liquid electrolyte-based PV cells, leading to poor cell stability. Solid-state HTM of spiro-OMeTAD replaces liquid electrolyte and increases efficiency to ~10% by resolving immediate perovskite instability [28]. Snaith et al. deposited a perovskite layer on an Al_2O_3 scaffold layer instead of a meso-TiO_2 layer, resulting in a high open circuit voltage and improved photovoltaic performance. Since Al_2O_3 scaffold has a large bandgap (insulator), it does not act as an electron collector. As a result, the electrons remain in the perovskite conduction band and are transferred into the compact TiO_2 layer and the FTO electrode. This kind of architecture is known as a meso-superstructure solar cell [29]. In a short period of time, this technology has resulted in an efficiency comparable to that of current silicon PV. As a result of these reports, to further improve device efficiency by optimizing device manufacturing process parameters.

12.6.1 Working Principle of PSC

The PSC working principle is divided into several steps: (i) carrier generation upon photon strikes (electron-hole pairs), (ii) the generated carriers separated by electron and hole are n-type and p-type semiconductors, and (iii) electrodes on both the front and back of the cell collect electrons and holes (see Figure 12.4a, b).

12.6.2 Perovskite Solar Cells Architecture

Generally, two kinds of PSC architecture such as mesoporous and planar structures have been developed. Among them, mesoporous structures are have some limitations. In mesoporous architecture, the n-type semiconductor needs to sinter at a high temperature, leading to longer sintering times and higher manufacturing costs. Planar structures are classified into two types (Figure 12.5):

i) Regular planar (n–i–p) structure
ii) Inverted planar (p–i–n) structure

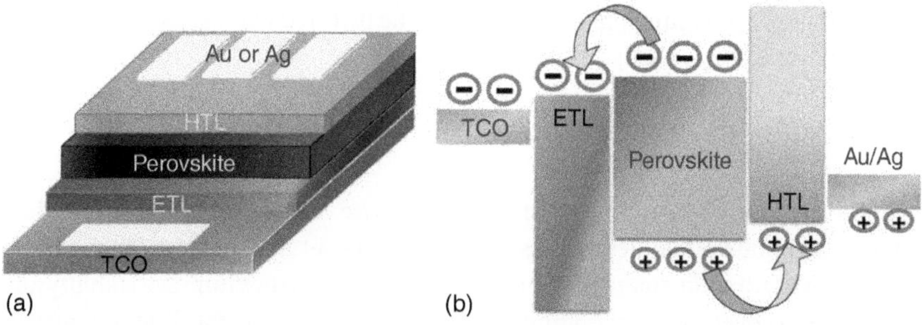

Figure 12.4 (a, b). Schematic architecture of a typical PSC and corresponding energy band diagram.

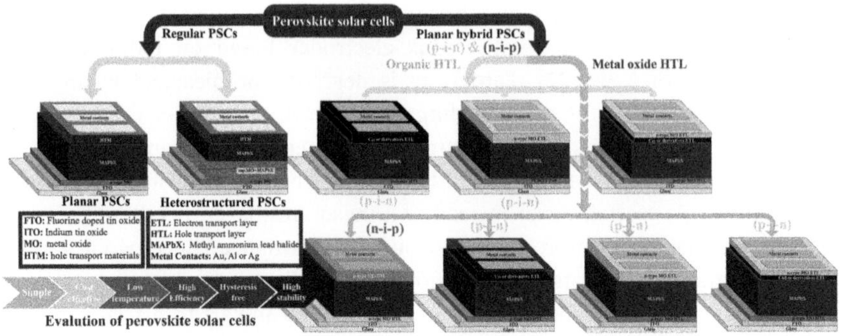

Figure 12.5 Evolution of perovskite solar cells. The various types of PSCs based on their device configurations. Source: Ref. [30] / with permission of Royal Society of Chemistry.

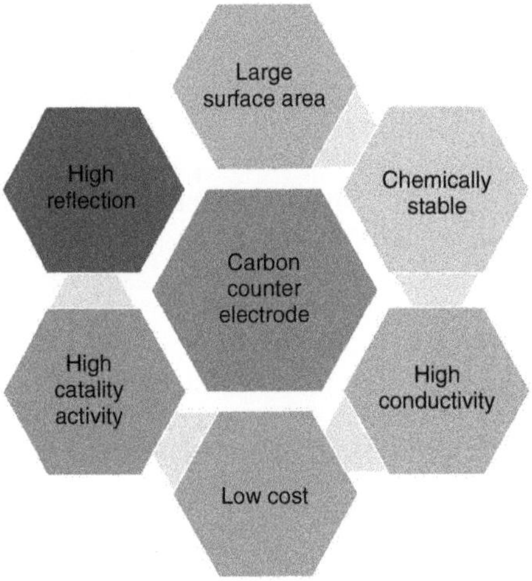

Figure 12.6 Properties of carbon counter electrode.

Choosing a type of electron transport and hole transport material (see Figure 12.6) results in a variety of architectural forms [30]. Solar cells based on inverted planar structures are very appealing since they have less hysteresis in current–voltage (J–V) characteristics than regular planar structures. The inverted PSC can be developed on flexible substrates very easily.

12.6.3 Hole Transport Material

Organic–inorganic hybrid layers degrade because of the electrolyte solution, making the photovoltaic device less effective. Graetzel and Park et al. employed spiro-OMeTAD, a solid-state hole transport material to further improve stability and efficiency. However, this organic hole transport material has merit and demerits. The main advantage of spiro-OMeTAD material has highly soluble in organic solvent, large hole mobility, annealing-free, and favorable electronic configuration.

Table 12.1 Type of biomass-derived carbon materials and its applications.

S.No.	Biomass carbon source	Component of materials	Synthesis method	Applications	Reference
1.	Coffee waste	Hierarchical porous Carbon	Carbonization method	DSSC	[35]
2.	Aloe peel	Honeycomb-like carbon	Hydrothermal method	DSSC	[36]
3.	Sunflower stalk	Layer-stacking activated carbon	Hydrothermal carbonization	Supercapacitors	[37]
4.	Pumpkin stem	Activated carbons	Carbonization method	DSSC	[38]
5.	Fallen leaves	Honeycomb-like porous carbon	Carbonization method	DSSC	[39]
6.	Rice husk	Hierarchical porous carbon	Carbonization method	DSSC	[40]
7.	Agriculture waste	Hierarchically porous nitrogen-doped carbon	Carbonization method	Supercapacitors	[41]
8.	Fallen pine cone flowers	Three-dimensional activated porous carbon	Carbonization method	DSSC	[42]
9	Filter paper and facial tissue	Carbon fiber	One-step pyrolysis	DSSC	[43]
10	Soybean dregs	Porous carbon	Carbonization process	PSC	[44]
11	Brewery residues	Activated biocarbon	Hydrothermally carbonized	DSSC	[45]

Its main drawbacks are poor crystallinity, possible degradation under environmental influences, needed doping strategies, and highly cost-effective and expensive due to the synthetic method to prepare high purity. An alternative p-type inorganic semiconductor material has been demonstrated in perovskite solar cells such as CuO, CuI, CuSCN, and NiO_x that satisfice the following properties: (i) a valence band characteristic of inorganic HTM matches perfectly the valence band (or work function) of organic–inorganic halide perovskites. (ii) Good hole mobility increases fill factor and reduces resistance. (iii) Ideally, it should transmit visible light. (iv) High thermal and photochemical stability [31].

12.7 Biomass-Derived Carbon Counter Electrode for DSSC

Carbon-based materials, such as carbon nanotubes, graphene, activated carbon, and highly porous carbon, have sparked a lot of interest due to their exceptional thermal, electrical, and mechanical properties. Wasted materials can be turned into usable materials by recycling biomass with a simple catalyst approach. The recent development of green methods for synthesizing porous carbon-based catalysts using biomass is important to several applications especially solar cells, lithium-ion batteries, and batteries for electric vehicles, etc. [32–34]. Carbon is the most prevalent substance used by humans and other living things on the planet. The carbon atom is composed of several different allotropes, including diamond, graphene, C60 (fullerene or Buckyball), C70, amorphous carbon, and carbon nanotubes.

The carbon in nature is plentiful, inexpensive, and environmentally friendly. Both directly and indirectly, they are used in many things. A number of factors influence their industrial uses, including manufacturing ease and cost of materials. The following attributes are needed for an effective carbon counter electrode as seen in Figure 12.6.

12.7.1 Performance of DSSC with Counter Electrode via Bio-derived Carbon

A nitrogen-doped porous carbon material was prepared in an experiment by H. Jing and colleagues out of coffee waste (CW). A combination of CW and $ZnCl_2$ was heated into carbon in one step in an inert environment, as illustrated in Figure 12.7. The treated carbon materials show high porous nature and high electrocatalytic activity than the outperforms of existing commercial counter electrode. The CW-based solar cells generated 8.32% efficiency, which was greater than the efficiency of a solar cell based on Pt [46].

Figure 12.7 Schematic representation of coffee waste-derived nitrogen-doped porous carbon synthesis. Source: Ref. [35] / with permission of American Chemical Society.

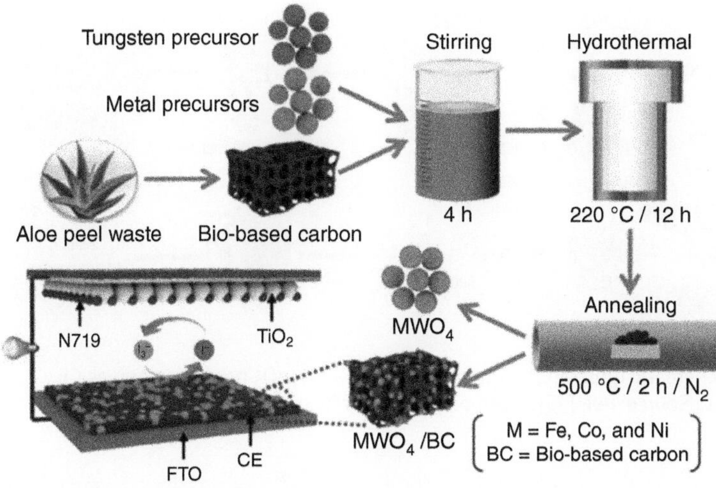

Figure 12.8 Schematic diagram of the synthesis of tungsten-based bimetal oxide with bio-carbon counter electrode and biocarbon. Source: Ref. [47] / with permission of Elsevier.

The chemical composition of tungsten-based bimetal oxide supports new bio-carbon-based electrodes for dye-sensitized solar cells proposed by Zhang et al. [47]. By hydrothermal and annealing methods, the researchers added porous bio-carbon into tungsten-based bimetal oxide to enhance its electrocatalytic properties. The synergistic effects of tungsten-based bimetal oxide and bio-based carbon on the electrode performance and charge-transfer resistance of these devices are extensive. Figure 12.8 depicts a graphical depiction of the synthesis material. This composited carbon electrode in DSSC exhibits an efficiency of 7.08%.

Three heteroatom-doped porous carbon was prepared from food waste (fish waste) and used as the counter electrode in a DSSC. During triiodide reduction, naturally doped three heteroatoms sulfur, nitrogen, and phosphorus demonstrate sterling electrochemical stability. An electrode with triple heteroatom doping shows a power conversion efficiency of 7.83% [48]. A. Arshad et al. proposed activated carbon obtained from aloe vera peel is incorporated into manganese-based and cobalt/manganese-based oxides to form highly electrocatalytic nanohybrids. Co/MnO_2 porous carbon formed in DSSC is shown schematically in Figure 12.9. Solar cells with bio-based carbon counter electrodes are significantly more efficient than platinum electrode-based solar cells.

12.7.2 Biomass-Derived Carbon as a Counter Electrode for Perovskite Solar Cells

A biochar was made by burning *Eichhornia crassipes* (EC) plant stems and leaves in an antique oil lamp. EC stems and leaves are being used to make charcoal [50]. EC-GC4, EC-GC8, and EC-GC10 were annealed at 450, 850, and 1000 °C, respectively, to obtain porous graphitic carbon. The manufacturing process provides

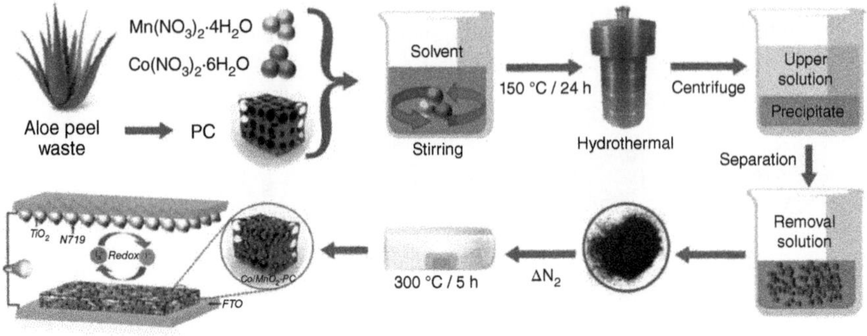

Figure 12.9 Schematic illustration of the synthesis of Co/MnO_2 porous carbon as CE material in DSSC. Source: Ref. [49] / with permission of Elsevier.

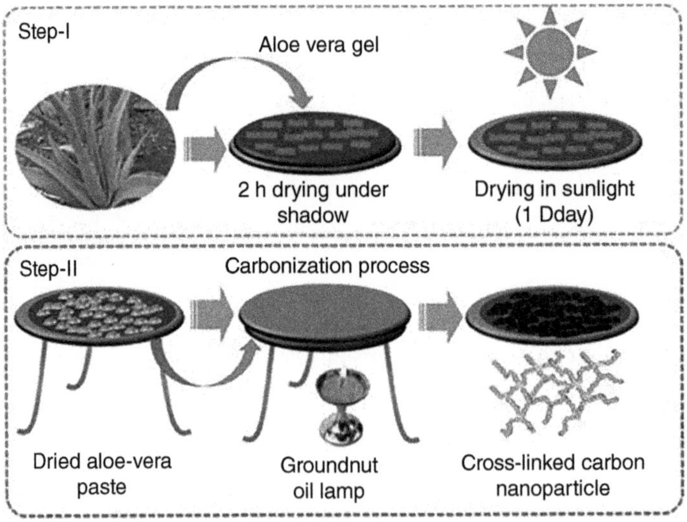

Figure 12.10 Pictorial representation of aloe-vera processed cross-linked carbon nanoparticles via ancient Indian method. Source: Ref. [51] / with permission of American Chemical Society.

low-cost carbon compounds to be utilized in hole transport materials and counter electrodes in perovskite solar cells. The device provides an efficiency of 8.52%, good moisture and air stability, and can operate for 1000 hours in any conditions. New ecofriendly carbon nanoparticles (aloe-vera plant) are used as hole transport materials (C-HTL) in carbon-based printable mesoscopic perovskite solar cells (C-PSCs), produced by using an old Indian technique (see Figure 12.10). This plant produced cross-linked carbon nanoparticles that proved effective at converting energy into electricity at a maximum of 12.50% [51].

For perovskite solar cells, L. Gao and colleagues created a number of low-cost bio-carbon counter electrodes. A variety of bio-masses were used to create bio-carbons, including corn stalk, peanut shell, phragmites australis, and bamboo chopsticks. The interfacial electron, work function sheet resistance, crystallinity,

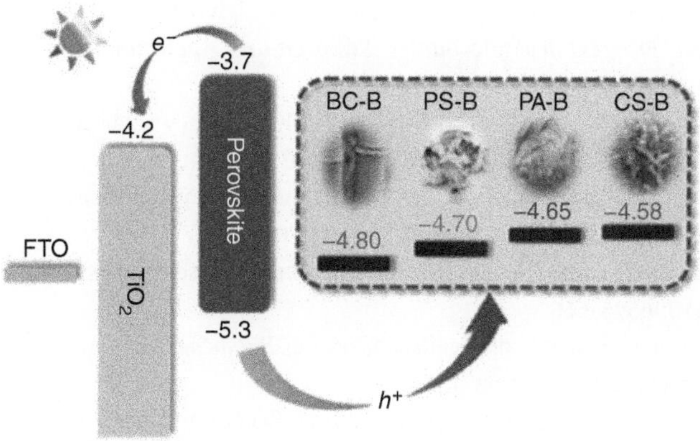

Figure 12.11 The energy-level diagram of the PSC device based on a different bio-carbon electrode. Source: Ref. [52] / with permission of Elsevier.

and shape of bio-carbons were used to assess the performance of photovoltaic devices depending on various bio-carbon electrodes. (see Figure 12.11). Based on bamboo chopsticks as the electrodes, the bio-carbon counter electrode of perovskite solar cells can attain a maximum efficiency of 12.82%, making it the best in the world so far. Bio-carbon electrodes are more stable than conventional electrodes [52]. The stability of PSC device retains 87% over 2000 hours in room temperature.

12.8 Conclusion and Future Perspectives

The biomass-derived carbon electrode has been found to have superior electrocatalytic activity, high surface porosity, excellent conductivity, and inexpensive. Carbon electrodes have been made using facile, cost-effective, and scalable techniques without the aid of toxic precursor chemicals. Moreover, the conventional device architecture has noble metal electrode and it required high vacuum deposition techniques, which is replaced by an alternative source that is carbon electrode, and it is feasible cheap and gives better performance. The biomass-processed carbon electrode may be commonly used in the manufacture using screen printing and inkjet printing processes, making it an ideal for massive manufacturing and long-term durability. Biomass-derived carbon is a crucial strategy to commercializing better performance photovoltaic technology at a cheap cost.

References

1 Green, M.A. (2009). The path to 25% silicon solar cell efficiency: history of silicon cell evolution. *Progress in Photovoltaics: Research and Applications* 17: 183–189.

2 Saga, T. (2010). Advances in crystalline silicon solar cell technology for industrial mass production. *NPG Asia Materials* 2: 96–102.

3 Green, M.A. (2001). Third generation photovoltaics: ultra-high conversion efficiency at low cost. *Progress in Photovoltaics: Research and Applications* 9: 123–135.

4 O'Regan, B., Grätzel, M., and Low-Cost, A. (1991). High-efficiency solar cell based on dye-sensitized colloidal TiO_2 films. *Nature* 353: 737–740.

5 Kong, F.-T., Dai, S.-Y., and Wang, K.-J. (2007). Review of recent progress in dye-sensitized solar cells. *Advances in OptoElectronics* 2007.

6 Hagfeldt, A., Boschloo, G., Sun, L. et al. (2010). Dye-sensitized solar cells. *Chemical Reviews* 110: 6595–6663.

7 Grätzel, M. (2003). Dye-sensitized solar cells. *Journal of Photochemistry and Photobiology C: Photochemistry Reviews* 4: 145–153.

8 Mehmood, U., Rahman, S.-u., Harrabi, K. et al. (2014). Recent advances in dye sensitized solar cells. *Advances in Materials Science and Engineering* 2014: 12.

9 Redmond, G., Fitzmaurice, D., and Graetzel, M. (1994). Visible light sensitization by cis-bis (thiocyanato) bis (2, 2′-bipyridyl-4, 4′-dicarboxylato) ruthenium (II) of a transparent nanocrystalline ZnO film prepared by sol-gel techniques. *Chemistry of Materials* 6: 686–691.

10 Dinh, N.N., Bernard, M.-C., Hugot-Le Goff, A. et al. (2006). Photoelectrochemical solar cells based on SnO_2 nanocrystalline films. *Comptes Rendus Chimie* 9: 676–683.

11 Ou, J.Z., Rani, R.A., Ham, M.-H. et al. (2012). Elevated temperature anodized Nb_2O_5: a photoanode material with exceptionally large photoconversion efficiencies. *ACS Nano* 6: 4045–4053.

12 Gholamrezaei, S., Niasari, M.S., Dadkhah, M., and Sarkhosh, B. (2016). New modified sol–gel method for preparation $SrTiO_3$ nanostructures and their application in dye-sensitized solar cells. *Journal of Materials Science: Materials in Electronics* 27: 118–125.

13 Tan, B., Toman, E., Li, Y., and Wu, Y. (2007). Zinc stannate (Zn2SnO4) dye-sensitized solar cells. *Journal of the American Chemical Society* 129: 4162–4163.

14 Hara, K., Zhao, Z.-G., Cui, Y. et al. (2011). Nanocrystalline electrodes based on nanoporous-walled WO_3 nanotubes for organic-dye-sensitized solar cells. *Langmuir* 27: 12730–12736.

15 Nazeeruddin, M.K., Pechy, P., Renouard, T. et al. (2001). Engineering of efficient panchromatic sensitizers for nanocrystalline TiO_2-based solar cells. *Journal of the American Chemical Society* 123: 1613–1624.

16 Wu, J., Lan, Z., Hao, S. et al. (2008). Progress on the electrolytes for dye-sensitized solar cells. *Pure and Applied Chemistry* 80: 2241–2258.

17 Wu, J., Lan, Z., Lin, J. et al. (2017). Counter electrodes in dye-sensitized solar cells. *Chemical Society Reviews* 46: 5975–6023.

18 Korshunova, K., Winterfeld, L., Beenken, W.J., and Runge, E. (2016). Thermodynamic stability of mixed Pb: Sn methyl-ammonium halide perovskites. *Physica Status Solidi B* 253: 1907–1915.

19 Goldschmidt, V. (1927). Crystal structure and chemical correlation. *Berichte der Deutschen Chemischen Gesellschaft* 60: 1263–1296.

20 Li, C., Lu, X., Ding, W. et al. (2008). Formability of abx3 (x= f, cl, br, i) halide perovskites. *Acta Crystallographica Section B: Structural Science* 64: 702–707.

21 Kieslich, G., Sun, S., and Cheetham, A.K. (2014). Solid-state principles applied to organic–inorganic perovskites: new tricks for an old dog. *Chemical Science* 5: 4712–4715.

22 Eperon, G.E., Stranks, S.D., Menelaou, C. et al. (2014). Formamidinium lead trihalide: a broadly tunable perovskite for efficient planar heterojunction solar cells. *Energy & Environmental Science* 7: 982–988.

23 Ball, J.M., Lee, M.M., Hey, A., and Snaith, H.J. (2013). Low-temperature processed meso-superstructured to thin-film perovskite solar cells. *Energy & Environmental Science* 6: 1739–1743.

24 Bailie, C.D., Christoforo, M.G., Mailoa, J.P. et al. (2015). Semi-transparent perovskite solar cells for tandems with silicon and CIGS. *Energy & Environmental Science* 8: 956–963.

25 Ye, Y., Run, X., Hai-Tao, X. et al. (2015). Nature of the band gap of halide perovskites ABX3 (A= CH3NH3, Cs; B= Sn, Pb; X= Cl, Br, I): First-principles calculations. *Chinese Physics B* 24: 116302.

26 Kojima, A., Teshima, K., Shirai, Y., and Miyasaka, T. (2009). Organometal halide perovskites as visible-light sensitizers for photovoltaic cells. *Journal of the American Chemical Society* 131: 6050–6051.

27 Im, J.-H., Lee, C.-R., Lee, J.-W. et al. (2011). 6.5% efficient perovskite quantum-dot-sensitized solar cell. *Nanoscale* 3: 4088–4093.

28 Kim, H.-S., Lee, C.-R., Im, J.-H. et al. (2012). Lead iodide perovskite sensitized all-solid-state submicron thin film mesoscopic solar cell with efficiency exceeding 9%. *Scientific Reports* 2: 1–7.

29 Snaith, H.J. (2013). Perovskites: the emergence of a new era for low-cost, high-efficiency solar cells. *The Journal of Physical Chemistry Letters* 4: 3623–3630.

30 Mali, S.S. and Hong, C.K. (2016). pin/nip type planar hybrid structure of highly efficient perovskite solar cells towards improved air stability: synthetic strategies and the role of p-type hole transport layer (HTL) and n-type electron transport layer (ETL) metal oxides. *Nanoscale* 8: 10528–10540.

31 Rajeswari, R., Mrinalini, M., Prasanthkumar, S., and Giribabu, L. (2017). Emerging of inorganic hole transporting materials for perovskite solar cells. *The Chemical Record* 17: 681–699.

32 Li, G., Xue, R., Chen, L., and Huang, Y.-Z. (1995). Carbon electrode materials for lithium-ion batteries. *Journal of Power Sources* 54: 271–275.

33 Zhou, C. and Lin, S. (2020). Carbon-electrode based perovskite solar cells: effect of bulk engineering and interface engineering on the power conversion properties. *Solar RRL* 4: 1900190.

34 Ghosh, S., Santhosh, R., Jeniffer, S. et al. (2019). Natural biomass derived hard carbon and activated carbons as electrochemical supercapacitor electrodes. *Scientific Reports* 9: 1–15.

35 Chung, D.Y., Son, Y.J., Yoo, J.M. et al. (2017). Coffee waste-derived hierarchical porous carbon as a highly active and durable electrocatalyst for electrochemical energy applications. *ACS Applied Materials & Interfaces* 9: 41303–41313.

36 Wang, Z., Yun, S., Wang, X. et al. (2019). Aloe peel-derived honeycomb-like bio-based carbon with controllable morphology and its superior electrochemical properties for new energy devices. *Ceramics International* 45: 4208–4218.

37 Wang, X., Yun, S., Fang, W. et al. (2018). Layer-stacking activated carbon derived from sunflower stalk as electrode materials for high-performance supercapacitors. *ACS Sustainable Chemistry & Engineering* 6: 11397–11407.

38 Madhu, R., Veeramani, V., Chen, S.-M. et al. (2014). Pumpkin stem-derived activated carbons as counter electrodes for dye-sensitized solar cells. *RSC Advances* 4: 63917–63921.

39 Cha, S.M., Nagaraju, G., Sekhar, S.C. et al. (2018). Fallen leaves derived honeycomb-like porous carbon as a metal-free and low-cost counter electrode for dye-sensitized solar cells with excellent tri-iodide reduction. *Journal of Colloid and Interface Science* 513: 843–851.

40 Wang, G., Wang, D., Kuang, S. et al. (2014). Hierarchical porous carbon derived from rice husk as a low-cost counter electrode of dye-sensitized solar cells. *Renewable Energy* 63: 708–714.

41 Zou, K., Deng, Y., Chen, J. et al. (2018). Hierarchically porous nitrogen-doped carbon derived from the activation of agriculture waste by potassium hydroxide and urea for high-performance supercapacitors. *Journal of Power Sources* 378: 579–588.

42 Nagaraju, G., Lim, J.H., Cha, S.M., and Yu, J.S. (2017). Three-dimensional activated porous carbon with meso/macropore structures derived from fallen pine cone flowers: a low-cost counter electrode material in dye-sensitized solar cells. *Journal of Alloys and Compounds* 693: 1297–1304.

43 Xu, S. (2017). One-step fabrication of carbon fiber derived from waste paper and its application for catalyzing tri-iodide reduction. In: *IOP Conference Series: Earth and Environmental Science*, 012014. IOP Publishing.

44 Liu, H., Xie, Y., Wei, P. et al. (2020). Interface optimization of hole-conductor free perovskite solar cells using porous carbon materials derived from biomass soybean dregs as a cathode. *Journal of Alloys and Compounds* 842: 155851.

45 Tiihonen, A., Siipola, V., Lahtinen, K. et al. (2021). Biocarbon from brewery residues as a counter electrode catalyst in dye solar cells. *Electrochimica Acta* 368: 137583.

46 Jing, H., Shi, Y., Wu, D. et al. (2018). Well-defined heteroatom-rich porous carbon electrocatalyst derived from biowaste for high-performance counter electrode in dye-sensitized solar cells. *Electrochimica Acta* 281: 646–653.

47 Zhang, Y., Yun, S., Wang, C. et al. (2019). Bio-based carbon-enhanced tungsten-based bimetal oxides as counter electrodes for dye-sensitized solar cells. *Journal of Power Sources* 423: 339–348.

48 Ma, P., Lu, W., Yan, X. et al. (2018). Heteroatom tri-doped porous carbon derived from waste biomass as Pt-free counter electrode in dye-sensitized solar cells. *RSC Advances* 8: 18427–18433.

49 Arshad, A., Yun, S., Si, Y. et al. (2020). Aloe vera-peel derived porous carbon integrated Co/Mn-oxide based nano-hybrids: an efficient electrocatalyst in advanced photovoltaics. *Journal of Power Sources* 451: 227731.

50 Pitchaiya, S., Eswaramoorthy, N., Natarajan, M. et al. (2020). Perovskite solar cells: a porous graphitic carbon based Hole transporter/counter electrode material extracted from an invasive plant Species Eichhornia Crassipes. *Scientific Reports* 10: 1–16.

51 Mali, S.S., Kim, H., Patil, J.V., and Hong, C.K. (2018). Bio-inspired carbon hole transporting layer derived from aloe vera plant for cost-effective fully printable mesoscopic carbon perovskite solar cells. *ACS Applied Materials & Interfaces* 10: 31280–31290.

52 Gao, L., Zhou, Y., Meng, F. et al. (2020). Several economical and eco-friendly bio-carbon electrodes for highly efficient perovskite solar cells. *Carbon* 162: 267–272.

13

Recent Advances of Biomass-Derived Porous Carbon Materials in Catalytic Conversion of Organic Compounds

N. Mahendar Reddy[1], *D. Saritha*[1], *Naveen K. Dandu*[2], *Ch.G. Chandaluri*[3], *and Gubbala V. Ramesh*[1]

[1] *Chaitanya Bharathi Institute of Technology (A), Department of Chemistry, Gandipet, Hyderabad 500075, Telangana, India*
[2] *Material Science Division, Argonne National Lab, Argonne, 60439, IL USA*
[3] *Faculty of Chemistry, Humanities and Sciences Division Indian Institute of Petroleum and Energy, Visakhapatnam 530003, Andhra Pradesh, India*

13.1 Introduction

The advancement of anthropic actions has affected a vast energy necessity and expenditure of assets that outcome the discharge of waste in the diverse vicinity such as water, land, biota, air, and aquatic sediments. As a consequence, it influenced the ecosystems and biodiversity of planets. The liveliness desires of the enhanced world population have further stressed the crisis. Recycling of biomass is vital to diminish the hazard effect on human health. Approximately 8 million tons of plastic wastes are thrown into the ocean every year [1]. Biological wastes include domestic and sewage wastes, food wastes, and scums from forestry, agriculture, and fisheries [2]. 2.2 billion tons of urban hard wastes formed per annum according to World Bank anticipation by 2025. The production of waste generation is doubled since the subsequent two decades in rising countries [3]. The familiar dumping process of biomass includes microbial decay under both aerobic or anaerobic thermal degradation circumstances and transfer to landfill [3]. Educational and industrial sectors seek out to identify approaches to convert waste into novel resources and products.

Enhancement in solid waste production is the most important challenge faced by society as a consequence of the growing economy and urban residents. The round financial system has achieved interest as a major perception for emergent closed-loop technical and biological cycles [4]. In this perception, materials are recycled or come back to the usual ecology with no spoil to the atmosphere [5]. Conversion of biomass into valuable products is obligatory without spoiling the normal ecology to realize sustainable growth. Recognizing procedures for the valorization of biomasss are a tough task to visualize the sort of outcome products.

Biomass-Derived Carbon Materials: Production and Applications, First Edition.
Edited by Alagarsamy Pandikumar, Perumal Rameshkumar, and Pitchaimani Veerakumar.

The earlier-mentioned procedure will resolve the crisis of municipal waste management and the controlled accessibility of resources. Sustainable progress can be achieved by the connection between the waste and fabrication sectors.

Porous carbon (PC) has been extensively exploited for a variety of purposes such as cleansing of drinking water, healing of exhaust gas, curing of wastewater, catalyst, gas storage, and capacitor appliances, etc. PC is essentially mentioned as carbonaceous materials [6]. Physicochemical stability [7], superior adsorptive capacity [8], fine porosity [9], especially fine surface reactivity with enormous surface area [10], and good mechanical strength are the distinctive benefits of PC. Fabrication of active carbons from renewable precursors is highly recommended due to their accessibility, abundancy, and low cost [11].

The generation of active PC nanoparticles (NPs) via biomass materials is the waste to the prosperity approach process, and this process encourages green chemistry [12]. The anticipated benefits of this process are effortless, secure, ecological, and price-effective [13, 14]. Hydrothermal synthesis, combustion, and chemical vapor deposition are the accessible processes for the production of PC NPs [15]. Activated PC NPs can be formed from carbon NPs by employing physical and chemical methods [16–21]. In a physical method, carbon NPs are produced from the carbonization of carbon source at high temperatures under inert circumstances and then activated by treating with activating agents for instance air, steam, and CO_2 to form activated carbon (AC) NPs [22]. In a chemical method, carbon precursor can be activated by employing chemical agents for instance acids, bases, and salts to form AC NPs [22]. The PC NPs formed from the physical procedures have a superior surface area when compared to chemical procedures [23]. Modern perceptions also recommended that the active carbons connected with heteroatom, for instance, sulfur, hydrogen, halogens, oxygen, and nitrogen, is the key in the form of carboxyl, carbonyls, phenols, and lactones [24–29]. However, the majority of these carbon resources are resulting from nonrenewable. Currently, it is obligatory to accomplish renewable active carbon materials resulting from natural biomass materials that can be used as electrode materials for energy storage devices to achieve sustainability (to resolve ecological issues). Wood, peat, coal, lignite, and petroleum residues are the costly and nonrenewable precursors to produce commercial ACs [30, 31]. PCs are produced from various biomasss such as melon seeds [32], coconut coir [33], coir pith [34], bamboo dust, coconut shell, orange peels, groundnut shell, rice husk, and straw [35], corncob [36], palm shell [37], coconut shells [38, 39], almond shell [40], and tomato waste. Consequently, other studies have shown a spotlight in the preparation of AC-supported from agricultural wastes and lignocelluloses resources which are valuable, efficient, and low cost [41, 42].

Activated PC or AC can be used in various fields such as chemical industries, petroleum industries, wastewater treatment plants, catalysis, energy storage, power stations, pharmaceutical industries, and food industries [43, 44]. AC has been employed to adsorb drug and risky chemicals in the medicine sector, gold recovery from leached liquors, and also in gas cleaning purposes [45, 46].

This chapter mainly focuses on the recent advances in the production of biomass-derived PC materials. The prepared materials can be utilized in several

Figure 13.1 Schematic diagram of commonly available biomass and its applications. Source: Highwaystarz / Adobe Stock; Green / Wkimedia Commons / CC BY-SA 3.0; Picture Partners / Adobe Stock; kariphoto / Adobe Stock ; Alexander Raths / Adobe Stock.

fields, for instance, catalysis, electronics, tissue engineering, and drug delivery. On the other hand, several such examinations are at an initial phase and necessitate to analyze both scientific and socio-ecological aspects (Figure 13.1).

13.2 Synthesis Procedures

Different routes have been developed to synthesize AC with desired properties. Most common method to prepare AC is carbonization of biomass material or biomass-derived material followed by activation. The properties of the synthesized AC depend on the parameters like nature of the precursor, temperature, heating time, and activating agent. Herein, we described their synthesis procedure in two parts: (i) carbonization and (ii) activation.

Carbonization procedure is performed to increase biomass's carbon content in the material, while decreasing the percentages of other atoms. Carbonaceous material, rich in carbon, is produced with high-temperature treatment; however, it is not sufficient for catalytic applications due to lack of sufficient porosity. Carbon material with high porosity, that is, high surface area, provides a capability for desired application, specifically in catalysis. High porosity is achieved by physical or chemical or physicochemical activation process. Heteroatom doping also enhances the properties of the AC. Steps involved in the AC synthesis process are shown in the schematic representation (Figure 13.2) [47]. Reported figure shows biomass as a starting feed; however, similar procedures are followed for the biomass feed as well. In the following sections, a brief description about different synthesis routes that influence the AC's porosity distribution and nature is explained.

13.2.1 Carbonization

The most common methodologies used for carbonization are hydrothermal treatment and pyrolysis. Hydrothermal process generally carried in an autoclave under

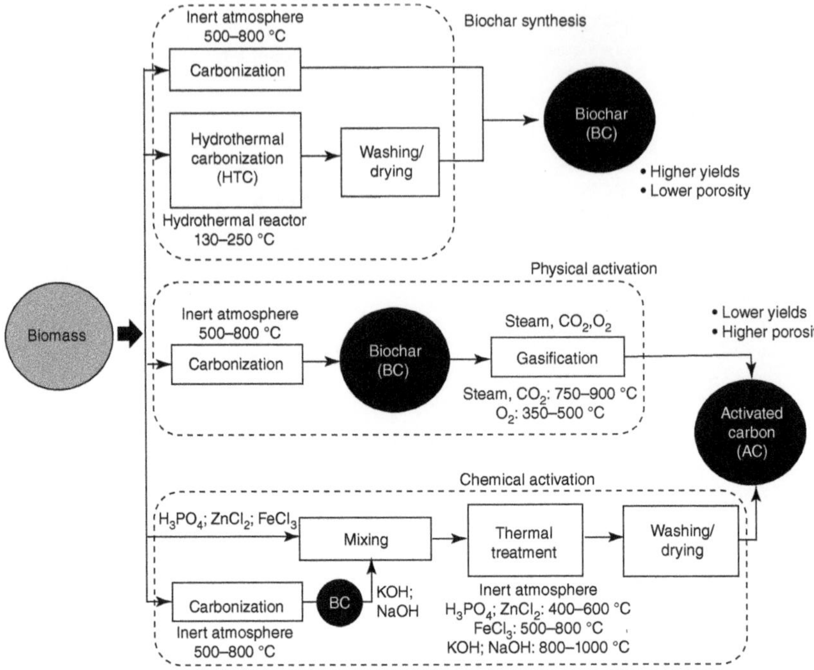

Figure 13.2 Synthesis methods used to produce activated PC from biomass. Source: Ref. [47] / MDPI / CC BY 4.0.

autogenous pressure generated in the presence of solvent maintained above boiling temperature of the solvent. Pyrolysis is done at a very high temperature in inert atmosphere. Both processes are explained briefly in the following sections along with few examples from published reports.

13.2.1.1 Hydrothermal Carbonization (HTC)

Generally, for any carbonization process, biomass (precursor) was dried and screened for different sizes using meshes sieve. Collected particular size was dispersed in solvent usually in deionized water or water with 1% of H_2SO_4 and sealed in a hydrothermal reactor. This reactor was heated to 120–300 °C in an autoclave for few hours [48–50]. In this process, obtained hydrochar (char obtained from hydrothermal process) converted to PC by an activation process. Generally, hydrothermal carbonization (HTC) followed by the activation is preferred in many reports; however, P. Chen et al. group used KOH in the HTC process to reduce the time of overall reaction [51].

Many biomass sources, especially that of agricultural waste, are used as a starting feed for the HTC process to synthesize the PC materials [52]. Agricultural waste contains cellulose, hemicellulose, and lignin, which are the main sources for carbon content. Cellulose and hemicellulose decompose below 250 °C and lignin required higher temperature. Zenghua Xu et al. reported a synthesis of AC derived from rice straw using hydrothermal treatment followed by chemical activation [48]. The obtained AC depends not only on the HTC process but also on the biomass feed, temperature, and other parameters. However, the HTC processes narrow down to the differences among different biomass sources. In a different study, Zhengang Liu et al. showed how waste biomass feed and hydrothermal temperature causes variation in the quality of the fuel [53]. The biochar obtained through HTC is a potential candidate for substitution of coal that has been used for heat production. In some cases, hydrothermal treatment was not only used for the production of hydrochar but also for the functionalization of activated PC. For example, α-MnO_2/f-AC functional composites were prepared using HTC for higher specific capacitance and salt removal and regeneration applications [54]. Varying the parameters such as temperature, synthesis time, and nature of the precursor in HTC process will affect the morphology, porosity, aromaticity, polarity, and stability of the nanoPC [52]. Zhu et al. reported that increasing peak temperature and retention time causes increased thermal stability and aromaticity with decreased polarity of hydrochar. Contrary to that, low peak temperature and retention time produced high porous activated PC [55].

13.2.1.2 Pyrolysis

Pyrolysis is a process in which organic materials are decomposed at very high temperature in an inert atmosphere to get the carbon content from the precursor. This process has been utilized for the preparation of functional PC materials and many other forms of carbon. During pyrolysis, biomass decomposes in different paths namely fragmentation, depolymerization, and char formation. Detailed pyrolysis process and mechanism are explained in the review by Collard and Blin [56]. The following section describes the next step after pyrolysis or HTC process, that is, the activation of the produced carbonaceous material.

13.2.2 Activation

Researchers have been working on to produce activated PC to provide better and safe environment. Properties of the char for various applications can be enhanced using activation process. As mentioned earlier, AC properties depend on morphology, porosity, and surface functionality. Steps involved in the activation process are discussed in the following subsections.

13.2.2.1 Physical Activation

Generally, char formed from the carbonization process possess locked pores, due to the presence of organic decomposition and inorganic metal content. Physical activation leads in recovery of the porous nature by creating new pores and increasing the existing pore size. In this method, carbonization followed by activation with oxidizing gases like CO_2 and steam at high temperature (600–1100 °C) for few hours was performed. Steam reacts faster with biochar compared to CO_2. Activation can be controlled with CO_2, as it is less reactive compared to steam. Moreover, another advantage of CO_2 is ease in handling, thus preferred in industrial scale. These oxidizing gases react with carbon and form activated PC as mentioned in the following reaction schemes.

Reaction with CO_2:

$$C + CO_2 \rightarrow 2CO$$

Reaction with H_2O:

$$C + H_2O \rightarrow H_2 + CO$$

$$CO + H_2O \rightarrow H_2 + CO_2$$

$$C + 2H_2 \rightarrow CH_4$$

The reaction of carbon with water and CO_2 not only enhances the porosity but also increases existing pore size of the char, and hence their surface area.

13.2.2.2 Chemical Activation

It involves the treatment of carbonized material with strongly dehydrating and oxidizing agents like alkali (NaOH, KOH), acids (H_3PO_4, H_2SO_4), and slats ($ZnCl_2$, $AlCl_3$) followed by heating under inert atmosphere. The advantage of the chemical activation involves a single step and requires low-temperature conditions. However, one should remember about the corrosive nature of the chemicals involved in this process that raised difficulties for industrial scaling. It may also require more steps as shown in Figure 13.2 for improved product quality.

There are two types of reported procedures: one type is where impregnation with activating agent is done before carbonization and the other type is where activating agent is added after carbonization. Activating agent acts as both dehydrant and as an oxidant at the same time. It helps in inhibiting the deposition of organic matter in the pores and enhances the surface area of the produced AC. In the final step, washing treatment was used to remove the materials deposited inside the pore. This washing treatment also helps to recover certain amounts of activating agent.

Generally, chemical activation is performed at lower temperature (450–900 °C) than physical activation. The plausible reaction between KOH and carbon is as follows [57]: carbonates and hydrogen gas are produced below 750 °C. Carbon monoxide and carbon dioxide evolution from carbonates are observed at higher temperatures. Preparation of activated PC from various sources using KOH is very promising and reaches specific surface area up to 3000 $m^2\ g^{-1}$ [58]:

$$6KOH + 2C \leftrightarrow 2K + 3H_2 + 2K_2CO_3$$

Activating agent plays a crucial role in the production of AC with desired properties. Researchers have been working on the quantity of activating agent required in the activation process and capability of different activating agents for the production of highly activated PC. Islam et al. prepared AC using coconut shell waste through hydrothermal treatment followed by chemical activation for methylene blue adsorption. They have varied impregnation ratio (hydrochar:NaOH) to obtain an efficient AC and compared it with other reports and showed that NaOH is better activating agent for their process [59]. A. Ros et al. studied activation process with CO_2, H_3PO_4, NaOH, and KOH for the synthesis of AC from sewage sludge-based precursors. The efficiency of AC depends on hydroxide:precursor ratio and also on the mixing of the activating agent and precursor [23]. KOH is highly corrosive in nature and difficult to use for industrial scale. To overcome this problem, Zenghua Xu et al. used environmental-friendly activating agent $KHCO_3$ and achieved high yield compared to the KOH product [48]. AC porosity (2786.5 m^2 g^{-1}) and wettability also enhanced by using $KHCO_3$ with melamine (for N-doping) compared to KOH. Besides, activating agent temperature could change morphology of the AC and hence properties. Nanaji et al. synthesized porous graphene sheet-like materials and amorphous carbon from jute stick biomass using chemical activation method. Quantity of the product was controlled with the help of activation temperature. Increasing temperature yields high content of porous graphene sheets. Morphology was tuned with the temperature range from 800 to 1000 °C [60].

13.2.3 Physicochemical Activation

It involves both physical and chemical activation process. Two methods are known in these processes: chemical treatment followed by carbonization and carbonization followed by chemical treatment. Huanlei Wang et al. explained the advantage of this process. In their experiments, commercially available carbon is treated with CO_2 at activation temperature of 1223 K. Carbon temperature brought back to room temperature followed by mixing KOH with activated PC at activation temperature of 1023 K in argon atmosphere. The resulting AC exhibited high surface area of 3190 m^2 g^{-1} with high hydrogen storage capacity [61].

13.2.4 Microwave-based synthesis

Microwave-based activation is a novel method that has been investigated in the recent past to avoid drawbacks of the conventional activation methods. In this method, heating involves radiation energy transfer that rapidly converts into heat instead of heat transfer involved in the case of conventional activations [62]. Due to this, the activation is achieved much faster, thus saving energy or gas consumption during their synthesis process. Foo et al. [63] reported that activation from the sugarcane bagasse precursor when synthesized using microwave activation yielded AC with Brunauer–Emmett–Teller (BET) surface area of 1612 m^2 g^{-1}. Whereas, conventional-based activation reported by Azmi et al. produced AC from same precursor with BET surface area of only 100 m^2 g^{-1} [64].

Additionally, microwave-based activation shows several more advantages over conventional thermal activation. The advantages include enhanced efficiency, selective heating, instantaneous start and shutdown facility, fewer steps in synthesis, low activation temperature, and improved safety.

Moreover, the radiation time and power play an important role by affecting the AC yields and their surface functional groups [65]. Angin et al. showed that microwave-based synthesis produced ACs with better texture and pore distribution [66].

The methodologies mentioned earlier are used for different sources. Starting biomass materials composition is very important for the efficiency of the produced

AC and price of the source for commercial purpose. Animal, mineral, agriculture, and industries are the main sources for waste generation. Agricultural waste is one of the best precursors due to low cost, availability, and environment-friendly. Researchers should consider the parameters like carbon content, inorganic matter, volatile matter, and potential to produce high yield to choose the precursor. Synthesis of AC is an important task, yet modification of AC is also another important task for chemists, to get desired products with better performance for various applications. The following session provides a brief idea of different methods used for modification.

13.2.5 Functionalization/Doping/Composites of ACs

Functionalization of activated PC introduces new properties which results in a wider scope in their application. Carboxylic groups on the surface of activated PC are introduced by simple oxidation treatment with chemicals like HNO_3, H_2O_2, and $(NH_4)_2S_2O_8$ [67]. Oxidation with HNO_3 is highly efficient and can be controlled by varying concentration of acid and temperature. Activated PC was suspended in H_2SO_4 for 4 hours at 80 °C followed by washing treatment to get sulfonated material [68]. Xiqing Wang et al. reported covalent attachment of sulfonic acid using diazonium salt and hypophosphorous acid under mild condition instead of H_2SO_4 [69]. Oxidation and sulfonation produce hydrophilic activated PC. Reaction with F_2 gas produces fluorinated carbon on the surface and results in hydrophobic nature [70].

Andreas Stein et al. presented a review on functionalization of PC materials, in which grafting technique (Figure 13.3) was used in many reports for various surface functionalizations. Impregnation, nanocomposites, attachment of NPs via electrostatic interactions, and surface coating with polymers are also discussed (Figure 13.3) [71].

Xiaoling Zhong reported synthesis of nano-PC with rich functional groups by templating leaf extract with nano-$CaCO_3$. The advantage of nano-$CaCO_3$ is that it serves as a template and acts as a source for activating agent CO_2 during pyrolysis. Released CO_2 during pyrolysis produce micropores and mesopores and enlarges pore size [72]. Nano-PC prepared from tomato waste by using an activating agent $ZnCl_2$ was used for the elimination of anionic dye [73].

Heteroatom doping helps to develop ACs for large surface area, high electrical conductivity, and other applications. Doping is performed by introducing organic or inorganic molecule or metal oxides at the time of carbonization. Other approach is self-doped, where source itself contains the element. Schematic representation of doping is shown in Figure 13.4 [74].

A brief concept on the nitrogen doping and an example for co-doping and core–shell NPs are discussed. Organic molecule melamine is used in the nitrogen doping process because of high nitrogen content in it. Resulted nitrogen-doped porous activated PC shows much larger surface area and high capacitance compared to non-N-doped sample [75]. L. Niu et al. reported self-doped activated PC derived from waste pork, blackfish, and eel bone for supercapacitor applications. Waste

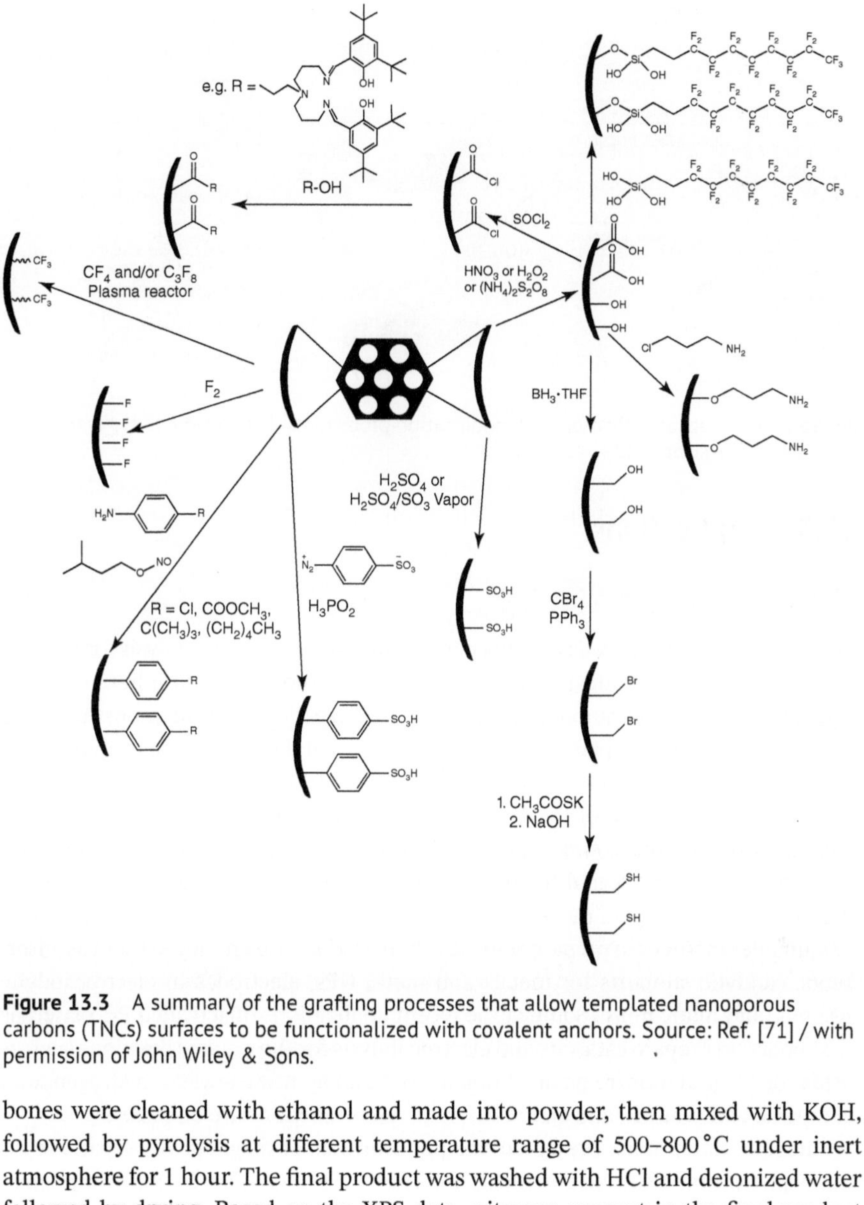

Figure 13.3 A summary of the grafting processes that allow templated nanoporous carbons (TNCs) surfaces to be functionalized with covalent anchors. Source: Ref. [71] / with permission of John Wiley & Sons.

bones were cleaned with ethanol and made into powder, then mixed with KOH, followed by pyrolysis at different temperature range of 500–800 °C under inert atmosphere for 1 hour. The final product was washed with HCl and deionized water followed by drying. Based on the XPS data, nitrogen present in the final product belonged to pyridinic-N, pyrrolic-N, and quaternary-N. The presence of lone pair of electrons on pyridinic and pyrrolic nitrogen enhances the electrochemical properties of the electrodes [76]. Similarly, N and P co-doped PC was synthesized from inexpensive oil seed waste using H_3PO_4 and used for electrocatalytic activity [77]. AC was doped with thiourea to obtain N and S co-doped AC to improve electrochemical performance [78]. Even well-defined shapes of core–shell nanostructures of activated PC with ZnO (AC on the surface of ZnO) were fabricated from silkworm cocoons and ZnO, which act as a dual-functional sensor for H_2 and UV [79].

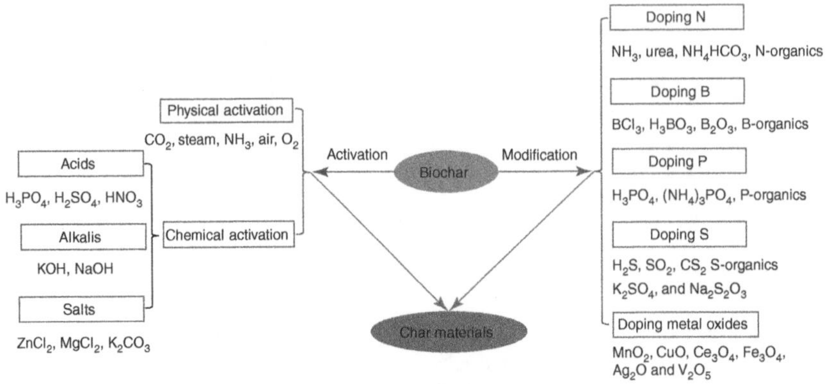

Figure 13.4 Biochar activation and modification procedures. Source: Ref. [74] / with permission of Elsevier.

13.3 Applications

13.3.1 Heterogeneous Catalysis

The heterogeneous catalysis reckons paramount importance in catalysis for the production of numerous industrial chemicals owing to the associated inherent advantages such as easy separation of the catalyst from reaction mixture, robustness, and minimum cost. The preparation of heterogeneous catalysts commands appropriate catalytic supports. Numerous materials such as AC, silica, and alumina are available, and the AC is one of the frequently used catalytic supports. The AC can be obtained from various sources, but the activated PCs derived from biowaste materials are the most preferred for the reason that they are abundant, low cost, and the ease in processing. In the recent past, many biowaste materials were explored as sustainable resources to prepare activated PC materials and employed them as adsorbents, catalytic supports for metals and metal NPs, electrodes in electrocatalytic systems, and many more. Some of the recent applications of activated PC materials as supports to prepare catalysts and electrocatalysts for the organic reactions such as oxidation/degradation/removal of organic pollutants in wastewater, hydrogenation of organic compounds, and electrocatalytic redox reactions are discussed.

Lincheng Zhou et al., presented one-pot synthesis of magnetically separable PC from agricultural waste peanut shells and $(NH_4)_3Fe(C_2O_4)_3$ via a simple impregnation, a combined activation and carbonization process. The crushed peanut shells and magnetic precursor ferric ammonium oxalate ($(NH_4)_3Fe(C_2O_4)_3$) solutions were prepared and dried in oven at 70 °C to get cylindrical precursors, which are carbonized in a tube furnace under the nitrogen flow. First, the sample was carbonized at 500 °C (rate of heating 5 °C min^{-1}) for 30 minutes, then at 800 °C (rate of heating 10 °C min^{-1}). The resulting catalyst was used as a heterogeneous Fenton catalyst for the removal/degradation of methylene blue from the wastewater in the presence of persulfate (PS). The study shows that the catalyst is exhibiting higher efficiency, can remove 90% of methylene blue within 30 minutes, and also can be recycled for seven times. [80] Another most abundant biomass, rice husk, was converted to nanoporous activated carbon (NPAC) by pre-carbonization and chemical activation method by Sekaran et al. The rice husk was pre-carbonized by heating at 400 °C for 4 hours under N_2 flow. The chemical activation was carried out by stirring the pre-carbonized rice husk with H_3PO_4 at 85 °C for 4 hours. The treatment with H_3PO_4 results in chemical changes and structural alterations in the pre-carbonized rice husk. Thus, obtained NPAC was treated with cobalt nitrate fallowed by sodium borohydride to get cobalt-impregnated NPAC (Co-NPAC) catalyst. The Co-NPAC catalyst was successfully used for the removal of organic dye chemicals from tannery industry wastewater by Fenton oxidation process [81].

Shanshan Tang et al. applied simple methods such as carbonization and chemical activation for the preparation of a PC material (PCCS) from corn straw, an agriculture waste. Typically, the carbonization is carried out at 500 °C for 60 minutes under nitrogen flow. The raw carbon straw material washed with deionized water and crushed to different particle sizes, and then carbonized in a horizontal tube furnace. Then, the 1 : 1 mixture of KOH and NaOH was used for the chemical activation process. The carbonized carbon straw was mixed with alkali mixture and heated to activation temperature (700 °C) for 60 minutes under nitrogen flow. The resulting PCCS was coated with Fe_3O_4 NPs and used in the adsorption of rhodamine B (RhB) subsequently by removing the dye from the wastewater. The study claims that the PCCS shows maximum surface area and dye adsorption capacity when the 1 : 1 alkali mixture is used as a chemical activator. The present study shows the advantages such as the separation of adsorbent from the mixture due to magnetic property of Fe_3O_4 NPs, and single activator is replaced with mixed activators to influence the surface area and adsorption capacity, resulting enhanced dye removal performance [82]. Wen–Da Oh et al. studied the preparation of N-doped PC from various biomass materials such as waste peels and dried leaves using thermal annealing method under inert atmosphere. These PCs were employed for the degradation of bisphenol A (BPA) in the presence of peroxymonosulfate (PMS) as an activator. It is observed that the extent of doping and surface area was influenced by the choice of biomass material. The N-doped PCs prepared from spent coffee ground and saw dust exhibited higher graphitic N (44–46%) content, specific surface area (>400 $m^2\ g^{-1}$), and better catalytic performance than those prepared from banana peel, orange peel, and dried leaves. The present study demonstrates the importance of biomasss as beneficial resources to produce PCs [83].

Zakaria Anfar et al. used almond shell as a source for the preparation of N-doped graphitic PC (N-GPC) using pyrolysis method. Initially, the almond shell was chemically activated by treating with NH_4OH at 60 °C for 12 hours. Then, the sample was pyrolyzed at 300 °C for 2 hours and then at 900 °C for 4 hours at a heating rate of 5 °C min^{-1} under nitrogen flow. The resulting N-GPC exhibited high surface area of 2054 $m^2\ g^{-1}$ and total pore volumes of 1.19 $cm^3\ g^{-1}$. The catalytic activity of N-GPC was evaluated for the degradation of dye pollutant Orange G in the presence of PS as an activator. The high surface area and generated charges from the surface of the prepared N-GPC are prominent factors for the better performance of N-GPC [84]. Ting Zhang et al. developed a one-pot pyrolysis technique for the synthesis of iron

NPs encapsulated within nitrogen and sulfur co-doped magnetic PC (Fe-N-S-MPC) with lignin obtained from black liquor as the primary precursor. A lignin solution was treated with sodium bicarbonate, L-cysteine, and iron nitrate. The material was then carbonized at 700 °C to produce Fe-N-S-MPC. The catalytic performance of Fe-N-S-MPC for the degradation of 1-naphtol in the presence of PMS was effectively assessed [85].

Zaeni et al. synthesized oxygen-functionalized PC via direct thermal treatment of biomass under inert atmosphere. The wood residue was taken in a sealed crucible and calcined at 800 °C for 2 hours in a furnace (carbolite) with heating rate of 10 °C min^{-1}. The characterization of obtained PC showed the surface area of 993 m^2 g^{-1} and amorphous in nature. The catalyst was used for the degradation of organic pollutants in the presence of PMS and found to be extremely favorable for catalysis due to its high specific surface area and abundant active sites (ketonic groups) (Figure 13.5) [86].

Biomass tea leaves were used as an economical and bio-friendly resource for the preparation of three-dimensional (TD) hierarchical tea leaf PC (TPS) by Jiafeng Wan et. al. First, the tea leaves were pretreated with KOH and then subjected to high-temperature calcination to get the TPC with a high surface

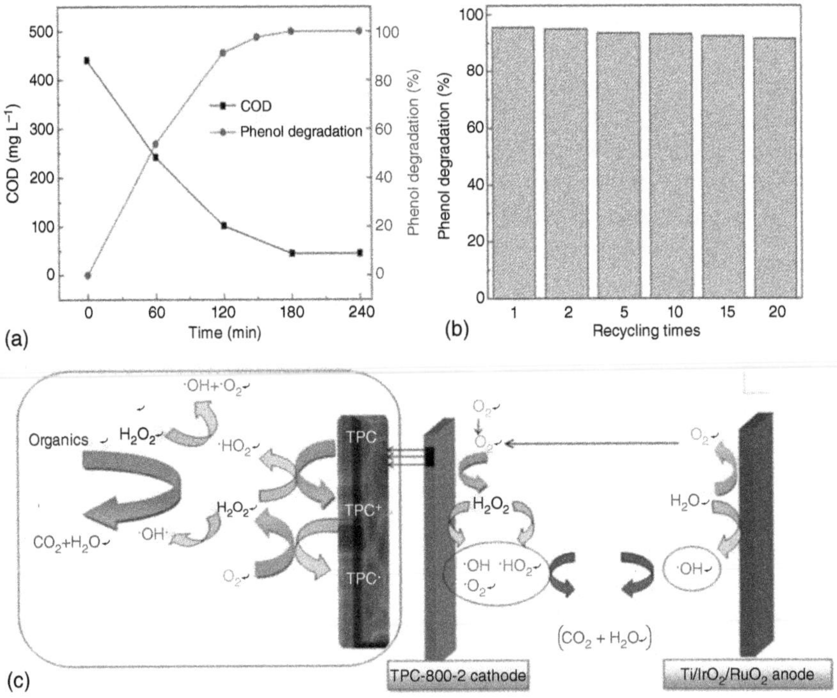

Figure 13.5 (a) Degradation of COD and phenol by the TPC-800-2 cathode over time. (b) Recyclability of the cathode in degradation of phenol. (c) The electrocatalytic mechanism of the TPC-800 cathode is depicted schematically. Source: Ref. [86] / with permission of American Chemical Society.

area of 1620.05 m^2 g^{-1}, high oxygen content of 15.51%, 3D multilayer pore structure, and uniform pore size. The TPC material was used to fabricate a cathode, and its electrocatalytic properties were evaluated for degradation phenol. The conclusions insinuate that the layered inter-connected microporous structure improves the adsorption and transport of organic pollutants, and larger specific surface area results in a greater number of active sites by enhancing the catalytic performance [86].

Jixian Yang et al. reported one-step pyrolysis method for the preparation of biomass-derived PC cathode (wood-derived cathode, WDC) from the recovered wood waste and used as a cathode for the electro-Fenton (EF) oxidation thereby the degradation of sulfathiazole (STZ) contaminant presents in the water bodies. The cut waste wood pieces were initially pre-carbonized at 260 °C for 6 hours in air (rate of heating at 10 °C min^{-1}). Then, the complete pyrolysis was performed at different temperatures (500, 700, and 900 °C) (rate of heating at 10 °C min^{-1}) in a quartz tubular furnace under continuous nitrogen flow (150 mL min^{-1}). It was finally stabilized for 2 hours. The WDC as a cathode, boron-doped diamond (BDD) as anode, and pyrophosphate (PP) as electrolyte were used in EF process at pH 8 for the treatment of STZ as an antibiotic contaminant in most water bodies. The WDC cathode pyrolyzed at 900 °C showed better results due to its larger electroactive surface area (28.81 cm^2) (Figure 13.6) [87].

Recently, one-pot pyrolysis method was used to prepare N-doped carbon from renewable biomass (chitin) and used as a catalyst support for metals. The chitin (biomass) and metal salt were used to prepare a N-doped carbon-supported reusable bimetallic catalyst Ru-MoO_x/CN_x by Cao et al. and used for the hydrogenation of furfural (FAL) to tetrahydrofurfuryl alcohol (THFA). The typical procedure is as follows: chitin, metal precursors ($H_3PMo_{12}O_{40}{\cdot}xH_2O$ and $RuCl_3{\cdot}xH_2O$) in required quantities were taken in ethanol and heated at 60 °C until complete drying to get a

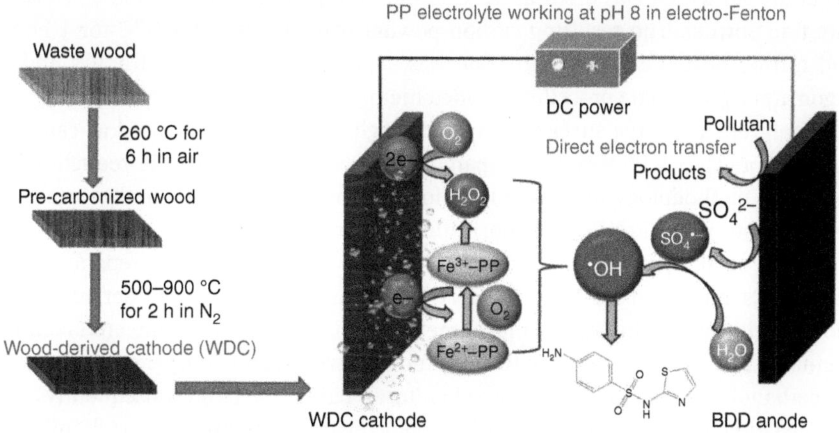

Figure 13.6 Graphical abstract, biochar cathode derived from waste wood and its use in electro-Fenton for sulfathiazole treatment at alkaline pH using pyrophosphate electrolyte. Source: Ref. [87] / with permission of Elsevier.

powder. The resulting powder was heated at 700 °C for 2 hours in a furnace. Then, the catalysts were calcined at 250 °C in a stream of air for 3 hours and in hydrogen flow at 350 °C for 2 hours. The characterization showed the BET surface area of 452.4 m^2 g^{-1} and pore volume of 0.322 cm^3 g^{-1} by these catalysts. The catalysts were evaluated for direct hydrogenation of FAL to THFA. The bimetallic Ru-MoO_x/CN catalyst showed much higher activity for the direct hydrogenation in comparison with corresponding monometallic Ru/CN catalyst. Further, the catalysts can be reused for five cycles without deactivation. Hence, the one-pot pyrolysis method can be a good alternative for the preparation of N-doped carbon-supported bimetallic catalyst using renewable chitin and metal salts [88].

Yangyang Zhu et al. reported the preparation of N-doped PC material (PNCM) using biomass soybean curd residue (SCR) as the precursor material using the carbonization method. In a typical procedure, the SCR powder was pyrolyzed at 300 °C in an inert atmosphere. After activating the carbon sample with KOH, it was oven-dried overnight in vacuum at 40 °C. Then, carbon sample was carbonized with a heating rate of 5 °C per minute at 700, 800, and 900 °C in a tube furnace for 2 hours under nitrogen atmosphere. The obtained PNCM samples were modified with Pd NPs to yield the Pd/PNCM catalyst. The catalyst was successfully used for the hydrogenation of phenol to cyclohexanone. Further, the catalyst can be used for at least 10 cycles. This strategy may capable of producing cost-effective N-doped PC materials from biomass and may offer new ways for the preparation of useful nanocatalysts [89]. Di Hu recently reported the preparation of PC from biomass waste sorghum and used as a catalyst support for Pd and Pt NPs. The catalyst was evaluated for the reduction FAL to get value-added chemicals. In a typical procedure, the sorghum was grounded to fine powder and mixed and fully solvated with ammonia solution. The solution was decanted into glass vial, and the vial was placed inside a Teflon liner vessel. The vessel was heated at 180 °C for 24 hours in a preheated oven to form gel. Upon cooling the system, the mixture was centrifuged to get the gel which was dried at 80 °C under an argon atmosphere and converted into fine powder. The resulting carbon powder was calcined at 350 °C for 1 hour and further heated at 550 °C (the rate was kept at 5 °C min^{-1}) for 3 hours under argon atmosphere. The procedure yielded hierarchically PC support with a number of thin layers and large surface area from sorghum. This PC was used as catalyst support for Pd and Pt NPs. So, prepared catalysts were used for the reduction of FAL. The methodology has a potential for the synthesis of highly active catalysts using various metallic NPs and valorization of biomass waste [90].

Advani et al. reported an attractive synthetic method for the preparation of nickel NPs supported on N-doped CNT nanocatalyst (Ni@N-CNT) derived from biomass chitosan. Initially, the nickel metal precursor and chitosan were taken in methanol and heated at 50 °C with continuous stirring. After complete evaporation of methanol, the resultant Ni-adsorbed chitosan flakes dried and subsequently carbonized directly in a tubular furnace by slowly raising the temperature (5 °C min^{-1}) at 800 °C under a constant flow of 5% H_2/N_2. The resulting catalyst Ni@N-CNT was used for the selective hydrogenation of nitroarenes. Initially, the nickel adsorbed on chitosan, and then the carbonization at higher temperatures led to the development

Table 13.1 Biomass-derived carbon material catalysts anc their uses are listed.

	Biowaste	Catalyst	Dopent content (%)	Surface area ($m^2\ g^{-1}$)	Application	Ref no.
1	Peanut shells	Peanut shell magnetic carbon	–	299–503	Removal of methylene blue	[80]
2	Rice husk	Cobalt oxide-supported PC	–	52	Removal of refractory organics	[81]
3	Corn straw	PC material (PCCS)@Fe3O4	–	1993–3467	Degradation of RhB	[82]
4	Waste peels and dried leaves	Nitrogen-doped PC	Graphitic N (44–46 at%)	>400	Degradation of bisphenol A	[83]
5	Almond shells	N-doped graphitic PC	N-2.03	2054	Degradation of Orange G	[84]
6	Black liquor (lignin)	Nitrogen and sulfur co-doped magnetic PC	N-1.85, S-0.98, Fe-0.54	663	Degrade 1-naphthol	[85]
7	Wood residue	Oxygen-functionalized PC	O-5.3--12.0	993	Degrade acid orange 7	[86]
8	Waste tea leaves	3D tea leaf PC	O-15.5	1620.05	Electrocatalytic degradation of phenol	[86]
9	Chitin	N-doped carbon	–	452.4	Hydrogenation of furfural	[88]
10	Soybean curd	N-doped PC	–	1351.6–1604.2	Phenol hydrogenation	[89]
11	Sorghum	Hierarchically PC	–	531.9	Hydrogenation of furfural	[90]
12	Chitosan	N-doped carbon	N-5.7%	–	Hydrogenation of nitroarenes	[91]
13	Chitosan	Co@Chitosan	N-2.53%	–	Hydrogenation of terminal and internal olefins	[92]

of a carbonaceous graphitized structure. These graphitic carbon layers move toward the Ni NPs during the heat treatment, resulting in the production of hollow N-doped CNTs with anisotropic and deformed patterns. These NPs are generated at the opening end of the hollow CNTs. The nitrogen that remains in the carbon network stabilizes the NPs, resulting in an improved NP dispersion [91]. Scharnagl et al. reported the synthesis of cobalt chitosan (Co@Chitosan-700) catalyst from biomass that used for the hydrogenation of olefins under mild conditions. The catalyst was prepared by stirring the cobalt acetate and chitosan in ethanol. After removing the solvent, the sample was crushed and pyrolyzed in between 700 and 1000 °C. The Co@Chitosan-700 catalyst exhibited better performance in the hydrogenation. The catalyst also used in the hydrogenation of some industrially important compounds such as diisobutene, fatty acids, and their esters. The catalyst can be recycled and found to be stable during several consecutive runs [92]. Some of the other materials are shown in Table 13.1.

Chi Van Nguyen et al. synthesized N-doped nano-PC (NNC) with a zeolitic-imidazole framework (ZIF-8). The ZIF-8 was prepared from the zinc nitrate and 2-methylimidazole. The subsequent carbonization of the ZIF-8 produced NNC with high nitrogen quantity, uniformity, and increased specific surface area. The resulting NNCs were successfully tested for metal-free aerobic oxidation of 5-hydroxymethylfurfural (HMF) to 2,5-furandicarboxylic acids (FDCA) with higher FDCA yields [93]. Yamei Lin et al. modified the nitrogen-doped PC with Co and used for the conversion of 5-HMF into 2,5-furandicarboxylicacid dimethyl ester using oxidative esterification method under mild conditions. The Co-modified N-doped graphite (Co@CN) catalysts were prepared by using the chitosan as a carbon source and the zinc was used as a sacrificial template to offer the catalyst with high specific surface area and catalytic activity. The presence of Zn was found to be useful in the regulation of acidic and basic sites, which rendered the increased activity of the catalyst and also the evaporation of Zn during the pyrolysis resulted in the increased specific surface area [94]. Vitaly L. Budarin et al. prepared a mesoporous carbon material from the starch, a renewable and more abundant polysaccharide, and used it as a catalytic support. Pd NPs were fixed on this mesoporous material and used in organic reactions (e.g. Heck, Sonogashira, etc.) which results in C—C bonds. The starch-derived mesoporous materials in combination with Pd NPs were exhibiting higher activity and stability and also can be reused multiple times [95].

13.4 Conclusion and Future Challenges

This chapter discusses recent advances in synthesizing activated PC from biomass and their applications in catalysis. A most common method of synthesis takes place in two steps, that is, carbonization and activation. Carbonization is performed either by hydrothermal treatment or pyrolysis. The resulting carbonaceous material is finally activated, using a physical, or a chemical, or a physicochemical method, to make activated PC. Physical activation is performed at high temperature, using

CO_2 or steam as an activating agent, whereas chemical activation is done by adding a strong dehydrating or an oxidizing agent, in the form of base or acid or salt under inert atmosphere and heat. Apart from these methods, a novel method, which is a microwave-based activation, has also been in practice since recent times, which has been reported to be more advantageous over conventional methods. Although there are different synthesis methods, each of the processes produces a different AC material with different porosity and texture. Thus, one should make a note that the resulting AC depends on several parameters including nature of precursor, type of activating agent, temperature, and heating time. All these parameters play a crucial role in producing an AC with desired properties with better performance for various applications. Aside from correct biomass precursor selection, the preparation process is also a key aspect in achieving carbon compounds with outstanding physicochemical qualities. As a result, additional efforts must be made to create greener synthetic methods in order to build biomass-derived carbon compounds with the needed properties for their catalysis-related applications of future importance. This chapter mainly focuses on the synthesis of ACs from biomass and AC-supported materials for applications in the area of catalysis including hydrogenation, oxygen reduction reaction, etc.

References

1 Weedmark, D. (2018). Human activities that affect the ecosystem. https://sciencing.com/human-activities-affect-ecosystem-9189.html (accessed 20 August 2020).

2 Tun, M.M., Juchelkova, D., Raclavska, H. et al. (2018). Utilization of biodegradable wastes as a clean energy source in the developing countries: a case study in Myanmar. *Energies* 11 (11): 1–20.

3 Hoornweg, D. and Bhada-Tata, P. (2012). *What a Waste: A Global Review of Solid Waste Management*, Urban Development Series knowledge papers no. 15. Washington, DC: The World Bank Group © World Bank. https://openknowledge.worldbank.org/handle/10986/17388 License: CC BY 3.0 IGO.

4 Bocken, N.M.P., Olivetti, E.A., Cullen, J.M. et al. (2017). Taking the circularity to the next level: a special issue on the circular economy. *J. Ind. Ecol.* 21 (3): 476–482.

5 McDonough, W. and Braungart, M. (2002). *Cradle to Cradle: Remaking The Way We Make Things*. North Point Press.

6 Cuhadaroglu, D. and Uygun, O.A. (2008). Production and characterization of activated carbon from a bituminous coal by chemical activation. *Afr. J. Biotechnol.* 7 (20): 3706–3713.

7 Zhu, Z.L., Li, A.M., Xia, M.F. et al. (2008). Preparation and characterization of polymer-based spherical activated carbons. *Chin. J.Polym. Sci. (English Edition)* 26 (5): 645–651.

8 Hu, Z. and Srinivasan, M.P. (2001). Mesoporous high-surface-area activated carbon. *Microporous Mesoporous Mater.* 43 (3): 267–275.

9 Yacob, A.R., Abdul Majid, Z., Sari, D.D.R. et al. (2008). Comparison of various sources of high surface area carbon prepared by different types of activation. *Malaysian J. Anal. Sci.* 12 (1): 264–271.

10 Idris, S. (2012). Kinetic study of utilizing groundnut shell as an adsorbent in removing chromium and nickel from dye effluent. *Am. Chem. Sci. J.* 2 (1): 12–24.

11 Bouhamed, F., Elouear, Z., and Bouzid, J. (2012). Adsorptive removal of copper (II) from aqueous solutions on activated carbon prepared from Tunisian date stones : Equilibrium, kinetics and thermodynamics. *J. Taiwan Inst. Chem. Eng.* 43 (5): 741–749.

12 Jain, A., Balasubramanian, R., and Srinivasan, M.P. (2016). Hydrothermal conversion of biomass waste to activated carbon with high porosity : a review. *Chem. Eng. J.* 283: 789–805.

13 Manaf, S.A.A., Roy, P., Sharma, K.V. et al. (2015). Catalyst-free synthesis of carbon nanospheres for potential biomedical applications: waste to wealth approach. *RSC Adv.* 5 (31): 24528–24533.

14 Dizaj, S.M., Mennati, A., Jafari, S. et al. (2015). Antimicrobial activity of carbon-based nanoparticles. *Adv. Pharm. Bull.* 5 (1): 19–23.

15 Marsh, H. and Reinoso, R.F. (2006). *Activated Carbon*, 1ee. Elsevier.

16 Sivamani, S. and Leena, G.B. (2009). Removal of dyes from wastewater using adsorption – a review. *Int. J. BioSci. Technol.* 2 (4): 47–51.

17 Gupta, V.K., Mittal, A., Jain, R. et al. (2006). Adsorption of Safranin-T from wastewater using waste materials – activated carbon and activated rice husks. *J. Colloid Interface Sci.* 303 (1): 80–86.

18 Manoj, P., Reddy, K., Verma, P. et al. (2016). Bio-waste derived adsorbent material for methylene blue adsorption. *J. Taiwan Inst. Chem. Eng.* 58: 500–508.

19 Manoj, K.R., Krushnamurty, P., Mahammadunnisa, K. et al. (2015). Preparation of activated carbons from bio-waste: effect of surface functional groups on methylene blue adsorption. *Int. J. Environ. Sci. Technol.* 12 (4): 1363–1372.

20 Kristianto, H., Putra, C.D., Arie, A.A. et al. (2015). Synthesis and characterization of carbon nanospheres using cooking palm oil as natural precursors onto activated carbon support. *Procedia Chem.* 16: 328–333.

21 Abdul, H.A., Anuar, K., Zulkarnain, Z. et al. (2001). Preparation and characterization of activated carbon from Gelam Wood Bark (Melaleuca cajuputi). *Malaysian J. Anal. Sci.* 7 (1): 65–68.

22 Nor, N.M., Chung, L.L., Teong, L.K. et al. (2013). Journal of environmental chemical engineering synthesis of activated carbon from lignocellulosic biomass and its applications in air pollution control – a review. *Biochem. Pharmacol.* 1 (4): 658–666.

23 Ros, A., Lillo-Rodenas, M.A., Fuente, E. et al. (2006). High surface area materials prepared from sewage sludge-based precursors. *Chemosphere* 65 (1): 132–140.

24 Chiang, H.L., Huang, C.P., and Chiang, P.C. (2002). The surface characteristics of activated carbon as affected by ozone and alkaline treatment. *Chemosphere* 47 (3): 257–265.

25 Adib, M., Al-qodah, Z., and Ngah, C.W.Z. (2015). Agricultural bio-waste materials as potential sustainable precursors used for activated carbon production: a review. *Renew. Sustain. Energy Rev.* 46: 218–235.

26 Yahya, M.A., Al-Qodah, Z., Ngah, C.W.Z.C.W. et al. (2015). Preparation and characterization of activated carbon from desiccated coconut residue by potassium hydroxide. *Asian J. Chem.* 27 (6): 2331–2336.

27 Allwar, A. (2012). Characteristics of pore structures and surface chemistry of activated carbons by Physisorption, Ftir and Boehm methods. *IOSR J. Appl. Chem.* 2 (1): 9–15.

28 Patil, B.S. and Kulkarn, K.S. (2012). Development of high surface area activated carbon from waste material. *Int. J. Adv. Eng. Res. Stud.* 1: 109–113.

29 Belhachemi, M., Rios, R.V.R.A., Addoun, F. et al. (2009). Preparation of activated carbon from date pits: effect of the activation agent and liquid phase oxidation. *J.Anal. Appl. Pyrol.* 86 (1): 168–172.

30 Altenor, S., Carene, B., Emmanuel, E. et al. (2009). Adsorption studies of methylene blue and phenol onto vetiver roots activated carbon prepared by chemical activation. *J. Hazard.Mater.* 165 (1–3): 1029–1039.

31 Ahmedna, M., Marshall, W.E., and Rao, R.M. (2000). Production of granular activated carbons from select agricultural by-products and evaluation of their physical, chemical and adsorption properties. *Bioresour. Technol.* 71 (2): 113–123.

32 Djilani, C., Zaghdoudi, R., Modarressi, A. et al. (2012). Elimination of organic micropollutants by adsorption on activated carbon prepared from agricultural waste. *Chem. Eng. J.* 189–190: 203–212.

33 Sharma, Y.C., Uma, and Upadhyay, S.N. (2009). Removal of a cationic dye from wastewaters by adsorption on activated carbon developed from coconut coir. *Energy Fuel* 23 (6): 2983–2988.

34 Namasivayam, C. and Kavitha, D. (2002). Removal of Congo Red from water by adsorption onto activated carbon prepared from coir pith, an agricultural solid waste. *Dyes Pigm.* 54: 47–58.

35 Kannan, N. and Meenakshisundaram, M. (2002). Adsorption of Congo Red on various activated carbon A comparative study. *Water Air Soil Pollut.* 138 (1–4): 289–305.

36 Preethi, S., Sivasamy, A., Sivanesan, S. et al. (2006). Removal of safranin basic dye from aqueous solutions by adsorption onto corncob activated carbon. *Ind. Eng. Chem. Res.* 45 (22): 7627–7632.

37 Mohammadi, M., Hassani, A.J., Mohamed, A.R. et al. (2010). Removal of Rhodamine B from aqueous solution using palm shell-based activated carbon: adsorption and kinetic studies. *J. Chem. Eng. Data* 55: 5777–5785.

38 Cazetta, A.L., Vargas, A.M.M., Nogami, E.M. et al. (2011). NaOH-activated carbon of high surface area produced from coconut shell: kinetics and equilibrium studies from the methylene blue adsorption. *Chem. Eng. J.* 174 (1): 117–125.

39 Afrane, G. and Achaw, O.W. (2008). Effect of the concentration of inherent mineral elements on the adsorption capacity of coconut shell-based activated carbons. *Bioresour. Technol.* 99 (14): 6678–6682.

40 Nabais, J.M.V., Gomes, J.A., Suhas et al. (2009). Phenol removal onto novel activated carbons made from lignocellulosic precursors: influence of surface properties. *J. Hazard.Mater.* 167 (1–3): 904–910.

41 Sugumaran, P., Susan, V.P., Ravichandran, P. et al. (2012). Production and characterization of activated carbon from Banana Empty Fruit Bunch and Delonix regia Fruit Pod. *J. Sustain. Energy Environ.* 3: 125–132.

42 Buasri, A., Chaiyut, N., Loryuenyong, V. et al. (2012). Transesterification of waste frying oil for synthesizing biodiesel by KOH supported on coconut shell activated carbon in packed bed reactor. *Sci. Asia* 38 (3): 283–288.

43 Tawallebeh, M., Allawzi, M.A., and Kandah, M.I. (2005). Production of activated carbon from Jojoba seed residue by chemical activation residue using a static bed reactor. *J. Appl. Sci.* 5 (3): 482–487.

44 Sirichote, O., Innajitara, W., Chuenchom, L. et al. (2002). Adsorption of Iron (III) ion on activated carbons obtained from Bagasse, pericarp of rubber fruit and coconut shell. *Songklanakarin J. Sci. Technol.* 24 (2): 235–242.

45 Qureshi, K., Bhatti, I., Kazi, R. et al. (2007). Physical and chemical analysis of activated carbon prepared from sugarcane bagasse and use for sugar decolorisation. *Eng. Technol.* 34: 194–198.

46 Yusufu, M.I. (2012). Production and characterization of activated carbon from selected local raw materials. *Afr. J. Pure. Appl. Chem.* 6 (9): 123–131.

47 Bedia, J., Penas-Garzon, M., Gomez-Aviles, A. et al. (2018). A review on the synthesis and characterization of biomass-derived carbons for adsorption of emerging contaminants from water. *C. J. Carbon Res.* 4 (63): 1–53.

48 Xu, Z., Zhang, X., Liang, Y. et al. (2020). Green synthesis of nitrogen-doped porous carbon derived from Rice Straw for high-performance supercapacitor application. *Energy Fuel* 34: 8966–8976.

49 Huang, G., Wang, Y., Zhang, T. et al. (2019). High-performance hierarchical N-doped porous carbons from hydrothermally carbonized bamboo shoot shells for symmetric supercapacitors. *J. Taiwan Inst. Chem. Eng.* 96: 672–680.

50 Sun, W., Lipka, S.M., Swartz, C. et al. (2016). Hemp-derived activated porous carbons for supercapacitors. *Carbon* 103: 181–192.

51 Chen, P., Zang, J., Zhou, S. et al. (2019). N-doped 3D porous carbon catalyst derived from biomassTriarrhenasacchariflora panicle for oxygen reduction reaction. *Carbon* 146: 70–77.

52 Yahya, M.A., Al-Qodah, Z., and Ngah, C.W.Z. (2015). Agricultural bio-waste materials as potential sustainable precursors used for activated carbon production: a review. *Renew. Sustain. Energ. Rev.* 46: 218–235.

53 Liu, Z., Quek, A., Hoekman, S.K., and Balasubramanian, R. (2013). Production of solid biochar fuel from waste biomass by hydrothermal carbonization. *Fuel* 103: 943–949.

54 Govindan, B., Alhseinat, E., Darawsheh, I.F.F. et al. (2020). Activated porous carbon derived from *Phoenix dactylifera* (PalmTree) and decorated with MnO_2 nanoparticles for enhanced hybrid capacitive deionization electrodes. *Chem. Select* 5: 3248–3256.

55 Zhu, X., Liu, Y., Qian, F. et al. (2015). Role of hydrochar properties on the porosity of hydrochar-based porous carbon for their sustainable application. *ACS Sustain. Chem. Eng.* 3: 833–840.

56 Collard, F.-X. and Blin, J. (2014). A review on pyrolysis of biomass constituents: mechanisms and composition of the products obtained from the conversion of cellulose, hemicelluloses and lignin. *Renew. Sustain. Energ. Rev.* 38: 594–608.

57 Lillo-Rodenas, M.A., Juan-Juan, J., Cazorla-Amoros et al. (2004). Aboutreactions occurring during chemical activationwith hydroxides. *Carbon* 42: 1371–1375.

58 Wang, J. and Kaskel, S. (2012). KOH activation of carbon-based materials for energy storage. *J. Mater. Chem.* 22: 23710–23725.

59 Islam, M.A., Ahmed, M.J., Khanday, W.A. et al. (2017). Mesoporous activated coconut shell-derived hydrochar prepared viahydrothermal carbonization-NaOH activation for methylene blueadsorption. *J. Environ. Manag.* 203: 237–244.

60 Nanaji, K., Upadhyayula, V., Rao, T.N. et al. (2019). Robust, environmentally benign synthesis of nanoporous graphene sheets from biomass for ultrafast supercapacitor application. *ACS Sustain. Chem. Eng.* 7: 2516–2529.

61 Wang, H., Gao, Q., and Hu, J. (2009). High hydrogen storage capacity of porous carbons prepared by using activated porous carbon. *J. Am. Chem. Soc.* 131: 7016–7022.

62 Haque, K.E. (1999). Microwave energy for mineral treatment processes-a brief review. *Int. J. Miner.Process.* 57 (1): 1–24.

63 Foo, K.Y., Lee, L.K., and Hameed, B.H. (2013). Preparation of activated carbon from sugarcane bagasse by microwave assisted activation for the remediation of semi-aerobic landfill leachate. *Bioresour. Technol.* 134: 166–172.

64 Azmi, N.H., Ali, U.F.M., Muhammad Ridwan, F. et al. (2016). Preparation of activated carbon using sea mango (Cerberaodollam) with microwave-assisted technique for the removal of methyl orange from textile wastewater. *Desalination Water Treat.* 57 (60): 29143–29152.

65 Canales-Flores, R.A. and Prieto-Garcia, F. (2016). Activation methods of carbonaceous materials obtained from agricultural waste. *Chem. Biodivers.* 13: 261–268.

66 Angin, D., Altintig, E., and Kose, T.E. (2013). Influence of process parameters on the surface and chemical properties of activated carbon obtained from biochar by chemical activation. *Bioresour. Technol.* 148: 542–549.

67 El-Hendawy, A.A. (2003). Influence of HNO3 oxidation on the structure and adsorptive properties of corncob-based activated carbon. *Carbon* 41: 713–722.

68 Budarin, V.L., Clark, J.H., Luque, R. et al. (2007). Versatile mesoporous carbonaceous materials for acid catalysis. *Chem. Commun.* 634–636.

69 Wang, X., Liu, R., Waje, M.M. et al. (2007). Sulfonated ordered mesoporous carbon as a stable and highly active protonic acid catalyst. *Chem. Mater.* 19: 2395–2397.

70 Li, Z., Cul, G.D.D., Yan, W. et al. (2004). Fluorinated carbon with ordered mesoporous structure. *J. Am. Chem. Soc.* 126: 12782–12783.

71 Stein, A., Wang, Z., and Fierke, M.A. (2009). Functionalization of porous carbon materials with designed pore architecture. *Adv. Mater.* 21: 265–293.

72 Zhong, X., Yuan, W., Kang, Y. et al. (2016). Biomass-derived hierarchical nanoporous carbon with rich functional groups for direct-electron-transfer-based glucose sensing. *ChemElectroChem* 3: 144–151.

73 Guzel, F., Saygili, H., Saygili, G.A. et al. (2014). Elimination of anionic dye by using nanoporous carbon prepared from an industrial biomass. *J. Mol. Liq.* 194: 130–140.

74 Chen, Y., Zhang, X., Chen, W. et al. (2017). The structure evolution of biochar from biomass pyrolysis and its correlation with gas pollutant adsorption performance. *Bioresour. Technol.* 246: 101–109.

75 Han, X., Jiang, H., Zhou, Y. et al. (2018). A high performance nitrogen-doped porous activated porous carbon for super capacitor derived from pueraria. *J. Alloys Compd.* 744: 544–551.

76 Niu, L., Shen, C., Yan, L. et al. (2019). Waste bones derived nitrogen–doped carbon with high micropore ratio towards supercapacitor applications. *J. Colloid Interface Sci.* 547: 92–101.

77 Konwar, L.J., Sugano, Y., Chutia, R.S. et al. (2016). Sustainable synthesis of N and P co-doped porous amorphous carbon using oil seed processing wastes. *Mater. Lett.* 173: 145–148.

78 Liu, Y., Xiao, Z., Liu, Y. et al. (2018). Biomass-derived 3D honeycomb-like porous carbon with binary-heteroatom doping for high performance flexible solid-state supercapacitors. *J. Mater. Chem. A* 6: 160–166.

79 Saravanan, A., Huang, B.-R., Kathiravan, D. et al. (2017). Natural biomass-cocoon-derived granular activated porous carbon-coated ZnO nanorods: a simple route to synthesizing a core–shell structure and its highly enhanced UV and hydrogen sensing properties. *ACS Appl. Mater. Interfaces* 9: 39771–39780.

80 Zhou, L., Ma, J., Zhang, H. et al. (2015). Fabrication of magnetic carbon composites from peanut shells and its application as a heterogeneous Fenton catalyst in removal of methylene blue. *Appl. Surf. Sci.* 324: 490–498.

81 Karthikeyan, S., Boopathy, R., and Sekaran, G. (2015). Insitu generation of hydroxyl radical by cobalt oxide supported porous carbon enhance removal of refractory organics in tannery dyeing wastewater. *J. Colloid Interface Sci.* 448: 163–174.

82 Chen, S., Chen, G., Chen, H. et al. (2019). Preparation of porous carbon-based material from corn straw via mixed alkali and its application for removal of dye. *Colloids Surf. A* 568: 173–183.

83 Oh, W.-D., Veksha, A., Chen, X. et al. (2019). Catalytically active nitrogen–doped porous carbon derived from biomasss for organics removal via peroxymonosulfate activation. *Chem. Eng. J.* 374: 947–957.

84 Anfar, Z., Fakir, A.A.E., Ahsaine, H.A. et al. (2020). Nitrogen doped graphitic porous carbon from Almond Shell as an efficient persulfate activator for organic compounds degradation. *New J. Chem.* 44: 9391–9401.

85 Zhang, T., Li, C., Sun, X. et al. (2020). Iron nanoparticles encapsulated within nitrogen and sulfur co-doped magnetic porous carbon as an efficient peroxymonosulfate activator to degrade 1-naphthol. *Sci. Total Environ.* 739: 139896.

86 Zaeni, J.R.J., Adnan, R., Lim, J.-W. et al. (2020). One-pot synthesis of oxygen-functionalized porous carbon from biomass for rapid organics removal via peroxymonosulfate activation. *Mater. Today Chem.* 17: 100314.

87 Deng, F., Olvera-Vargas, H., Garcia-Rodriguez, O. et al. (2019). Waste-wood-derived biochar cathode and its application in electro-Fenton for sulfathiazole treatment at alkaline pH with pyrophosphate electrolyte. *J. Hazard.Mater.* 377: 249–258.

88 Cao, Y., Zhang, H., Liu, K. et al. (2019). A biomass-derived bimetallic Ru-MoOx catalyst for the direct hydrogenation of furfural to tetrahydrofurfuryl alcohol. *ACS Sustain. Chem. Eng.* 7 (15): 12858–12866.

89 Zhu, Y., Yu, G., Yang, J. et al. (2019). Biomass soybean curd residue-derived Pd/nitrogen-doped porous carbon with excellent catalytic performance for phenol hydrogenation. *J. Colloid Interface Sci.* 533: 259–267.

90 Hu, D., Xu, H., Yi, Z. et al. (2019). Green CO2-assisted synthesis of mono- and bimetallic Pd/Pt nanoparticles on porous carbon fabricated from Sorghum for highly selective hydrogenation of Furfural. *ACS Sustain. Chem. Eng.* 7 (18): 15339–15345.

91 Advani, J.H., Ravi, K., Naikwadi, D.R. et al. (2020). Bio-waste chitosan-derived N-doped CNT supported nanoparticles for selective hydrogenation of nitroarenes. *Dalton Trans.* 49: 10431–10440.

92 Scharnagl, F.K., Hertrich, M.F., Ferretti, F. et al. (2018). Hydrogenation of terminal and internal olefins using a biomass-derived heterogeneous cobalt catalyst. *Sci. Adv.* 4: 1–9.

93 Nguyen, C.V., Liao, Y., Kang, T. et al. (2016). A metal-free, high nitrogen-doped nanoporous graphitic carbon catalyst for an effective aerobic HMF-to-FDCA conversion. *Green Chem.* 18: 5957–5961. https://doi.org/10.1039/C6GC02118B.

94 Lin, Y., Guo-Ping, L., Zhao, X. et al. (2020). Porous cobalt@N-doped carbon derived from chitosan for oxidative esterification of 5-hydroxymethylfurfural: the roles of zinc in the synthetic and catalytic process. *Molecular Catal.* 482: 110695. https://doi.org/10.1016/j.mcat.2019.110695.

95 Budarin, V.L., Clark, J.H., Luque, R. et al. (2008). Palladium nanoparticles on polysaccharide-derived mesoporous materials and their catalytic performance in C–C coupling reactions. *Green Chem.* 10: 382–387. https://doi.org/10.1039/B715508E.

96 Zhou, P., Wan, J., Wang, X. et al. (2019). Three-dimensional hierarchical porous carbon cathode derived from waste tea leaves for the electrocatalytic degradation of phenol. *Langmuir* 35: 12914–12926.

14

Summary on Properties of Bio-Derived Carbon Materials and their Relation with Applications

S. Vinodha[1], L. Vidhya[1], and T. Ramya[2]

[1] *Sethu Institute of Technology, Department of Chemical Engineering, Kariapatti-626115, Virudhunagar, Tamil Nadu, India*
[2] *Bharathiar University, Department of Environmental Science, Coimbatore-641046, Tamil Nadu, India*

Biosphere (universe) comprises a huge volume of biowastes from the major source of agricultural sectors, food sectors, and industrial sectors. Most of them are highly desirable traits in terms of degradability, and effectively making use of these biowastes is being a prominent one. A few examples of sustainable transformations of waste into a commodity (carbonaceous materials) are presented in the first chapter. Carbonaceous materials can be prepared from the abundant resources of carbon includes biowaste like coconut shells, wood, rice husk, etc., via different synthesis approaches such as pyrolysis, microwave hydrothermal, chemical vapor deposition, thermal combustion, hydrothermal/solvothermal, arc discharge, and other biological methods. Recycling waste materials into high-valued products are markedly feasible which have inspired researchers to synthesize carbon-based nanomaterials from wastes resources for various potential applications. Biowaste refers to the organic materials that are derived from living organisms (plants or animals). Biowaste can be divided into two major groups: virgin biowaste (crops, trees, fruits, and vegetables) and aquatic biowaste (water plants and algae). It comprises the waste from major resources of agricultural, municipal, food, and industrial sectors. The synthesized carbonaceous materials are applied in diverse field like water treatment, biosensing, electrochemical sensing, solar cells, catalyst, super capacitors, antimicrobial products, imaging, drug delivery, and other biomedical applications. Microbicidal also refers to as antimicrobials agents that eradicate or suppress the spread of microorganisms. Continuous developments have been made in the development of antimicrobial agents in numerous aspects to attain improved pharmacodynamics with better oral absorption drugs and distribution in inflammation, and antimicrobial chemotherapy also recognized drug safety organisms including viruses, bacteria, fungi (mold), parasites, protozoans, and mildew. The antimicrobial products can be regulated into two categories: (i) drug/antiseptics (used to treat or avoid diseases on peoples, animals, and other living beings) and (ii) pesticides (used in objects like toys, countertops,

Biomass-Derived Carbon Materials: Production and Applications, First Edition.
Edited by Alagarsamy Pandikumar, Perumal Rameshkumar, and Pitchaimani Veerakumar.

grocery carts, and hospital equipment). The major factors that lead to microbicidal resistance are unnecessary antimicrobials used in agriculture, self-medicating and overprescription of the antibiotics, poor infection control in clinics and hospitals, and poor hygiene and sanitation practices. Microbicidal-resistant threats became questionable for the cure and hindrance of a broad range of infections caused by microorganisms. Due to the evolving crisis of microbicidal resistance in global health, it is an urgent need for developing new classes of antibiotics which leads the scientists and researchers to develop various types of alternatives. The carbon-based materials are primarily engaged due to their physicochemical properties for the inhibition of microbes. Carbon-based materials such as graphene, carbon nanoparticles, graphene quantum dots, etc., are derived from the biowastes such as rice husk, grape seed oil, and sugarcane bagasse, respectively. In the recent years, several studies have been done based upon carbon dots (CDs) because of their high efficacy toward antimicrobial activities. Even though there are numerous experimental proofs emphasizing the leading part of ROS-dependent oxidative stress about the antimicrobial action of carbonaceous materials, certain researchers have demanded that the carbon-based materials can be able to induce antimicrobial effect by ROS-independent oxidative stress also. There are various reports has been given for supporting ROS-independent oxidative stress-based antimicrobial mechanism for graphene. Numerous studies that were observed the wrapping and isolation of microbes can also inhibit microbial cell membrane perturbation to some extent with minor variations. Carbon-based materials can work as photosensitizers to kill microbial cells by photo-dependent ROS production with the presence of photoexcitation. Graphene and microbial cell interactions are more complicated than originally predicted ones. Killing microorganisms through lipid extraction is a new mechanism compared to the others. Researchers have found the strong hydrophobic interaction of graphene and lipid molecules known as "nanoscale dewetting" other than the van der Waals interaction, to give a driving force for cell wall damage. In the recent years, AgNPs-based wound bandages/dressings exposed some of the clinical success, specifically by controlling the infections and treating superficial wounds. Additionally, with the help of histological analysis, it proved that Ag–graphene composite hydrogel with higher ratio has its potential application toward wound treatment. Sun et al. [1] reported the efficiency of GQD/CDs bandages in wound disinfection by in vivo studies using mice. The antimicrobial properties of carbonaceous materials are favorable for dental applications. An early study has established a substantial reduction of cariogenic species like *Streptococcus mutans* in ZnO-GM-coated biofilm for acrylic teeth implants without any toxic effects. Taking advantage of absorbing light irradiation using GO in nanocomposites, effective photothermal activity is achieved for cellular inhibition toward the microbes. It is found that the carbon-based materials tend to induce ineligible toxic effects in mammalian systems by damaging the plasma membrane, lysosome dysfunction, or mitochondrial disruption, etc. Nowadays, preparing carbonaceous materials from biowastes has been trending in the research field as it can solve both of our problems in a single way. Since the antimicrobial efficacy of the carbonaceous materials varied according to its physicochemical properties, the mechanism of action, and also functionalizing with other nanomaterials like a metal ion/oxide nanoparticles, the detailed summary of the same has been reviewed with respective sections to provide an overview for best understanding. Furthermore, the biomedical perspective toward the prepared carbonaceous materials and their biosafety in clinical trials has also been incorporated.

An in-depth discussion is made in Chapter 2 with the biowastes that can be used as resources for producing carbon nanomaterials, that is, various methods of carbon nanomaterials synthesis are followed by discussing different catalytic applications. Carbon is a major component of living beings. Carbon is one of the basic building blocks for organic life together with nitrogen and oxygen. Carbon is an old and new catalytic material. Active carbons (ACs) are used commercially in many catalytic formulations, particularly for hydrogenation catalysts, for the excellent properties of dispersion of metal particles (particularly based on noble metals), and absence or (limited) presence of sites, which may catalyze side reactions. Carbon quantum dots (CQDs) have attracted remarkable attention as a novel and promising fluorescent carbon material with their unique optical properties, good water solubility, high stability, low toxicity, excellent biocompatibility, and low environmental impact. In the past few years, there have been numerous reports of natural biomass being used as a carbon source for CQD synthesis as an alternative to chemical carbon sources. In the recent years, in order to improve the efficiency of the fuel cell, a lot of studies are carried out from different ways. In order to improve the efficiency of the fuel cell, the catalyst is needed. The noble metal Pt has the good catalytic efficiency. However, Pt is expensive, which is a huge obstacle for fuel cell to be commercialized. Graphene nanosheets as an ideal alternative compared with the traditional carbon support materials have high electrical activity of the catalyst and superior durability than the commercial Pt/C catalyst. The use of CNPs as photosensitizers displays several advantages such as abundance, nontoxicity, stability, and sustainability. Nanocomposites based on transition metal oxides and carbon nanostructures with low cost, high electrical activity, and good stability are promising catalysts toward electrochemical water oxidation which is desirable but remains challenging.

CDs are a novel sort of carbon-based zero-dimensional material and obligate drawn substantial attention due to their excellent photostability, favorable biocompatibility, low toxicity, outstanding water solubility, high sensitivity and excellent selectivity for target analytes, tunable fluorescence emission and excitation, high quantum yield, and broad stokes shift. Carbon-based nanomaterials were deliberated as a milestone in the field of sensors and their environmental solicitations remaining to their exclusive physical and chemical properties. The fluorescence properties of CDs are essentially allied with surface functional groups, and these are highly sensitive to the adjoining environment subsequently to make further robust interaction with analytes. Metal ions can be efficiently interacting with CDs over surface bonding, consequently ensuing in the tuning of CDs properties. The CDs consequent from each orange juice, *Coccinia indica*, and barley were developed as nanosensors for the detection of Hg^{2+}. Recently, iron (Fe^{3+}) ions are one of the utmost collective described metal ions by fluorescent CDs detection. Pb^{2+} ions have strong affinity with surface functional groups of CDs to enable the electron transfer from the excited CDs to Pb^{2+} leading to the fluorescence quenching. The oxygen and amino functional groups of CDs have a high binding affinity with Pb^{2+} ions that induce proximity between them which facilitates effective electron transfer. Cu^{2+} ion was proficiently detected using several biomass-derived CDs as a fluorescent sensing probe. The CDs organized from tulsi leaves and groundnuts were initiated to have good selectivity and sensitivity toward the revealing of Cr(VI) based on the fluorescence turn-off via the IFE mechanism. Likewise, N/S co-doped CDs were fabricated to the sensing of Ag^{+} with high sensitivity and selectivity. The detection of As^{3+} ion by using CDs prepared from quince fruit has been found in previous literature. Hence, numerous metal ions can quench the biomass-derived CDs fluorescence. Rose-derived CDs can be used to detect tetracycline (TC) based on the interactions between TC and CDs. Grape skin has been used to synthesize biomass CDs by a hydrothermal method. There is a dehydration condensation reaction between the phenolic hydroxyl groups on the picric acid and the carboxyl groups on the CDs. The advance of novel biomaterial schemes to deliver a prodigious prospect aimed at the effective clinical translation of nanomaterials for adapted biomedicine to benefit patients. Many studies propose that graphene and its derivatives as bioimaging agents. Probable challenges and stance for the impending advance of mesoporous carbon in biomedical fields oblige been conferred in aspect. Carbon nanotube and CDs entice widespread attention for their capable enactment in the biomedical arena comprising drug delivery system (DDS). Biocompatible inorganic mesoporous provisions have fascinated substantial consideration in biomedicine over the earlier spans. The nanosized mesopores in mesoporous carbon nanoparticles (MCNs) can similarly diminish the drug release rate, presenting sustained releasing behavior. MCNs retain lesser density, higher porosity, and robust adsorption facility, bequeathing them obligate an advanced drug-loading ability. MCNs revealed an excellent loading capacity of doxorubicin (DOX) owing to the hydrophobic interactions and the supramolecular p–p stacking amid DOX and MCNs and exhibited a sustained release. Amid innumerable photothermal agents, MCNs are considered by durable optical absorption in the near-infrared region. Tumor targeting peptide-conjugated core–shell graphitic carbon@silica nanospheres through dual-ordered mesopores (MMPS) remained magnificently contrived. Mesoporous silica shell might be amended to confirm hydrophilicity and targeted DDS. Semi-graphitized carbon was familiarized in situ to the mesopores of MSNs for contemporary photothermal therapy and DDS. MCN is advanced for operational dual-triggered synergistic cancer therapy. Self-assembly of CDs foremost to aggregates is frequently convoyed by the fluorescence quenching that delivers the origin intended for the advance of biosensing methods. The predictable sensing system has been admirably used for the assay of glucose in human serum. Conflicting to the aggregation-induced fluorescence quenching, a portent of aggregation-induced emission enhancement (AIEE) effect for CDs was freshly revealed and used for sensing applications. Pt-containing mesoporous carbon improves the electron relocation and redox capability of glucose oxidase. Biomolecules show vital roles in entirely lifespan advances embracing disease enhancement, so the detailed acquaintance of biomolecules is unsafe to disease analysis and treatment. The graphene-based ingredients necessitate endured used to paradigm numerous biosensors based on diverse sensing mechanisms encompassing optical and electrochemical signaling. Magnetic resonance (MR) imaging is an extensively used biomedical utensil that is proficient to noninvasively attain anatomic evidence through high spatial and temporal resolution. Emerging an intelligent stimuli-responsive nanosystem is established on MCNs to advance the resolution and specificity of investigative imaging and improve the therapeutic efficiency for cancer management. MCNs holds prodigious probable as a drug carrier to curb drug release and apprehend spatial-temporal DDS owing to a large surface area and pore volume, amendable pore structure, and easily modified surface. Owing to their distinctive mesoporous structure, carbonaceous conformation, and high biocompatibility, MCNs spectacle high concert in photothermal therapy, synergistic therapy, fluorescent labeling, bioadsorption of toxic pathogenic constituents, peptide parting, and biosensing. Furthermore, due to the capacity of MCNs to adapt NIR light to heat, the combinational PTT and chemotherapy could be accomplished to advance the clinical therapeutic effect. Surface modification remnants inspiring contemporary oxidization of MCNs can bequeath the carrier with some precise organic groups for additional alteration. MCNs are anticipated to spectacle high recital in advance biomedical regions due to their exceptional structure, composition, and physicochemical properties. Systematic biosafety assessments of MCNs are obligatory to promise their clinical translation presently, and these assessments will intensely be contingent on the expansion of approaches to attain anticipated MCNs. Biosafety evaluations should focus on the biodistribution, biodegradation, excretion, and other precise toxicities such as neurotoxicity, reproductive toxicity, and embryonic toxicity. Radiolabeling of MCNs potency solves this concern for imminent biosafety assessment.

Several toxic chemicals such as heavy metals, organic dyes, and phenolic compounds evacuated from the industries are hazardous and create major havoc to the environment. Various treatment technologies have been developed for the removal of toxic chemicals from water and wastewater. Some of the important methods are reduction and precipitation, ion exchange, reverse osmosis, evaporation, electrodialysis, and adsorption. Of these technologies, adsorption proved more economical, efficient, and eco-friendly. Recently, mesoporous material derived from biowaste has developed a greater impact and has led to several innovations. This includes biomass materials for the removal of toxic chemicals like organic dyes, heavy metal, and phenolic compounds. The mesoporous carbon derived from biomass materials will have great development and practical applications in the near future in various fields such as removal of toxic chemical, energy conversion, and storage application.

14.1 Removal of Toxic Chemicals

The mesoporous carbon material derived from biowaste exhibits an excellent adsorption capacity with respect to their nature of surface area, porosity, and functional group. Mesoporous carbon materials lessen the pollution in the environment and management-related problem of the wastes. This provides a justifiable substitute method to generate lucrative value-added materials for the advantage of society. Due to their aromatic character present on the surface of the mesoporous carbon, the dyes and phenolic compound carry their main interaction through π–π interactions. The kinetics and adsorption models of all the dyes, heavy metals, and phenolic compounds directly influence the structural characteristics of the synthesized carbon materials. The system of adsorbent-adsorbate is unique and mandatory to note. Consequently to make a safe comparison, the adsorption conditions must be precisely the same. Otherwise, the comparison will be defective. The mesoporous carbon materials are synthesized from cheap and naturally occurring biodegradable resources. Recently, it is a trend in new materials in the field of research owing to its inexpensive cost and eco-friendly impact on the environment. The synthesis of mesoporous carbon materials from biowastes can easily be prepared by physical, chemical, physicochemical, and microwave-assisted activation. Using the activating agent and activation method, the micro-, meso-, and macroporous materials are synthesized. These mesoporous carbon materials are used for all applications include removal of dyes, heavy metal ions, and phenolic compounds. The mesoporous carbon materials synthesized from biowastes resources have more microporous structure and substantial mesoporous structure. The amalgamation of pure mesoporous carbon and ordered mesoporous carbon from biowaste resources is still a great task. The chemical activation technique plays a vital role in producing a mesoporous structure and increasing the surface area. This also leads to the release of unwanted materials into the aquatic ecosystem. Henceforth, a more dependable, lucrative, and efficient physical activation method should be developed.

14.2 Electrode Materials for Batteries

Electrochemical energy storage devices are in great need of the energy research circle with the intention of confronting the energy shortage in the upcoming years. The application of these devices generally relies on the electrode materials used. Among lots of batteries, carbon materials are found to be the most attractive applicants owing to their abundant resource, low cost, good constancy, nontoxicity, and high security. These carbon materials which are utilized as electrode materials can be derived from the biomass and, in turn, will ameliorate the accomplishment of the batteries. The biowaste-derived carbons as electrode materials are utilized in various types of batteries such as lithium-ion batteries (LIBs), sodium-ion batteries, lithium-sulfur batteries, and Zn-air batteries. Furthermore, the association with the structural and morphological characteristics of biowaste-derived carbon materials with the enhancement of electrochemical behavior will help further the rational design of biomass-derived carbon material for energy storage devices.

In accordance with the energy research aspect, the rapid development in advanced energy technology and the rise of high-quality energy techniques have revolutionized the world. Henceforth, the utilization of conventional fossil fuels for fulfilling the global energy demand has become a huge alarm. During the past several decades, the usage of conventional fossil fuels, such as coal, petroleum, and natural gas, led to rapid exhaustion and also created a great hazard to the environment. The modern era has focused its view on renewable resources and also specified more significance to satisfy the depletion of nonrenewable fossil fuels. Green energy technologies and the rapid development of renewable energy resources are supposed to be the great potential for future energy generation, storage, and their usage to humanity. Recently, electrical energy storage devices are becoming progressively significant to develop clean energy technologies. In particular, the grid-scale storage of the energy derived from renewable energy resources can address the problem of a continuous supply of energy to the world. Among the electrical energy storage devices, electrochemical energy storage devices and electrochemical capacitors have received remarkable attention due to their high energy conversion efficiency. These have created precise interest to the energy researchers in order to challenge the energy shortage in the future years. However, the supercapacitors possess a high power density and scalability, and they have limited energy density. Furthermore, the rechargeable batteries are known for their high energy density and have the expense of charge–discharge rate and cycle life. Therefore, the investigation of new and maintainable electrochemically active materials with higher energy density to accommodate the demands of next-generation electronic devices is unavoidable.

14.3 Electrochemical Sensor Applications

The carbon material derived from biomass serves as an exclusive host pattern for the innovative electrodes which can ameliorate the performance of sensors. Most recently, we are in need of an inexpensive feasible generation of robust, movable, supersensitive, and selective electrochemical sensing devices for a biosensor and point-of-care device. This can be achieved easily by bio-derived carbon structures that are produced by simple steps, mainly from inexpensive and plentifully accessible renewable biomass. Based on the properties of these novel materials, they can be incorporated with the devices, and various technologies can be used for detection, where optical and electrochemical detection are the most prevalent. This technique is used to enhance the surface properties in order to increase the electroanalytical behavior of working electrodes. Sensors are devices which are used to detect phenomena of interest by interpreting their state and variations into an analytical signal. The "sensing element" is the most significant part of a sensor which it "senses" or detects the preferred phenomenon. However, the nature of the sensing element for similar applications may differ by varying for different sensor techniques and technologies. Further electrochemical sensors are classified as voltammetric, potentiometric, chemically sensitized field effect transistor, and potentiometric solid electrolyte gas sensors. An electrochemical reaction occurs in an electrochemical sensor, and it works between the substance/phenomena to be observed and the electrode and, in turn, produces an electric current corresponding to the chemical excitation. Thereby, it produces the analytical signal. It is convenient to use inexpensive device for quantitative and qualitative determination of electroactive species.

Nanomaterials give an abundant porosity and surface area for chemical reactions to occur, and also they are very popular as catalysts. A biological precursor uses two common ways in preparing the carbon nanomaterial. They are biomass pyrolysis and hydrothermal carbonization (HTC). Several types of research were carried out by pyrolysis as it offers wider flexibility in the selection of biomass precursors and requires only mild reaction conditions in an overall low-cost procedure. Generally, electrochemical sensors have found widespread applications in environmental monitoring and biosensing applications. Due to their low cost, they are widely used in large-scale environmental monitoring efforts and other industrial applications.

14.4 Fuel Cell Applications

The prepared sensor enables to detect the compounds in real samples. For example, in human urine, serum, it provides satisfactory recovery. This, in turn, points toward its great potential for detection in human body fluids for the clinical diagnostics. According to the electrochemical investigation analysis, the prepared bio-derived electrode shows more active electrocatalytic properties toward the oxidation of various compounds, and the field is explored very less. Modern civilization requires a power source, and it imparts the need for the life of humanity. The supply of fossil fuel resources-based fuels like coal, oil, and natural gas is still not balanced because of increasing world economic expansion and population growth. Novel and advanced electrocatalyst materials are the fundamental requirements for the sustainability of electrochemical energy conversion and storage devices. The carbon-based nanomaterials play a vital role as a catalyst and catalyst support for the fuel cells. The specific characteristics of nanocarbon materials such as electronic conductivity, chemical inertness, porosity, surface area, performance, and durability nurture a significant role in electrochemical storage energy devices. During energy production, the waste materials released are efficiently recycled to manufacture nanocarbon porous materials for energy storage and conversion devices, such as batteries, fuel cells, solar cells, and super capacitors. The biomass-derived carbon materials prove an effective sustainable green material, thus safeguarding the environment.

The modern era needs fuel cells, solar cells, supercapacitors, and LIBs. Of all these cells, fuel cells have a significant role in commercial devices because of their performance, durability, and cost. The energy device used in fuel cells should use high-performance materials with coveted properties and should be perfect for fuel cell applications. Furthermore, these materials must be simple for access, earth copious, environment-friendly, and inexpensive. In spite of various materials tested in fuel cell applications, carbon materials are found to be more potential owing to their exceptional properties such as electronic conductivity, surface area, porosity, chemical stability, and durability. Generally, it is observed that the annual global biomass production is estimated as 105 pentagrams of carbon. The biomass-derived carbon materials possess numerous nanopores that will facilitate the diffusion of electrolyte and will result in efficient ion transfer. Advanced technologies, employed in synthesizing biomass-derived carbon will exhibit high surface area, high porous morphology, enhanced electrical conductivity, and good mechanical properties. This is basic requirement for the application of all electrochemical storage devices. The HTC method produces rich carbon precursors for fuel cell applications. The surface morphology characteristics of biomass-derived carbons are graded porosity and high surface area, and colloidal synthesis method is highly considerable in electrochemical devices. Furthermore, other methods like ion thermal carbonization (ITC) and molten salt carbonization methods are also used to synthesize energy-efficient carbon materials. The features of biomass-derived carbon nanomaterials like high electronic conductivity, inexpensive, readily available, and eco-friendly influence more potential in electrochemical energy storage devices. This various synthesis methods of mesoporous carbon materials derived from biowastes have good significance of oxidation-reduction reaction (ORR) for fuel cell applications. The extensive research conducted on different types of ORR electrocatalysts synthesis currently depicts greater selectivity, high stability, low cost, and excellent electrocatalytic properties for increasing the general execution of the fuel cell system.

Porous carbon materials are applied in a wide range of functions, for instance, conversion of energy and biological applications, storage devices, photocatalysis, etc. This is attributed to their excellent physiochemical properties such as diffusion, accessible active sites, mass transport, high surface area, etc. It is interesting to note that porous carbon materials are being considered as futuristic substitutes which have the potential to replace traditional and costly Pt-based electrocatalysts for oxygen reduction reaction in fuel cells. When biomass-based sources are utilized in the synthesis of porous carbon materials, one can reap multifaceted benefits such as viability, ecofriendly characteristic, abundant quantity, and sustainable nature. The current summary is aimed at helping researchers to gain an in-depth understanding about the biowaste sources and their constituents, methods of synthesis, different parameters that decide the synthesis methods, templating methods, and role of dopants and transition metals in influencing electrocatalytic activity. The current review helps in understanding the importance of enabling biomass-based PCM as electrocatalysts in future energy devices.

The development of novel and efficient methodologies for the conversion of biomass into useful materials is still an unaddressed and important challenge. Biomass is under use for a long period of time, in the form of feedstock, when preparing porous-derived carbonaceous materials via thermal composition, i.e. pyrolysis. There are various activation methods being followed, for instance thermal and acid/base of steam-provided activated carbon, in which the latter can be used as a catalyst in adsorption, electrochemistry, chromatography, etc. However, it is highly challenging to design the carbonaceous materials with functional groups and tunable surfaces following such established thermal activation processes. There are still some examples available, whereas the original biomass macrostructure can be found and more or less controlled. In such scenario, the synthesis of porous carbonaceous structures is completely a new area and is gaining strong recognition in the recent times due to its wide application and cost-effective strategies. In sustainability perspectives, the porous carbonaceous structures enact an important and heavy role in future applications and implementation. This is due to the versatile nature of systems designed for advanced applications.

Prior to unlocking the potentials of renewable resources, fossil fuel remained the major source in the synthesis of carbon materials. The synthesis of carbon materials was performed under harsh conditions such as laser ablation, direct pyrolysis of organic compounds, and chemical vapor deposition. In such scenarios, the sustainability and accessibility of these procedures remained limited. HTC process has been explored to in-depth level in the past 20 years. Various researchers upgraded the process which increased its efficiency and helped in the transformation of biomass into carbonaceous products under control. According to a general classification based on its diameters, the porous carbon materials are of three types such as microporous (pore size <2 nm), mesoporous (2 nm $<$ pore size <50 nm), and/or macroporous (pore size >50 nm). The traditionally used porous carbonaceous materials like carbon molecular sieves and activated carbon contain broad pore size distributions in both micropore and macropore ranges. The synthesis of such carbon materials is mostly by pyrolysis and physical or chemical activation of organic precursors such as coal, wood, fruit shell, or polymers at elevated temperatures.

The studies conducted earlier reported numerous cost-effective biomass sources such as sucrose, glucose, cyclodextrins, cellulose, starch, or biochar. One can perform carbonization and active waste-derived biomass to yield microporous carbon with pore size distribution ranged between 0.77 and 0.91 nm. High surface area ($3000\,m^2\,g^{-1}$) and large micropore volumes ($1.7\,cm^3\,g^{-1}$) are exhibited in general among biomass-derived microporous carbons. It is possible to improve the applications of porous carbons alike mesoporous silica, through different processes such as surface functionalization and fine-tune interactions with guest molecules. Further, when bulk and interfacial properties of the materials can be optimized, it results in advanced applications and ocean of possibilities. Counter electrode (CE) is one of the important components in dye-sensitized solar cells (DSSCs). It also plays a vital role in holistic efficiency and cost mitigation of a device. In spite of the fact that platinum is the preferred and efficient metal as a CE in DSSC, it has serious disadvantages such as high cost, unavailability, and low stability in I^-/I^{3-} redox couple. These drawbacks inhibit the metal from large-scale applications. The current chapter provides an overview of porous carbon materials in the form of excellent metal-free CE for DSSCs. DSSCs and their importance along with the working principle are described in the chapter in a precise manner. Further, they also highlight the impact of CE upon photovoltaic (PV) performance of DSSCs. Various methods of synthesis, precursors of porous carbon materials, and their efficiency in DSSCs have also been discussed together with an evidence about the characterization and evaluation of device performance with the help of porous carbon materials using CE. Overview of porous carbon material as CEs in DSSCs has also been summarized. DSSC is the latest generation of solar cells and proved to be a promising alternative to silicon solar cell. Being a third general solar cell, DSSC has various advantages such as mechanical robustness, good plasticity, easy assembly procedure, transparency, low light, and environment-friendly nature, and it possess the ability to work at wide angles. In 1991, a significant breakthrough occurred in the efficiency of photoelectric conversion (7.1–7.9%) by DSSCs. In this event, mesoporous film of TiO_2 nanocrystalline was introduced by O'Regan and Grätzel [2] research group to adsorb dye instead of planar semiconductor electrode. After 1991, DSSCs triggered a series of research investigations in the next two decades and underwent significant changes to achieve an increased efficiency of ca. 14%. Carbon materials can replace platinum in CE, since the former has various advantages such as high catalytic activity, good corrosion resistance against iodine, low cost, environment-friendly nature, high availability, high surface area, high electrical conductivity, high reactivity for triiodide reduction, and high thermal stability. From then, strenuous research investigations have been conducted in this research arena upon carbonaceous materials such as carbon black, mesoporous carbon, graphite, graphene, carbon nanotubes, and carbon nanofibers. All these efforts deemed to be successful, i.e. in being employed as CEs.

Carbonization methods which are generally followed in the conversion of biomass carbon are of various types such as hydrothermal, pyrolysis, carbonization, etc. The activation methods are also being followed, for instance physical and chemical activation methods. Carbonization process keeps few parameters under control such as surface properties, time, chemical reagents, carbon materials, temperature, morphologies, and costs. Pyrolysis and hydrothermal processes are the two common methods used in the carbonization of biomass. Pyrolysis is performed at elevated temperatures with limited volume of oxygen. HTC is a method in which the biomass is converted into carbonaceous materials by following thermos-chemical process. Few parameters such as catalyst, temperature, temperature ramping rate, and size of the particles decide the type of important transmutation materials extracted from the biomass. Either with or without catalyst, the hydrothermal combustion process is performed at low temperature range, i.e. 120–250 °C. Within short period of time and high reaction rate, the biomass becomes natural coal. Various review articles were published in the recent years focusing hydrothermal conversion of biomass. There are various parameters considered in thermochemical methods such as concentration of precursor, time, temperature, and catalyst. Subcritical water is used in the biomass conversion to carbonaceous products which contributes to best hydrolysis rate and dehydrates the precursors. These processes results in the bestowing of hydrochars with high and tunable oxygen-containing functional groups. With the help of dopant containing precursors, one can introduce other practicalities, for instance nitrogen-containing groups, into hydrochars. In general, pyrolysis is performed before physical activation. During pyrolysis, high temperatures upto 1200 °C is also used, whereas the temperature for chemical activation lies in the range of 450–900 °C. KOH, NaOH, $ZnCl_2$, $FeCl_3$, H_3PO_4, and K_2CO_3 are the chemicals used in chemical activation process.

With advancements in solar and wind energy, high-end energy storage systems are developed, for instance batteries, supercapacitors, etc. The need for high-performing electronic portable devices increased in the recent years which demand high-end energy storage systems. There are various advantages present in novel rechargeable batteries such as long life cycle, high energy storage, and huge power density. The authors has detailed about device configuration and the mechanism behind energy storage in hybrid electric devices, smart grids, etc. Due to increasing popularity of automatic start–stop technology coupled with proven track record of high reliability in spite of severe climate conditions, the SCs are being preferred these days in trams, trucks, trains, buses, and metros. The supercapacitors capture power from regenerative braking system, and the energy is released to help in acceleration. In addition to this, the supercapacitors are able to provide rapid storage and excellent energy delivery for heavy duty applications irrespective of harsh conditions. For example, the supercapacitors hold burst power and good low-temperature operation performance which make them use extensively in hybrid forklift and cranes. There has been research conducted in the recent times to amalgamate PVs and supercapacitors so that the latter can function both as an energy reservoir and power buffer. Being a green energy producer, the PV solar cells are predicted to lead the stage for global sustainable energy development in the future. But the energy converted from solar cell is an interim one, which can be overcome by supercapacitors, i.e. the latter can get rid of the intermittence and make the sustainable green energy available throughout the day and night and even on cloudy days as well. In the previous decade, the integration system is packed as a parallel combination of PVs and supercapacitors, while the photo-supercapacitor experienced tremendous advancements in laboratory scale. IWS (Intelligent Wireless Sensor) is one of the interesting applications of supercapacitors. At present, the systems being used can fetch different modules of data with the help of battery-powered or wired sensors. The future systems are expected to be installed with wireless sensors in remote areas or in some special structures. The energy required for IWS can be harvested from thermal energy, radiofrequency, vibration, and electromagnetic energy. In order to store the harvested energy for IWS, the energy storage system should meet the criteria, i.e. low self-discharge, high energy density, and enough count of charge and discharge cycles without any deterioration of lifetime of the device. SCs have a key challenge, i.e. low energy density. The commercially available SCs can provide energy density of a mere $10\,\mathrm{W\,h\,kg^{-1}}$, a value even lesser than LIBs, i.e. the latter can supply energy density of $>180\,\mathrm{W\,h\,kg^{-1}}$. Both organizations and research/academia are highly interested in increasing the energy density of SCs with the induction of new electrode materials, ingenious device design, and novel electrolytes with wide operation voltage window.

The current summary classifies different porous carbons derived from biomass, organic materials, and carbides into different groups. This section focuses on some of the new-age porous carbons and its novel synthesis methods from energy storage perspectives. The authors have discussed the practical implementation of porous carbon in activated carbon, biowaste, biowaste valorization, carbonization, and its different applications. The segments detail about the importance, benefit, and significance of energy storage in modern society. Further, it also discusses about the complex structures involved in a wide range of energy storage systems. Porous carbon possesses unique and vibrant physiochemical properties which make it an important stakeholder in energy storage domain. In spite of the presence of different attractive features, some specific features are given attention in porous carbon materials, for instance, surface area which has a close association with energy storage performance. In porous carbon, mesopores and macropores function as reservoirs, while the micropores save all the ions that affect the delivery of energy. Porosity of the materials should be taken into account based on the application. For instance, when micropores are present in porous carbon, it increases its ion storage capacity which in turn boosts the capacitance. This remains crucial for SCs since they possess deadly low energy density. Further, in terms of energy storage, the wettability and conductivity of porous carbons are important, and these features can be customized by heteroatoms doping. Porous carbon has been widely applied in the recent years where its growth and applications are remarkable as partially listed earlier. However, since the porous carbon is getting lighter in the due course, it reduces the tap density of carbon-based electrodes which in turn inhibits its commercialization. Few approaches have been proposed by the researchers such as capillary compression, mechanical compression, and self-assembly which are quite promising in terms of outcome. However, either the structure or the electrochemical performance should be compromised to achieve optimal output. The behaviors of lithium-ion storage containing amorphous and crystalline carbons are quite different. The development of advanced technology helps in the visualization of production mechanism behind porous carbon. This helps in the development of facile preparation of 2D/3D porous carbon with excellent accuracy. Care should be taken to overcome few challenges so that porous carbon can be applied in a broad range of energy storage applications.

References

1 Sun, H., Gao, N., Dong, K. et al. (2014). Graphene quantum dots-band-aids used for wound disinfection. *ACS Nano* 8: 6202–6210.

2 O'regan, B. and Gratzel, M. (1991). A low-cost, high-efficiency solar cellbased on dye-sensitized colloidal TiO_2 films. *Nature* 353: 737–740.

Index

a
absorption-catalytic graphitization-oxidation process 187
activated carbon materials 2, 149, 163
activated carbons (ACs) 1, 4, 19, 27, 28, 34, 42, 64, 70, 76, 93, 147–164, 184, 217–219, 222, 231, 236, 257, 259–261, 283, 285, 294, 303, 319, 325, 326, 328
activated PC 294, 296–302, 308
active carbons 294
active nanofibers 17
adsorbent–adsorbate 165, 322
adsorption intensity 150
Ag/AgCl electrode 222
Ag–graphene composite hydrogel 80, 318
aggregation-induced emission enhancement (AIEE) effect 142, 320
Ag nanoparticles 64
agro-residues 3
amalgamate PVs 328
amino dextran-coated Fe_3O_4 nanoparticles 142
ammonium molybdate tetrahydrate (AMM) 186
animal-based biomass 7–8
antimicrobial agents 64, 67, 68, 79, 80, 84, 317
antimicrobial nanoagents 64
antimicrobial nanomedicine 64
antimicrobial properties 71, 73, 74, 76, 77, 81, 82, 318
antimicrobial resources 63
antimicrobial therapeutic drugs 67
artificial neural network (ANN) model 153

b
bimetallic sulfides 188, 189
bio-carbon electrodes 287
biochar 4, 5, 7, 8, 11, 13, 147, 158, 160, 188, 223, 285, 297, 302, 305, 326
bio-derived carbon-based nanomaterials 94, 218
bioengineering, carbonaceous materials 79
  biosafety 83–84
  nanoantibiotic formulations 82–83
  surface modifications (coating) on medical devices 81–82
  wound dressing 80–81
biological precursor 215, 324
biomacromolecules HA 135
biomass, composition 3–4
biomass carbon sources 113, 188, 283
biomass CDs (BCDs) 113, 320
biomass-derived carbon
  activation methods
    chemical activation 13, 14
    combination of, physical and chemical activation 14
    modification and structure control of 14–17
    physical activation 11–13
  batteries characteristics 172–174
  batteries classification 172
  biomass resources and composition
    animal-based biomass 7, 8
    fruit-based biomass 5–7

Biomass-Derived Carbon Materials: Production and Applications, First Edition.
Edited by Alagarsamy Pandikumar, Perumal Rameshkumar, and Pitchaimani Veerakumar.

biomass-derived carbon (*contd.*)
microorganism-based biomass 7
plant-based biomass 4, 5
carbonaceous materials 179–181
carbonization
hydrothermal carbonisation 9, 10
pyrolysis 10, 11
cost analysis 19
electrochemical performances
LIBs 181–189
Li-S batteries 195–199
SIBs 189–195
Zn-air batteries 199–201
heteroatom-doped carbon electrodes
dual-heteroatom-doped 204–206
single-heteroatom-doped 202–204
mechanism
LIBs 174, 175
Li-S 175–176
SIBs 176–178
ZAB 178–179
precursor selection of, biomass-derived carbon 8
production methods of 8, 9
production process description 17, 18
biomass-derived catalysts 236
biomass-derived ORR catalyst 238–245
biomass-derived PC materials 294
biomass materials 64
carbon and its derivatives 65–66
synthesis of CNTs 99
biomass precursor 3–5, 8, 71, 149, 182, 184, 187, 190, 194, 199, 215, 217, 218, 222–224, 309, 324
biomass pyrolysis 8, 10, 215, 324, 327
biomass, recycling of 293
biomass, resources 3–4
biosafety 83–85, 143, 319, 321
bio-wasted carbon 262, 266
bisphenol A (BPA) 303
bizarre microstructures 1
boron-doped diamond (BDD) 305
Bougainvillea spectabilis 222, 264
Brunauer–Emmert–Teller (BET) 30, 32, 40–43, 47, 50, 153, 162, 183, 187, 198, 243, 244, 259, 268, 299, 306
1-butyl-3-methylimidazolium hexafluorophosphate 142
butyrylcholinesterase (BChE) 123

c
Candida albicans 68
capacitor 171, 193, 253–258, 269, 294, 317, 323, 324
carbohydrates 3, 7
carbonaceous materials 17, 20, 33, 37, 64, 66, 69–85, 149, 151, 175, 179–182, 190, 202, 206, 231, 237, 294, 295, 297, 308, 317–319, 325–327
carbon-based multifunctional DDS strategy 131
carbon-based nanostructures 70, 117
carbon black (CB) 2, 4, 18, 238, 262, 327
carbon-dependent catalysts 201
carbon dots (CDs)
application of
anion sensors 122, 123
metal ion sensing 117–122
miscellaneous molecules 123
characterization of 114, 115
optical properties 115
synthesis of
chemical oxidation method 116, 117
hydrothermal carbonization method 116
microwave method 116
pyrolysis 117
carbon fibers 2, 34, 152, 176, 179, 182, 183, 191, 206, 218, 263, 283
carbonization, hydrothermal carbonization 9, 10
carbon materials 1–20, 27, 30, 35, 39, 42, 45, 47, 48, 79, 93, 94, 102, 104, 115, 116, 148–150, 157, 158, 161–163, 165, 176, 181, 184, 187, 188, 190, 194, 195, 201–203, 206, 207, 216, 224, 230–232, 236, 239, 242, 243, 245, 258, 259, 262, 264, 265, 267, 269, 283, 284, 293–309, 317–329
carbon monoliths 152
carbon nanomaterials 129, 215
carbon nanoplates 222
carbon nanotubes (CNT) 1, 2, 17, 64, 65, 70, 77, 94, 99–102, 104, 113, 129, 147, 176, 200, 218, 220, 223, 224, 230, 257, 262, 283, 284, 320, 327
carbon powder 3, 181, 191, 306
carbon quantum dots (CQDs) 65, 74, 102–104, 113–124, 216, 319

catalyst support materials 234–236
cationic radical intermediate 219
CD44 receptors 131, 138
cellulose/carbon nanotube nanocomposite fibers 17
chemical demineralization 8
chemical oxidation method 116–117
chemophotothermal synergistic therapy 129
chitin 3, 7, 8, 187, 305, 306
chitin-catecholamine 7
cobalt-impregnated NPAC (Co-NPAC) catalyst 303
Coccinia indica 118, 319
coconut shell (CSC) 12, 17, 27, 64, 154, 204, 205, 218, 259, 294, 298, 317
coffee waste (CW) 159, 244, 245, 283, 284
CO_2 gasification 14, 29
coir pith 159, 184, 185, 191, 196, 199, 294
coir pith-derived carbon (CPC) 184, 193, 196
commercial-grade activated carbon (CAC) 152, 155
conduction band minimum (CBM) 280
confocal laser scanning microscopy (CLSM) 138
congo red (CR) 151
controlled hierarchical porous carbon (HPC) 158
copolymer-templated nitrogen-doped mesoporous carbon (CTNC) 139
copper (Cu^{2+}) ions 118, 120–122, 157, 223, 320
cornstalk-derived carbon (CSC) electrodes 186
correlation coefficient (R^2) 150, 152
counter electrode (CE) 222, 262, 275, 277, 278, 282–287, 326
crude fiber 7
cucumber peel (CP) 153
cyclic voltammetry (CV) 220, 236, 239, 255, 262, 264, 267
cyclic voltammogram (CV) 184, 185

d

degradation mechanism 234
deproteinization 8
dichloro-dihydro-fluorescein diacetate (DCFH-DA) 73
2,4-dichlorophenol (2,4-DCP) 162
N,N-dimethylformamide 218
disorganized carbons 11
dopamine 142, 218, 222
dopamine's fortitude (DA) 142
doping, B-d-CMs 15
doxorubicin (DOX) 130, 138, 320
dried raw biowaste precursor 149
drug delivery systems (DDS) 129–131, 320
dual-heteroatom-doped carbon electrodes 204–206
dye-sensitized and perovskite solar cells
- biomass-derive carbon counter electrode 283–287
- development of
  - hole transport material 282–283
  - perovskite solar cells architecture 281–282
  - working principle 281
- DSSC components
  - counter electrode 278
  - dye sensitizer 277
  - electrolyte 278
  - photoanode 277
  - TCO 277
- DSSC working principle 276
- perovskite solar cells 278–280
- tunability of bandgap energy 280

dye-sensitized solar cells (DSSCs) 280–287, 326
dye sensitizer 277, 278

e

eco-friendly synthesis approach 216
edible fungi residue (EFR) 153
Eichhornia crassipes (EC) 201, 223, 285
electrical double-layer capacitor (EDLS)
- activated carbon electrodes 257
- carbon nanotubes and graphene electrodes 257

electrical double-layer supercapacitor 256
electrocatalytic activity 47, 48, 107, 142, 200, 201, 217, 222, 224, 236–240, 242, 243, 284, 287, 301, 325
electrochemical active surface area (ECSA) 234
electrochemical capacitor 171, 253, 323
electrochemical detection 217, 218, 222, 223, 323

electrochemical energy storage devices 3, 171, 237, 243, 245, 259, 269, 322, 323, 325
electrochemical energy storing system applications 14
electrochemical impedance spectroscopy (EIS) 262
electrochemical method 74, 175
electrochemical (redox) reactions 172
electrochemical sensors 215–225, 323–324
electrochemical storage 3, 15, 17, 19, 20, 231, 324, 325
electrode materials for batteries 171–207, 322–323
electrode reactions 177
electro-Fenton (EF) process 305
electrolyte, DSSC 255, 278
electrolyte's iodide molecule 276
electrolytic oxidation 237
electron transfer reaction 234, 235
elemental sulfur (S) 175, 176, 195, 203
embedded fluorescent CDs 138
energy density 171–174, 178, 190, 195, 206, 207, 254–256, 258, 263, 264, 269, 323, 328, 329
environmental monitoring 216–218
Environmental Protection Agency (EPA) 160
Eragrostis plana Nees (EPN) biomass 163, 164

f

fatal bacterial infections 68
fibrous structure 33, 263
fluorescence intensity 117, 119, 122
fluorescent carbon nanoparticles 113
fluorescent nitrogen-doped carbon nanoparticles (FNCPs) 118
fluorescent porous carbon nanocapsules (FPC-NCs) 138
fluorine doped tin oxide (FTO) 277
folic acid-conjugated polydopamine (PDA) 77
fossil fuels 3, 171, 229, 230, 257, 259, 322–324, 326
Freundlich model 150
fruit-based biomass 5–7
fuel cells
  applications 324–329
  classification of 233
  theory and fundamentals 233
2,5-furandicarboxylic acids (FDCA) 308

g

glassy carbon electrode 216–218, 220, 222–224
glutathione disulfide (GSSG) 73
glycosaminoglycans (GAGs) 141
GO-based wound dressing 80
Goldschmidt tolerance factor (t) 279
graphene electrodes 257
graphene nanocomposities 220
graphene nanosheets 72, 76, 104, 319
graphene oxide (GO) 65, 72, 76, 94, 95, 147, 223, 224
graphene quantum dots 74, 79, 95, 223, 318
graphite compound (LiC_6) 177, 202
graphitic carbon capsules (FeS@GCC) 189, 192
graphitic carbon fibers 182
graphitic shells like carbon nano onions (GS-CNOs) 243

h

halide p-state orbital 280
halide perovskite 278–280, 283
hard carbon microtubes (HCT) 194
heavy metal adsorption 158
heavy metals removal 159
hematoxylin and eosin (H&E) 135
hemicellulose 4, 5, 7, 182, 186, 296
heteroatom doping 14, 15, 43, 47–48, 103, 116, 202, 238, 285, 295, 300
heterogeneous catalysis, applications 302–306, 308
heterogeneous Fenton catalyst 302
highly pathogenic human coronavirus (HCoV) 81
high-resolution transmission electron microscopy (HRTEM) 100, 114
hollow carbon nanospheres (HCSs) 132
hollow mesoporous carbon (HMC) 130
HSF-1 protein homotrimers 132
human hair (HHC) 3, 203, 204, 259
hyaluronic acid (HA) 131, 135
hybrid capacitors 258
hybrid MnO_2/CNT electrode 16, 17
hydrochar material 9
hydrogenation of furfural (FAL) 305, 307
hydrophobic contacts 130

hydrophobic graphitic mesoporous carbon core 132
hydrothermal carbonization (HTC) 8–10, 65, 116, 181, 182, 191, 215, 231, 296, 297, 324, 325
hydrothermal dealing 116
5-hydroxymethylfurfural (HMF) 308

i
identical mesoporous carbon spheres (UMCS) 135
indium-doped tin oxide (ITO) 277
inhibitory concentration (IC50) 139
insertion mode 71
internal rate of return (IRR) 17
intracellular hyaluronidase-1 131, 135
in vivo real-time imaging 129, 130
ion thermal carbonization (ITC) 325
ionic liquid 95, 142, 255
ionothermal (ITC) 232
iron (Fe^{3+}) ions 119, 319
iron (Fe^{3+}) sensor 119–120
iron sulfide (FeS) 189
isoniazid drugs 68

j
jamun seed-derived carbon (JSC) 199

k
kapok fibers (KFCNTs) 199
kiwi peel (KP) 153
KOH activation 14, 30, 34, 162, 182, 184, 191, 196–198
KOH-activated carbons 184

l
Langmuir adsorption isotherm 150
Li^+ intercalation/deintercalation anode reaction 175
linear sweep voltammetry (LSV) 220, 239
Li-S batteries (Li-S) 175, 176, 195–199, 204, 206, 207
Li-S discharge processes 176
lithium-ion batteries (LIBs) 20, 173–175, 181, 229, 263, 284, 322, 328
  biomass-derived carbon nanocomposite electrodes 186–189
  biomass-derived undoped carbon electrodes 181–186
lithium metal anode 175
lithium sulfide (Li_2S) 176
lithium-sulfur batteries (Li-S) 173, 322
lysosomal membrane permeation (LMP) 132, 133
lysosome dysfunction 83, 318

m
magnetic resonance (MR) imaging 142, 143, 321
malachite green (MG) 105, 151, 152
material physicochemical properties
  oxidative stress
    extraction of lipid 78, 79
    metabolic inhibitory effect 79
    photothermal effect 77, 78
    ROS-dependent oxidative stress 73–75
    ROS-independent oxidative stress 75–76
    wrapping effect 76, 77
  structural destruction 70–73
maximum adsorption amounts (Qmax) 163
mercury (Hg^{2+}) ions 118
mesocrystals 268–269
mesoporous activated carbon (MAC) 76, 151
  dyes removal
    activated carbon from cattail biomass (CAC) 152
    activated carbon from Rattan waste 152
    coconut coir dust 154
    Corozo oleifera shell 154
    edible fungi residue (EFR-AC) 153
    GWAC adsorbent 150, 151
    macadamia nutshell waste (MNS), for MB 155
    meso/micropore-controlled hierarchical porous carbon 158
    Neobalanocarpus heimii wood saw dust (WSAC) 155
    novel magnetized activated carbon, for heavy metal ions sorption 157
    plant wastes for, methylene blue (MB) 154
    RH derived mesoporous activated carbon (AC) 151, 152
    wood sawdust waste activated carbon (WACF-P) 152, 153
  metal ions removal 155–157
  phenolic compounds removal 158–164

mesoporous carbon 148–149
mesoporous carbon nanoparticles (MCN)
  controlled/targeted DDS 131
  immediate-released DDS 130
  sustained-release DDS 130–131
mesoporous materials 148, 308
mesoporous silica shell 132, 133, 320
metabolic inhibitory effect 79
metal ion sensing 117
  copper (Cu^{2+}) sensor 120–122
  iron (Fe^{3+}) sensor 119–120
  lead (Pb^{2+}) sensor 120
  mercury (Hg^{2+}) sensor 118–119
  miscellaneous metal ions 122
metallic potassium 13, 30
metal s-state orbital 280
metanil yellow (MY) 150–151
methicillin-resistant *S. aureus* (MRSA) 68
methylene blue (MB) 103, 105, 150–155, 298, 302
microbial cells 67, 70, 72, 76–79, 318
microbial contamination 81
microbial diversity 63
microbicidal 66, 67
  material physicochemical properties 70
  mechanism of action 67
  resistance 68
microorganism-based biomass 7, 8
microwave-based synthesis 299, 300
microwave method 116
mitochondrial disruption 83, 318
molecular dynamics simulation 79
mushroom precursors 7

n
N-acetylcysteine (NAC) 73
Na-intercalated binary graphite compound (b-GIC) 177
Na-ion batteries (SIBs) 176–178, 189–190
nanochannel-confined graphene quantum dots 223
nanocomposite coatings 82, 318
nanoporous activated carbon (NPAC) 303
nanoporous graphene-like sheets 223
nanorod coated glassy carbon electrodes 224
nanoscale dewetting 79, 318
nanostructured carbon materials 2, 202, 216, 230
N-CNO catalyst 239, 241
N-CNT 200, 306
N-doped carbon 305
N-doped carbon microspheres (NCSs) 194
N-doped carbon nanotube aerogels (N-CNT-As) 200
N-doped nano PC (NNC) 308
N-doped PC material (PNCM) 306
N-doping 45, 193, 201, 202, 243, 299
near-infrared (NIR) 129, 277, 320
net present value (PPV) 17
β-nicotinamide adenine dinucleotide 222
$Ni/NiO/NiCo_2O_4$ nanoparticles (NPs) 200
nitrogen atmosphere 151, 152, 154, 162, 186, 266, 267, 306
nitrogen-doped biomass 7
nitrogen-doped fullerene-like carbon shell (NDCS) 201
nitrogen sorption isotherm pattern 150–152, 154
4-nitrophenol 218
N-methyl-2-pyrrolidone (NMP) 262

o
O_2^--independent oxidative stress 75
okara 223
one-step electrochemical method 74
organic molecule melamine 300
organic–inorganic hybrid layers 282
oxidation–reduction reaction (ORR) 104, 325
oxidative esterification method 308
oxygen evolution reaction (OER) 107, 178
oxygen reduction reaction (ORR) 230, 231, 236–238, 309, 325
oxygen reduction response (ORR) 178

p
paper electrode 222
p-chlorophenol (PCP) 163
penetration mode 71
penicillin 67, 68
peptidoglycan (PG) 79
peroxymonosulfate (PMS) 303, 304
persulfate (PS) 302, 303
phenol adsorption 161, 162
phosphoric acid modification 154
photoacoustic imaging 139–140
photoanode 275–278

photoexcitation 77, 318
photoluminescence (PL) stability 65
photoluminescent blue emission 65
photosensitizers 77, 105, 275, 318, 319
photo-supercapacitor 328
photothermal conversion capacity 129
photothermal effect 77–78, 131–134
photothermal therapy (PTT)
  biosensing 140–142
  cell labeling 135–139
  magnetic resonance imaging 142
  photoacoustic imaging 141–140
  synergistic therapy 135
  therapeutic biomolecule delivery 140
  toxic substances, removal of 139
  transmembrane delivery 139
picric acid (acceptor) 123
plant-based biomass 4–6
plasmodium genus 68
platinum (Pt) 235
Plectranthus amboinicus 224
p-nitrophenol 222
polyacrylonitrile-block-poly (*n*-butyl acrylate) (PAN-b-PBA) 139
polyatomic doping 245
polyethylene (PE) 73, 174
polypropylene (PP) 174
polyvinylidine fluoride (PVDF) 262
porous carbon 198
  materials 10, 13, 148, 149, 158, 162, 195, 202, 231, 236, 242, 243, 293–309, 325, 326, 329
  nanomaterials 2
  synthesis from different biomass 237–238
porous heteroatoms-rich carbon (PHC) 200
porous structure 10, 11, 14, 28, 29, 162, 187, 193, 202, 206, 238, 265–269
potassium carbonate 29, 30, 153, 155
potassium ferricyanide 218
potato peel (PP) 153
power density 132, 134–136, 171, 172, 174, 206, 254, 255, 257, 258, 262, 263, 267, 323, 328
p–p stacking interaction 142
precarbonization 151
pristine materials 202
progesterone 218, 219
propidium iodide (PI) 72, 134
Prosopis africana seed hulls (PASH-AC) 163
protein (glucose oxidase) 142
pseudo-second-order model 151–155, 162, 163
pseudocapacitor 257, 258
pyrolysis 4, 5, 7–11, 13, 28, 30, 32–34, 46, 47, 64, 79, 94, 97, 117, 162, 163, 181–183, 190, 193–195, 201–204, 215–217, 223, 231, 238–240, 242, 243, 295–297, 300, 301, 303, 305, 306, 308, 317, 324–327
pyrolysis carbonization method 239, 327
pyrophosphate (PP) 305

q

quantum hall effect 220

r

Raman spectrum 46, 74, 114, 182
rattan wastes 152
rechargeable batteries (RBs) 172–179, 323, 328
redox reaction 13, 172, 187, 257, 302
reduced graphene oxide (rGO) 65, 223, 224
return on investment (ROI) 17
revised half-cell test (RHT) 194
rhodamine B (RhB) 153, 303
rice husk or rice hull (RH) 150, 151
rifampicin 68
rose-derived CDs 123, 320

s

screen-printed carbon electrode (SPCE) method 224
self-doped nitrogen porous carbon nanosheets (NPCNS) 240, 241
separation factor (R_L) 150
sheet structure 263–265
shuttle effect 199
silver (Ag) nanoparticles 64
silver-based antimicrobial agents 64
silver-coated GO nanocomposites (GO-IONP-Ag) 82
simple hyacinth pyrolysis 201
simply payback period (SPP) 17
single-heteroatom doped carbon electrodes 202–204
single-walled carbon nanotubes (SWCNTs) 65, 99, 113
SnO_2 nanoparticles 74
sodium-ion batteries 173, 176, 177, 232, 322
solid-electrolyte interface (SEI) 189
solid waste production 293

soybean curd residue (SCR) 306
soybean root-derived carbons (SRC) 266, 267
sp2-based carbon nanomaterials 129
spherical structure 33, 263
spontaneous reaction 172
S protein through pseudolectin interactions 81
sulfathiazole (STZ) 305
supercapacitors 255
  activated carbon, from biomass 259
  battery 254
  capacitor 253, 254
  EDLS 256, 257
  electrochemical measurements 262
  electrode material 205
  energy density 255
  hybrid capacitors 258
  power density 255
  pseudocapacitor 257
  structural diversities of biomass 262
    fibrous structure 263
    mesocrystal structure 268–269
    porous structure 265–268
    sheet structure 263–265
    spherical structure 263
    tubular structure 263
supramolecular p–p stacking 129, 130, 320
surface incorporation 17
surface modification
  coating on medical devices 81–82
  heteroatom doping 15
synthesis procedures 295
  carbonization 295–296
  chemical activation 298–299
  functionalization/ doping/ composites, of ACs 300, 301
  hydrothermal carbonization 296–297
  mesoporous carbons 148, 149
  microwave based synthesis 299, 300
  physical activation 297–298
  physicochemical activation 299
  pyrolysis 297
synthesized carbonaceous materials 64, 317
synthetic methods 3, 115, 231, 232, 283, 306, 309

t
tea leaf PC (TPS) 304
ternary graphite intercalation compound (t-GIC) 177
tetracycline (TC) 67, 123, 320
theoretical lithium storage capacity 175
theranostic functions 130
theranostic utility 129
therapeutic biomolecule delivery 140
3D bacterial cellulose-derived carbon nanofibres 188
three-dimensional (3D) hybrid nanoaerogels 200
three-dimensional (TD) hierarchical 304
thrombin-binding aptamers 139
α-tocopherol 73
topical ointment formulation 64
toxic chemical removal 147–165, 321–322
transparent conducting substrate (TCO) 277
2,4,6-trichlorophenol 162
triple-phase boundary (TPB) 234, 235
tubular furnace 155, 305, 306
tubular structure 102, 263
tungsten-based bimetal oxide 285
2D hierarchical porous carbon nanosheets (HPNC-NS) 203
2D/3D porous carbon 329

u
ultracapacitor 253
uniform oxygen 10

v
van der walls interaction 79, 318
verapamil (VER) 138

w
water-wettability 65
wound dressing 80, 81
wrapping effect 76–77

z
zero-dimensional structures 216
zinc-air batteries (ZABs) 173, 178–179, 199–201, 322